WHOLE-ING

NON-JUDGMENT, WHOLENESS, AND PERSONAL FREEDOM IN OUR QUANTUM UNIVERSE

TIM CROSS

ONE RIVER PRESS

Austin, Texas

www.thearchitectureoffreedom.com

Also by **Tim Cross**

The Architecture of Freedom

A Path to Personal Freedom

Copyright © 2020 by Tim Cross

Published and distributed in the U.S. by
One River Press, Austin, Texas.

First print edition ISBN: 978-0-9888344-9-1

Dedication

This book is dedicated to my two amazing daughters: Emily and Rachel, and my loving and extremely patient wife Connie. They have all been such a big part of my journey. It is also dedicated to the Covid-19 crisis sweeping the planet that is so clearly highlighting our connectedness and past choices. All external events, including pandemics, only reflect our shadow world within. Our darkest shadows are always our best teachers and this crisis will expose our collective soul and help to heal many of our planet's deepest wounds. It will not be an easy path and many individuals will experience extreme difficulties because we have ignored the clear signals for far too many years. Every day we are being asked to be more conscious about where we place and focus our attention; we must always remind ourselves that "energy always goes to where our attention flows." Covid-19 presents us with an especially powerful opportunity to shape a different kind of world: one that we each imagine in the most sacred place in our hearts.

TABLE OF CONTENTS

Table of Contents

Table of Contents

Table of Contents

"We live under the impression that in order for something to be divine it has to be perfect. In fact, the exact opposite is true. To be divine is to be whole and to be whole is to be everything."
Debbie Ford

"Enlightenment is not imagining figures of light but making the darkness conscious."
Carl Jung

"As long as I am this or that, I am not all things."
Meister Eckhart

"Live life as if everything is rigged in your favor."
Rumi

"We don't see the world as it is. We see it as we are."
Anais Nin

"Opening to oneself fully is opening to the world."
Chögyam Trungpa

"What you see depends mainly on what you look for."
Words above entry to Tom's Burned Down Cafe

"When the soul wants to experience something, she throws out an image in front of her and then steps into it."
Meister Eckhart

"There are no actual things in our universe, instead there are fluctuations or disturbances of the energetic Quantum Field which only then appear to us as things."
Paraphrasing a significant number of today's Quantum physicists

"All the world's a stage, and all the men and women merely players. They have their exits and their entrances."
Shakespeare

PREFACE

Are we just tiny impotent beings attempting to negotiate an enormous, unfathomable, and often terrifying universe? This is the clear message of our senses. Despite this constant messaging from our bodies and its consistent reinforcement by our culture, it is now becoming apparent that our true nature might be quite different. Beginning about one hundred years ago, this rational, but quite numbing, perspective of our relationship to the universe began to be completely reevaluated because of two life-changing scientific discoveries: *relativity* and *quantum physics*. Through these two transformative revelations, we are now beginning to understand, as many traditions have long foretold, how the entire universe can also be found within everything in existence, including a grain of sand and each of us. Today, as more scientists discover surprising new patterns of interconnectedness throughout our universe, what seems to be emerging is the recognition of a deeply intertwined relationship that some scientists and philosophers describe as an expression of "oneness." While it may seem that we are only a small part of an enormous universe, today we are realizing that the truth may be just the opposite: everything that we once experienced as being outside or external to ourselves, originates from within. **We don't just live in our universe, the entire universe also lives within each of us.**

Our culture is relentless in its messaging about the need to improve ourselves, to constantly strive to become better at something or just better individuals. We find that no matter how hard we work at this, the difficulties we experience in our lives rarely seem to subside, at least for long. This onslaught of difficulties isn't because we aren't yet "good enough." It happens, instead, because our conscious awareness is incomplete and not integrated, leaving most of our activity to the large unconscious, or "dark," parts of our being. When we operate through this unconsciousness, as most of us do continuously, we just repeat our old familiar, but often disruptive, patterns. Growing beyond these dysfunctional patterns is not about focusing upon or choosing the "good" parts of ourselves and rejecting the parts we think of as "bad." It also does not require that we learn more or something new. Instead, to dissolve these old automatic patterns, we must first become more whole and integrated. We can only do this by bringing new illumination to the uncomfortable darkness hiding within all of us, a shadow world dividing our naturally whole being into many broken, isolated, and incommunicative fragments. Our most important journey involves a process of discovering, welcoming, and honoring all parts of our authentic and complete being, especially the vast areas of inner darkness that we fear and avoid the most.

While our new science may help us to understand the relevance of this inner journey, the only way to have a different type of experience is to become more aware and conscious of our inner shadow world. Because we are all so different, there are as many unique variations for this type of growth as there are individuals, yet each personalized path can only be discovered from within each individual's journey. This book describes one such path: the very successful process that radically changed my life through the discovery and conscious illumination of my personal shadows. Today, instead of life appearing to work against me, I experience a world that always works in joyful harmony with my entire being. I still see and observe the many great problems throughout our world, but now I recognize that they are only the messengers reminding me of the continuing need for this type of inner work. My life today is powerfully energized and filled with satisfaction and joy, but it has not always been this way; for most of my seventy years, my experience was the complete opposite.

In this book, I describe, as best as words allow, my healing journey and the new perspective that changed absolutely everything about my life. I now know that each of us, our universe, and life's dramas do not exist independently; instead, the entire spectacle that we call life is formed only as a secondary and changeable manifestation of an underlying but invisible *form* that physicists have named the *quantum field*. As awareness of this deeper truth about the nature of our true *being* grows, we each will be called to expand our bounds to become more whole and complete. As we do this, the specific way that our universe manifests for each of us will also change. *Energy* comes first, *form* only comes after. The solid-appearing nature of our world is illusionary but manifest so that we can more clearly observe our true nature. Happily, it is also an illusion that will quickly and naturally change as we evolve, for it is only our reflection.

Below the surface of our egos, life is always feverishly working for us, no matter how things appear externally. The old common belief that life works against us and is something to overcome is just a product of our incomplete, wounded, and agenda-driven minds. Once we fully understand and begin to integrate this new perspective, our lives become easier, less stressful, much more fun, and far more joyful. Today, many people that I meet permanently and easily reside in this peaceful place, living full lives of authentic self-expression. With additional experience, proper preparation, and the right amount of time, this joyful approach to life evolves into an unwavering bliss.

This is the gift of "living in the now," but this way of walking through life is far from a new idea. Instead, it is an ancient wisdom that is now being powerfully reinforced by our latest physics. Life, just as it appears, is an individualized, customized, and perfectly choreographed performance that is being presented to each of us entirely for our personal benefit. We are here to experience our lives fully, to witness and participate in all of our individual and collective creations, so that we may better "know" ourselves and understand our deeper calling to become the whole and transparent vessels that can better transmit the revealing light of consciousness. This book is about why this is so and how to prepare and maintain our individual but unique vessels so they can participate, contribute, and enjoy at their highest potential.

Deep inner growth requires time and direct experience, which is exactly what life on earth provides for each of us. We can't rush this natural unfolding; a certain amount of time is required for gaining each piece of necessary experience. We add to our experiences by simply living our lives; exploring and accumulating direct experience is a primary purpose of life. However, while we are gaining this necessary experience, we can also greatly improve the quality of our journey by reducing, and even ending, our inner suffering by engaging the conscious healing process. Committing to this process is not easy, at least initially because, for many, it may first appear to be both unnatural and extremely illogical. With all of our culturally driven attempts to do the "right thing" and become "better" individuals, we judge ourselves and others and this only further divides our wholeness. If instead, we learn how to illuminate the places of darkness that always divide our full expression of self, we become more integrated and this alone transforms our lives. Our lives suddenly begin to *flow* in ways that previously we might have imagined to be nothing short of miraculous. Those very aspects that we once rejected are instead discovered to be the critical missing parts of our now much more integrated and empowered being. The name that I have assigned to this internal healing process of discovering and illuminating our hidden shadows is *whole-ing*.

Our culture teaches us to strive to be this or that, to be special, different, or even "better," but whenever we engage this way, we are actually judging and then rejecting important aspects of our invisible yet much more complete *being*. When we judge and reject any aspect of ourselves or others, in our misguided and confused attempts to become better, we are actually shrinking our *being;* we become "less," not more. These fragments only appear to be problematic because they are operating alone in isolation, having become fully disconnected from the rest of our *being*. They have been separated from our wholeness by the *egoic* darkness that naturally and continuously arises within, simply from being born and living our everyday lives. These same aspects, once they are fully accepted, loved, and embraced then transform *(transmute)* from their old appearances to become integrated parts of a more whole *being* that, only now, is fully capable of experiencing the great joy, radiance, and love that is its birthright. Deep within all of us, there already exists a completely satisfying wholeness of *being*, but we only discover our ability to communicate and *flow* with these depths once we learn how to get out of our own way.

Whole-ing activates the *magical* flowing universe that I enthusiastically describe in all my books. The world is already perfect just as it exists and we participate in its unfolding, moment by moment, with our every thought, word, and action. However, an infinite number of other equally possible worlds also exist simultaneously, each perfect for whatever purpose is required and these are hidden within invisible "parallel universes." We eventually discover that we are awesome and powerful *beings* that already have the power to occupy and enjoy any of these already existing worlds, but only once we learn how to effectively "choose." We can't choose to change our world with just our thinking minds; we must "choose," instead, from a deeper, unexplored place: the non-thinking *resonance* of our entire *being* that is sometimes called our soul. **Because much, if not most of our life and being, is hidden from our awareness in the depths of our shadow world, to make better "choices" we must first illuminate our darkness and bring more of it into our conscious awareness.**

After seventy years of exploring life on Earth, years filled with wide and varied experiences, I have become deeply impressed by our awesome ability to shape and modify the appearance of our lives (and our universe) with every imagined thought, spoken word, and completed deed. Despite any impressions to the contrary, we are now discovering that we are extremely powerful and creative beings, but our *being* will always "create" and manifest the outer physical world only as a direct reflection of our deeper, but hidden, internal *energetic* state. Because the "outer" world only represents the nature of our inner world, to change or improve our "outer" experience we must first do our work within; it makes no sense to try to fix the outside world when we are all so fragmented within. Until we each heal ourselves in this way, our separated and divided parts will continue to express themselves through the drama of war, conflict, and chaos that we commonly witness throughout our external world.

Before we can experience peace without, we must first discover peace and non-division within. This is not just another platitude or call for "positive thinking," nor does it involve denial or avoidance of experiences that we might consider unpleasant. It is also not about finding some way to avoid life's often difficult personal lessons. Instead, this book explains why exactly the opposite approach is required. ***To live our lives most joyfully, instead of just focusing on the positive, each of us is being asked to take the deepest dive into our most frightening shadows and fears.***

Most of us live lives that are very different from the ideal world that we might imagine. Over and over, we automatically and unconsciously relive the very situations and patterns that disturb us the most. ***We do this again and again, without ever realizing that the chaotic life that we experience, and constantly complain about, is also our most perfect creation. We are continuously, but unconsciously, creating these patterns in our lives, not realizing that they are our best teachers because they always accurately reflect the truest nature of our deeper being.***

While several spiritual or religious traditions have always taught that the life we directly experience is only a temporary and illusionary expression of something that originates in a far deeper place, only recently has our contemporary science caught up. Our most profound science is now revealing that our unconscious shadows may be the real producers and directors of our lives. Our deepest shadows continue to appear before us in bold technicolor on the big screen of life through what may be the ultimate, but least understood example of an otherwise well-known process that filmmakers, mathematicians, and psychologists all refer to as *projection*. Our world's most horrid manifestations, such as war and incomprehensible human atrocities, rise directly from this, our unaddressed shadow world.

What follows is the journal of my *whole-ing* process: the third book of my series. The first, *The Architecture of Freedom*, was the enormous compilation of my lifetime of study. The second, *A Path to Personal Freedom*, presented that information in a shorter and easier to follow format. This third volume, *Whole-ing*, focuses on the integration of these often strange-sounding ideas into our everyday lives. I have been continuously writing about this subject matter for over ten years because this daily practice helps me clear and integrate my own constantly evolving awareness, and, for me, this daily exercise has also become an important form of meditation. Initially, I began journaling only to share my understanding with my two daughters, but now I publish to share my process with others who might be on a similar path.

Because my understanding is constantly evolving, some of these ideas may eventually be seen differently as my experience grows. Life is an evolutionary process and none of us can predict exactly how it will unfold.

As I will remind the reader, no one can comprehend the deepest truths because they lie beyond, or outside, all human words and concepts. *In fact, at some point, often quite early in the whole-ing process, words and concepts start to limit our ability to experience truth.* Because the deepest truths will remain completely indescribable, words will always fall short; the deepest truths must and will always remain unspoken. While words can be useful to point us in the general direction of the truth, ultimately, only deep personal experience and the eventual illumination of our inner shadows can reveal the next layer.

Whether or not we are aware of it, we are all, already, on this journey, a journey towards what some have called *enlightenment. Enlightenment* is usually badly misunderstood for it is too often seen as an end-goal. Rather than some final prize to be claimed, it is much better understood as a never-ending process of becoming or expanding. The longest and most critical stage of the entire *whole-ing* process is the gaining and integration of life experience, which is exactly what we all do every day. Whether or not this is recognized, little by little we are all becoming more *enlightened beings*. What is important is to recognize this expansion of our *being* as a continuous and ongoing process, and becoming more comfortable with this process helps us to be more comfortable with all of life.

For some reading this book there may have been a single and dramatic moment of recognition, or awareness, that forever shaped the rest of their lives: a sudden realization that the world was far different from what was previously thought. This first clear realization was possibly disturbing, and it likely faded as it became overshadowed by the pace and responsibilities of our day to day lives. Some readers might have, instead, experienced a series of smaller gradual incidents that together amounted to a profound realization, while others had that one big instant of a clear "knowing" followed by an extended series of additional insights. No matter how we have been introduced to this new vision, all types of realization require a full lifetime for integration and deepening; some traditions even teach that this process requires many lifetimes. Each time that we discover, examine and illuminate one or more of the shadows or concepts that block access to our full being, our lives incrementally become more enjoyable and we become able to experience the already existing sea of love. This book describes some of the methods and techniques that I utilized to dramatically change the way my life appears.

As a final note, I would like to remind the reader that any science included in my books is only there because it was a critical part of my growth process. For me, realizing the implications of our latest science helped to focus and streamline my process. If any technical parts of this book seem difficult, please feel free to skip them, because this book is about so much more than the science; the same *Truth* exists and our growth unfolds whether or not we reflect on the latest scientific discoveries.

INTRODUCTION

Whole-ing – The conscious illumination, acceptance, and integration of all that we are; all that we judge to be "good" along with all of the "bad," so that we may become more of a "whole" and complete being, allowing us to more fully and profoundly experience the sacred gift of physical life.

WHAT THIS BOOK IS ABOUT

Who are we? What is our purpose? Why do we often feel unhappy or incomplete? What is our universe? How do we best live our lives? Throughout our entire human history, similar thoughts and longings have been broadcast from every corner of our diverse and breathtaking planet. While questions like these have been the focus of human curiosity for as long as mankind has possessed awareness, language, and reason, satisfying answers have been surprisingly rare. Throughout human history, religions and spiritual systems have competed to provide meaningful responses, offering answers that occasionally assuage, but rarely satisfy.

However, recently this search for answers has changed dramatically because we are now able to re-visit these ancient questions from a new and expanded perspective. Our ancient search for good answers to these fundamental longings has been re-energized by our newest physics and technological tools, which have been producing completely unexpected insights at an astounding rate. *Quantum physics* and Einstein's *relativity,* together, have become two of the most paradigm-changing scientific breakthroughs since the flat-Earth days of Copernicus, half a millennium ago. Today, after an entire century of experiments and many awe-inspiring technological breakthroughs from these theories, we are finally arriving at a point where scientists and philosophers can relax their disbelief enough to speak more openly and honestly about just what these very consistent experimental and practical results might mean. **This new emergent vision is not just a clarification or expansion of older ideas; instead, it permanently and radically changes everything that we once believed about the manifestation, potential, and purpose of our small planet, the entire universe, and even the extremely mysterious process that we call life.** What is finally being acknowledged has already started to shift our understanding of our relationship to the universe, to each other, and most importantly, to ourselves. Once this new awareness is fully recognized, life on this planet will be forever transformed for the better.

After studying *quantum physics* at the university level, I eventually changed career paths and practiced Architecture professionally for over forty years; yet I never let go of my driving curiosity about the deeper meaning of this strange new physics that I encountered so early in my life. **Today, after a full lifetime of focused study and a wide range of varied experiences, I now understand, have integrated, and live my entire life based upon the most extreme and radical idea that sits at the very foundation of this new physics: the radically interconnected oneness of everything in existence.** Our latest and best science, most honored spiritual traditions, and most profound human experiences all lend powerful support to this extraordinary and radical vision – one that illuminates and completely reverses many of our old beliefs about life. Once I began to integrate this new perspective into my daily awareness, everything about my life seemed to change for the better.

I shared my full vision and its discovery in my first two books: *The Architecture of Freedom* and *A Path to Personal Freedom*. The focus of this new volume is less about my realization and more about how to help others integrate this wonderful new vision into their lives. **Because this integration changes everything about the way our everyday lives appear, in some ways, this is the ultimate self-help book. One of the profound realizations is that because of this deep oneness, nothing ever exists "outside" of us. While this has many implications, the most significant might be that everything, including all change, must begin from the only "real" place in all of existence: within.**

When viewed from the conceptual framework of our three-dimensional realm, there appear to be as many different possible paths to personal freedom as there are individuals; no single path works for everyone. My books describe only one of these paths, but it is a path that has worked extremely well for at least one person: myself. With the deepening of each of our journeys, all of our varied paths will begin to converge as the more profound and universally shared truths are discovered, explored, and then integrated to gradually reveal the deeper oneness of everything. We are each like the separate-appearing leaves of a tree, discovering our connectedness through a personal voyage that begins at our individual and separated leaf-like identities, and follows a path of exploration down through the tree's branches, only to converge deep within in the "trunk" that is common to all the individual appearing leaves. Only from within this shared "trunk" can we begin to understand the nature of the massive and profound "root" of our common being. The root we share has always been present, but it remains hidden beyond the limits of our conceptual thinking, beyond the horizon line of human awareness. As the roots of trees are hidden by the soil, our *conceptual horizon* hides the deeper source of our interconnection.

We are all on this evolutionary journey and there is but one common destination for all of our paths: the life-changing realization and integration of our profound interconnection with everything in existence. At the surface level, this means that our relationship with each other is similar to the relationship between the cells of our body or, as just discussed, leaves of a tree. *We all may appear separate, but ultimately we all function as tiny, but still extremely important, parts of a single spectacular organism.*

From a broader perspective, one which already includes a full awareness of this life-changing vision, it is understood that every living *being* is continuously evolving, deepening and expanding, no matter how their individual lives appear in any single moment. *One of the most profound implications of modern physics is that despite the random-appearing nature of everything, all parts of our universe are in constant and instantaneous communication with all other parts. This means that, despite our fears and training to the contrary, the universe always unfolds in perfect interactive harmony, even if almost all of this communication exists at levels that we can't see, hear, smell, taste, touch, or understand.* This also means that there can be no "wrong" paths, and the timing of all events will always be coordinated and perfect for every part of our collective *being*. *There are no accidents in nature. Our universe is, and always will be, the perfect vehicle for our personal and collective evolution.*

What we are also starting to realize from this recent physics is that our physical universe, including all of the individual things, organisms, and people that we think of as "real," are far from the solid fixed objects that we commonly perceive. Scientists are now discovering that all the physical things of our world seem to be only secondary *images* that are springing from a deeper, well-hidden source: *the quantum field.* *This radical idea of our physical existence being a secondary manifestation is certainly not a new one, for it has been spoken of for thousands of years.* However, it is only very recently that our science has been able to weigh in on this radical idea.

Mankind already has a host of different names for this type of omnipresent creative source; it has been called God, Allah, Shiva, or similar word symbols for this presence which no words can ever accurately describe. Descriptions of that which creates and destroys and then re-creates this solid-seeming world can be found in ancient religious texts. These speak to an omnipresent power that lies far outside of our physical realm, occupying a different kind of *space*, one that can't be reached by conventional forms of travel or adequately described using words and conceptual thinking. Hindu texts reference the *Maya:* the illusionary nature of our world and physical lives. **Now, after more than 100 years of confirming experimental evidence, some physicists are concluding the same – that there exists, at the deepest level of creation, an ineffable source from which all of our lives unfold. Recently named the quantum energy field (QEF), it appears to be the source and "creator" of everything we experience in the entire universe, and yet, somehow and in some way that we don't yet understand, this omnipresent creator of our entire universe is also not separate from each of us.** For myself and many others, the *quantum field* has quickly emerged as the most probable mechanism for incorporating, explaining, and unifying all of our life experiences, deepest questions, and spiritual traditions.

The impressions of our senses are undeniable, and it is from them that we construct our vision of reality. It is impossible not to become convinced that the world of *time, space,* and things that we experience is real, solid, and largely immutable. However, over one-hundred years ago, Einstein made it very clear that our sensory impressions are far from accurate. In his own words, our impression of a real and solid world unfolding within *time* is, instead, just a "persistent and convincing illusion."

Initially, this idea may sound like far-fetched nonsense, religious doctrine, or science fiction; but before automatically dismissing it, first, re-consider our long human history of discovery. Everyone on Earth once believed that our world was flat with an edge over which one could fall. Only 500 years ago, this worldview was considered basic common sense because it was the only clear and obvious reading from all of our senses. Many other things that we take for granted today also once defied our collective "common senses." How could just air hold up heavy airplanes, and how do many types of animals detect and know things that are completely invisible to our senses? The more we learn, the more we realize that our logic and senses have never been very reliable guides for our probing of reality.

Of course, that "flat-Earth" understanding ultimately fell away as scientists and explorers gradually uncovered and then convinced the rest of us that our Earth was, instead, a round sphere, moving and spinning through *space* at millions of miles-per-hour, on a cosmic path not unlike that of a roller-coaster. Largely because this dramatic physical motion through *space* couldn't be sensed, we required many generations to integrate this new vision into our cultural understanding. Because a direct understanding was beyond the ability of our sensory-driven logic, it was left to science and further exploration to reveal the actual truth about the Earth's geometry and motion.

Once again, we are facing a dramatically different vision for our existence that similarly defies our logic and senses. The underlying "oneness" that mysteriously connects absolutely everything in the universe appears to us, instead, as a multitude of separate things, spread throughout what seems to be an enormous *space*, where all actions unfold as divided and separated moments within *time* and *space*. Similar to *space, time* appears to us as "spread out," but it also appears to be relentlessly marching in only one direction: forward. This spreading out of both *time* and *space* is a glorious gift, for it is precisely what gives us the wondrous opportunity to experience, grow, and evolve.

The illusionary appearance of our separateness is not accidental, problematic, or unimportant, for it has a critical evolutionary purpose. However, when we believe that this physical expression of separation is the only form of reality, we act in ways that divide our being and continue to create unnecessary human suffering.

Our universe has been divided and spread out this way so that we may observe and interact with all its different parts. *Time* and *space* together are not unlike a deck of cards: it can be experienced as one *thing* or something that can be fanned out and divided for all to see. The separation of *time, space,* and *things* is only one way to observe and experience existence, but it is one that serves us particularly well by delivering a plethora of varied and rich individual experiences and relationships, along with lots of "room" for building deeper levels of understanding, integration, and collective evolution. Through this separation of *time, space,* and *things*, we are given the perfect opportunity to explore many aspects of our remarkable *beingness*.

This entire physical universe of division and separation, including dividing our collective *being* into its many individuated selves, can be thought of as the ultimate amusement park, a Disneyland-like *space* created for the benefit of all. Through the astonishing variety of experiences that life provides, we eventually will discover that we are both one thing and many. We always embody both at the same *time*, and to deny either expression of our *being* is to render ourselves incomplete.

Whole-ing is the practice of first recognizing and then integrating all these divided and often rejected parts of our greater selves. As we better recognize our deep connection to "all things," we begin to realize that this also means we must embrace and include everything that exists in our universe, all that we label good, as well as all that we label bad. When first introduced, this idea seems crazy. Why would anyone ever want to examine and embrace "all of the bad?"

We all react to aspects of life that we find difficult: we push away certain things and judge others as bad, dangerous, wrong, or negative. This is a completely normal and natural reflex. Eventually, after a wide variety of experiences, we start to realize that when we reject something, we are only isolating or shadowing some part of existence which is also some isolated piece of ourselves. We also discover that this division and judgment only leaves us feeling more incomplete and unfulfilled. What most of mankind fail to realize is that if instead we embrace and love these very same parts that we have been judging and negating, these once-rejected aspects then shift to transmute their entire character. They then become important, supporting, and completing parts of a new, integrated, and expanded self. *Our most feared shadows, once embraced, can and will shift character to become our greatest allies.*

We eventually discover that the only thing we have to do to fully enjoy and transform our lives is to "get out of our own way." *The primary way that we all block access to our fulfillment is through personal and cultural judgment, both conscious and unconscious.* This book describes what this means, why this is so, and how to move beyond our culturally inherited and learned patterns of judgment. If the reader already knows that they need to "get out of their own way" but doesn't know how to do this, or feels "stuck," then this book may help; it discusses several of the most formidable obstacles that impede us in our common journey towards freedom.

The most significant part of this journey hinges on a very challenging recognition: that the separate and disparate parts of existence, regardless of how we judge them, are all equal and necessary for creating wholeness. All aspects of our lives are essential contributors to this life-changing wholeness: the good, the bad, as well as the beautiful and the ugly. *When our individual being judges, resists, or rejects any part of this wholeness, especially those parts we reflexively judge as repulsive or evil, we suffer. Whole-ing is the process of gradually reclaiming our lost, isolated, and rejected parts as preparation for the very natural, but surprising, transmutation of individual suffering into profound and permanent joy.* The resulting feeling of wholeness, along with its accompanying joy, is our universal birthright.

Freedom is only possible if we are free to choose, so freedom also means that all possible options must be readily available. Happily, our universe is built upon this architecture of freedom. Our universe is perfectly designed for exercising choice and freedom because, within it, the full range of possibilities always remains present and available. It is a space for exercising complete freedom only because, as I explain later, our physical universe is built, and completely depends, on the availability, continuing existence, and balancing of ALL polarities.

Whole-ing presents a radically different process for healing ourselves and living our lives in freedom and joy; but before we can fully enjoy this new way of being, we must learn how to embrace all of life, exactly as it appears before us in each and every moment. This book explains why this is so and how to integrate this deeper understanding of our universe into our everyday being.

CONVENTIONS USED IN THIS BOOK

Since I must use words and language to describe something that is ultimately indescribable, I will by necessity be using words in ways that are purposefully different from conventional usage. Other than for actual quotes, when I use quotation marks around very common words, I do so for one, or both, of two reasons.

The first possibility is that I am suggesting a deeper re-evaluation of the meaning of this common word or phrase. Within our culture, we have formed "consensus understandings" about the meanings of certain words and concepts. These "consensus understandings" might be efficient for quick impressions, self-protection, and rapid communication, but they also can be extremely narrowing or limiting for deeper exploration and understanding. This book is always asking the reader to recognize and understand how certain learned meanings and concepts that we habitually take for granted will contribute to, and even pre-determine, our experiences. Only when we recognize the presence and power of these assumed and inherited word-based concepts can new definitions that expand upon the old meaning be seen, understood, and eventually integrated. It may seem like only a small shift in awareness, but a more careful examination of our chosen words and definitions can open us to an entirely new and different view of our world and lives. **When I desire the reader to re-examine the meaning of a particular word, I will surround that word with "quotation" marks.**

Sometimes, these quote marks are, instead, asking the reader to treat the word as the best familiar metaphor for a concept that can't be fully understood with our rational minds. For example, I might say "window," "thing," "images" or "shadows." Here, I am not referring to the common glass "window" but rather "an opening to another way of understanding something." "Shadows" refer to parts of our *being* that are not seen by the "light" of consciousness, and "images" refer to how *particles* or "things" that appear in our realm are formed in a deeper invisible realm (the *quantum field*), and only then cast or *projected* into our three-dimensional realm.

Often, the quote marks refer to both of these uses simultaneously. For example, in the last paragraph quotes around "thing" refers to the fact that, in the depths of the universe, there really are no things at all, only *energy* in different-appearing forms. The solid thing-like quality of "things" is only an *illusion* within our realm. Here, and in many other places, the quotation marks are asking the reader to consider both uses: re-examining the word's common meanings and recognizing that this word is only *being* used as an approximating three-dimensional metaphor for something that also exists outside of our three-dimensional realm.

I will use *italics* when referencing words that have specific scientific definitions or are new and important concepts that I will be referencing throughout this book. For many of these, I will provide some form of explanation or definition for these terms when they first appear. A glossary with my definitions for all *italicized* words can be found in the appendix.

Throughout this book, I also will be using *italics* while discussing familiar words describing concepts like *mind, self,* and *soul.* In these cases, the *italics* are used to remind the reader that these familiar words have historical definitions that typically support our cultural belief that we are all separate selves confined to these physical bodies. However, with the expansion of awareness that is the subject of this book, these commonly used words take on new meanings; they shift to better reflect our deeper interconnection. When I am referring to this new and expanded understanding, these same words then become capitalized (*Mind, Self, and Soul*). Our long-embraced cultural understandings are not being replaced by new definitions, they are, instead, being extended and expanded upon. Our older cultural understandings are never discarded or lost within this new vision; the definitions are just being enlarged so that they now allow better access to the deeper secrets of our remarkable and mysterious existence. These expanded definitions have been modified to include the newly recognized realm where there is no separation of *time, space,* and "things." This deeper structure is what allows everything in the universe to be so instantly interconnected that, as several modern physicists have said, "It is as if everything in the universe is only one thing." *Mind* (with a capital M) describes the mysterious and unexpected availability of extraordinary *information* that a single individual's *mind* could not have known or processed. Many writers, the first being Aldous Huxley, have referred to this as "mind-at-large." At its fullest expression, *Mind* is that which contains and holds all of the *information* within and about the universe. Because *Mind* operates outside of *time* and *space*, and because our brains only process *information* tagged within *time* and space, *Mind* is usually inaccessible from our "normal" concept-based existence. (Here, I am also asking the reader to reevaluate that which we consider normal.) *Soul* and *Self* have similar expanded definitions: the small letter *soul*, and *self*, refers to the separated and common single individual understanding of these terms, while the capitalized version describes the fully interconnected state that only begins to become available as we learn to live and "think" outside of *time, space* and our state of separation.

This book discusses how and why this deeper source of "knowing" is always available to everyone, but only once we have properly prepared (tuned) our minds and hearts. ***This preparing or "tuning" of mind and heart requires that we release all tension, judgment, and divisive concepts (all concepts) so that we can better flow within the deeper energetic river of life.***

Another convention that I will be using is repetition. Learning how to live this way is a skill and when we desire to integrate any skillset like learning how to surf or play a tuba, endless repetition is required. For this reason, I will constantly be reminding readers of important guiding principles and present these from as many different perspectives as I can imagine. Also, and especially in part five, I wish for each of these open discussions to be as self-contained as possible, so naturally, this will require some repetition of ideas presented earlier. Because this is a self-help book and not a novel, and much of this material is new to our lives, the reader should be prepared for a significant amount of repetition. Perhaps frustrating at times, this repetition is intentional to facilitate our deepening.

PART ONE – WHY WHOLE-ING?

Much of this part is a review or distillation of critical ideas from my earlier books. If my earlier books have been read and understood and the concepts are clear, then the reader may prefer to jump to the heart of this book: "Part Two: The Healing Journey. Once these basic ideas are understood, the book can be opened to any section, for ultimately all these separate-appearing ideas untangle and merge into a perfectly arranged oneness.

My Personal Journey

There are thousands of well-traveled paths to this integrated holistic awareness; many involve well-known religions, philosophies, practices, and methods that have been taught for centuries or millennia. New and effective paths are also being discovered and developed right now. This book is only a description of one path: my own personal and highly successful path. Because all eventually lead to the same destination: deeper truth, an open and skilled traveler can follow any path that appears before them. All paths eventually lead home and all types of experience serve our collective evolution.

Like most, I was trained for life through the culture into which I was born. Born into the American middle-class, this meant that I was constantly instilled with the idea that I was alone in a competitive world, and nothing was more important than being strong while firmly "establishing" myself and making my "mark" in our existing system. My father taught me that I must carve a path built upon "personal achievement" by striving to be the "best" at everything that I attempted. I was encouraged to "win" at life by "working hard" and to be "first" at everything by "beating" all the "others," and if I wasn't successful, it was only because I wasn't trying hard enough. Like many raised in the United States, there was little focus on, or opportunity for, self-discovery and the unearthing of my special gifts or joys. Instead of learning about myself by exploring different ways of *being*, I was constantly taught to fear the dire consequences of "not doing it right." This cultural "education" rendered me and my peers severely frightened and resistant in so many ways: some were recognizable, but most remained deeply hidden. As I grew up stubborn, narrow-minded, and emotionally closed, I also held a deep dissatisfaction with myself and the general human condition. Along with these emotional issues, and not coincidentally, I also suffered from a constant stream of allergies and illnesses.

Even as a pre-teen, I ached to better understand a world that seemed chaotic, illogical, and far too often cruel. When I was only thirteen, my first self-chosen research project was titled "The Secret of Life." In a multimedia presentation, built from interviews with local scientists and religious leaders, I explored and compared their differing views on the purpose and meaning of life: my first formal exploration of a topic that would dominate the rest of my life. In the late 1960s, along with hundreds of thousands of others facing the military draft and the Vietnam war, I was motivated to examine and ultimately resist our endless cycles of wars that only seemed to be driven by profit, power, and greed. I was deeply disturbed by not being able to comprehend how and why humans continued to create such profound self-suffering, again and again and again.

As an undergraduate physics major, what I was learning in my studies further shook my inherited world view; this strange new physics only confused and compounded my growing list of questions about our lives and culture. I could not know back then that my professors' inabilities to answer my many questions about meaning were because they either could not accept the deepest implications of *relativity* and *quantum physics* or they were simply not able to muster the professional or individual courage to speak openly about, what was, then, a completely radical and academic career-damaging idea.

After teaching high school physics for a short *time*, and then working as an environmental researcher – including a profound "real world" education as the official environmental advisor to the government of Fiji – I began to better understand the influences of power, money, fear, and other aspects that shape the complex human motivations driving many of our culture's most self-destructive decisions. It was also around that period of my life that an extended series of impossible-to-explain personal experiences began to rattle my inherited world view.

By then, I clearly understood that what I had been taught wasn't the only truth or even the most workable truth, so I sought to expand my search for ways to improve my understanding of the human condition. I dove headfirst into deeper studies of philosophy, spirituality, religion, yoga, *psychedelics*, and meditation. Later, while raising my children and practicing architecture professionally, I actively continued this line of inquiry by reading everything that I could find and participating in the nearly endless stream of workshops and events focused on alternative perspectives, new-age philosophies, natural healing, interpersonal energetics, new religions, and spiritual practices that were pouring through Austin Texas. My overriding desire was always to learn as much as possible about people, relationships, technologies, health, and different philosophies. While I was exposed to a lot of wonderful *information* and methods, integrating these often disconnected and random-seeming discoveries into my everyday life still seemed impossible. *At that time it would have been completely impossible to imagine that one day I would feel not just content, but tremendous joy about the wonderful perfection of our crazy and wildly divergent mixed-bag world full of competing opinions, ideas, and actions.* As I aged and learned more, many ideas started to congeal and deepen, yet confusion, lack of integration, and an inability to communicate my thoughts remained.

Immediately after successfully reversing a serious health crisis at sixty years of age, I suddenly found myself compelled to record my journey for the benefit of my two young daughters. Only through this process of daily journaling did my vision began to untangle, clear, and unify. The constant reflection activated by writing, revisiting, and then rewriting, allowed my jumbled, but continuously evolving, vision to gradually take a more consistent and communicable form within a safe, dependable, yet fluid and *infinitely* malleable, repository: a digital word-processing file. Because so much of what I have learned might also be valuable to others traveling similar paths, I have taken the substantial extra effort that is required to publish my work. In truth, I am being quite selfish because part of my desire to share this work with others is also to find and connect with more playmates who are actively participating in this wonderful new expanded vision of freedom and joy.

While my journaling began as a letter to my daughters, my primary reason for continuing to write about my journey into this magnificent place of *being* has become to deepen and clarify my understanding. For this, the process has worked beautifully. My path has been shaped by the merging of three personal and powerful interests that were dominant in my life: my scientific curiosity and formal training in physics, my fascination with yoga, meditation and spirituality, and my strong native desire to directly experience as much of life as possible. While the ideas presented in my books summarize only one individual's viewpoint, they do describe the process of a particularly committed explorer: one who has actively devoted his entire life to this particular subject.

Finally, today, as I approach seventy, and after many breakthroughs – some very painful, some completely joyful – I have found and integrated surprisingly self-satisfying and complete-feeling answers to almost all of my early questions. Now seeped in a deep and sustainable peace, I find myself energized and enjoying life far more than ever before. *I now clearly see and fully understand that this entire expression of life, just as it appears to each and every one of us, is a perfect reflection that allows us to experience and grow to "know" the true extent of our being. Despite the loud, fearful, and often destabilizing outward expression of our modern civilization, we are also developing a completely new understanding of our universe and the human condition; one that presents an extremely positive, promising, exciting, and uplifting view of our purpose and our collective future*

MY TRAUMATIC PAST

"The wound is the place where the light enters you."
Rumi

I was twenty years old and in school studying *quantum physics* when I received my first clues about the buried trauma and hidden darkness within my own life: a completely forgotten period of my childhood that involved long-term sexual abuse under the care of a trusted neighbor. *Those two years of abuse, which were completely hidden from my conscious memory, would eventually be addressed and come to be understood as one of my greatest blessings. This trauma was to become the primary entry point for my lifelong healing (or whole-ing) journey.*

By then, I had already recognized that my childhood memories contained enormous unexplained voids, but it still took me another forty years to fully discover, embrace and reclaim many of those wounded, isolated, and temporarily lost parts of my spirit. Without this trauma demanding my full attention so early in life, it would have been much more difficult to maintain the focus that helped me reach my current clear and deep level of understanding. This experience became, and remains, one of my greatest teachers and I am still discovering new and often surprising truths as its illumination continues to reveal the secrets of my shadow world.

BUILT ON PERSONAL EXPERIENCE

My perspective was not built upon any existing religion or spiritual philosophy, even though as it developed it began to parallel or include ideas that are integral parts of many different religions and spiritual traditions. Everything that has contributed to forming this vision initially arrived through my own direct experience of life; my interest and study of science and philosophy was a natural part of that experience.

As a young man, I immersed myself in science and quickly became a devout proponent of materialism, convinced that every aspect of life would be eventually understood and described with our remarkable scientific tools if given enough focus, *time*, and effort. However, my personal life experiences gradually eroded this deterministic view of our scientific edifice as I found myself lacking the ability to understand or explain many of these unusual, but very real, personal experiences. A few of the most important stories are related in this book, but my earlier works describe many more of these "out of the box" experiences. Largely because of these experiences, over *time*, I shifted from a completely materialistic view of life to a more metaphysical view of a universe that is infused with deep mystery, unknowable to the human mind but not to the human heart. *I now recognize that our physical world and our life dramas are only one visible aspect of what is a much deeper, broader, and often misunderstood energetic existence.*

Nothing that I am writing about is just information from others that I am simply repeating or rewording; it all springs from my own direct experience. However, the discoveries and work of others and our ancient and contemporary spiritual traditions have certainly provided many new entry points, insights, correlations, and reinforcement for my journey and growing awareness.

SCIENCE, SPIRITUALITY, AND EXPERIENCE

While my scientific bent helped to initialize, shape, and crystalize this vision, it does not belong to just the science. Instead, it can be found and recognized, again and again, in the fertile fields where our science, personal experience, and spiritual traditions intersect. Many of our oldest spiritual traditions directly parallel the surprising revelations of our newest physics. Quotes from *Advaita Vedanta*, an ancient sub-sect of *Hinduism*, often read as if a *quantum physicist* might be writing the words. The earliest *Gnostic Christian* writings – those that were scribed closest to the *time* of Christ and before the agenda-driven politics of the Church re-shaped the story – reveal that Christ spoke with a clear awareness of the "oneness" principle as he described the illusionary, dream-like nature of our physical lives. The mystical Christian literature throughout history makes references to *miracles* and transformation induced by direct and deep personal relationships with the energetic Christ spirit. The poetry of *Sufi Muslims* such as *Rumi* and *Hafez,* along with the words and writings of many of the *Buddhists* and the *Hindu* faiths, also speak directly to this same vision. *A Course in Miracles*, the contemporary Christian revelation, clearly parallels this revolutionary new view of our existence. Even though their descriptions and language were tied to their culture and times, visionaries throughout history were still able to receive, embrace, and share a similar expanded vision for our universe. *One has to stop and notice the universal and timeless nature of this ancient but repeating vision.*

WRITING ABOUT THE INDESCRIBABLE

As my vision unfolds, I need to make one thing very clear. Words can never describe the actual machinery or operation of our universe because, deeper within, it operates at levels that are completely incomprehensible to the human mind. *Instead, what I am sharing is that if you picture the universe working like I am describing and then live your life accordingly, your entire life will be transformed for the better in completely unexpected, and even miraculous, ways.*

I am not able to use words to describe exactly how the universe and our lives work at the deepest levels; no one can. The most fundamental *truths* are formed within realms that lie far beyond human thought, logic, words, and concepts; they unfold outside of our awareness, far outside the current limits of the brain's processing ability. There exists no such thing as a *true* thought; thinking will never be able to describe *truth*. However, we all still have a full connection to these depths, for our entire *being* is in constant and direct communication with all parts of existence, including its deep unknowable source. *While we are not able to understand these depths through thinking, all of us still have direct access through our hearts. However, for this communication to be understandable and make "sense," our hearts must first be open, clear, free, and whole. If we have not learned how to locate and clear our resistances, then the heart's message is always filtered through our ego's unexamined agenda and with this filtering, a crazy kind of indecipherable confusion emerges. Whole-ing is my name for the process of preparing our hearts to receive this universal communication with clarity.*

Because this model for our universe works so much better for me than anything imagined before, I have become fully committed and now devote my life to the full exploration of this wondrous path. As mentioned, my first exploration of this subject began as a thirteen-year-old with a presentation to my social studies class, yet my first full book on the topic, *The Architecture of Freedom* only began when I turned sixty. At its inception, it was just intended to be a short letter to my daughters, one that could be read later when it would have more meaning. Since publishing that first book, scientists have made many discoveries, but thus far they all only served to reinforce the plausibility of my often wild and strange-sounding conclusions.

Today, we are witnessing a slow but steady shift in our awareness, one that is quickly becoming apparent everywhere. More and more, people from all walks of life are exploring, writing about, speaking of, and demonstrating their experience and integration of these ideas, much of it freshly approached from unique individual perspectives. Poets, novelists, musicians, painters, scientists, researchers, theologians, bloggers, doctors, psychics, archaeologists, and others are busy exploring and explaining this and similar visions from many interesting perspectives. However, it is not just these artists, harbingers, and educators, but so many more everyday people, especially the less fearful youth and the more experienced elders, that are now living wonderfully expansive lives, fully influenced and guided by these ancient principles. What was once only the realm of a few philosophers, poets and mystics has gradually emerged to become common, and even in some circles, very normal. We learn from our culture, so by now, several generations have some level of exposure to these once radical ideas.

While all of my books build upon this same vision, with each book I attempt to simplify the presentation and expand upon the ideas and range. With each publication, I strive to better communicate the message, amplify different aspects and implications, and reach out to connect with other co-explorers around the planet as I simultaneously enhance my integration.

My earlier books described the multidimensional structure of our universe, how it works for our benefit, and what this means for living our lives. *In this, my third book of the series, I focus on the personal journey towards our self-healing, or whole-ing.* While this book is designed to be self-contained and present ideas in a logical order, the reader should also feel free to jump around the book, as inspired in each moment. Such free-form exploration is a very useful practice for living in the *flow* of the "now." *However, if the reader has not read either of my earlier books, a quick detour to the science section in the appendix of this volume is highly recommended.*

Because the workings of our universe mostly lie outside the realm of reason, language, and our conceptual thinking, they can't be fully explained using language alone; direct experience is always required. A full life of witnessing, understanding, honoring, and expressing our uniqueness is the only way to uncover the deeper truth of "who we really are."

BEYOND THE VEIL

Our newest scientific understandings have already changed and expanded our collective vision of the universe. As we unravel the bigger picture and discover deeper truths, there is a rapidly-growing recognition that we and everything in existence are all so much more interconnected than we ever could have imagined. While our invisible ties to each other and all things throughout *space* and *time* are profound, now they are also becoming so clear that some *quantum* scientists are explaining their experimental results by musing, "it is as if all these separate-appearing things are actually one thing."

Who are we, really, and what is life actually about? In our familiar three dimensions, we appear as separate entities, growing, evolving, and transforming in every moment of marching *time*, while outside or beyond our spatial and temporal "box," we seem to appear as "one thing." What is very clear is that our dominant cultural-shaping concept about ourselves is still entirely built upon a 400-year-old Newtonian determinism that we know is riddled with mistaken understandings, approximations, and assumptions. This old mechanistic world view, which once promised to unlock and reveal the secrets of our universe, has been showing its age by having been repeatedly found to be incomplete, limited, and sometimes even wrong. It continues to do a good job when we are trying to describe three-dimensional interactions and motion for objects and things in our approximate size range, but it quickly falls short when trying to describe things that are much bigger, much smaller or much faster than that which humans can normally see or sense. In addition, we have discovered that our old concepts also lack the breadth or depth to be able to describe that great part of existence that we now know resides outside of our three-dimensional realm – a new emergent landscape for scientific exploration that completely dwarfs our known cosmos.

All concepts, scientific or otherwise – even these new ones – will fall impossibly short in this regard because they always must be built upon our three-dimensional reasoning. Boxed-in by rational reasoning, our concepts will always lack the flexibility to reach beyond. The greater *truth* of "who we are" can't accurately be described through any human concepts, for this level of *truth*, by definition, lies beyond all words and language, as well as human ideas. Learning how to *flow* within this unfamiliar but *infinitely* greater landscape requires that we no longer be tethered and restricted by our old concepts or anchors. ***Working with, instead of against, this deeper truth requires that we become comfortable and functional in psychic spaces***

that can be experienced and felt, while not able to be described or understood.

Despite this, our words and concepts can be useful for guiding us within this journey. They can still be used to describe and inform us about "what we are not." They can guide us by illustrating the holes, breaks, and gaps in our logic and our automatic default responses. We can use our words and concepts to guide us to the trail's end of our logical reasoning and, once there, point us in the general direction of what lies beyond.

This is precisely what our most sensational and adventurous science of the last 100 years has done. This newest science has finally caught up to the ancient spiritual traditions that instructed us to look beyond our logic, towards the incomprehensible mystery. One of the key discoveries of the last 100 years (since Einstein's *relativity*) is that *time* is not at all what we typically imagine; it does not march in evenly marked steps, relentlessly pushing us forward. Instead, it is now understood to be part of a completely malleable and changeable medium where all kinds of illogical ordering and relationships are possible depending on a particular *observer's* perspective. We now understand that many of our concepts and ideas that have been built around our old understanding of *time* only create illusions, artifacts, and artificial limits for our thoughts and actions.

Even though our lives often appear to be chaotic, every person on this planet is performing their role perfectly; nothing ever needs to change. Simply by living and experiencing our world, we are fully participating in life, and this participation in all of its forms is exactly what our lives are for. Despite all of our judgments and personal agendas, our lives are perfect instruments for our growth and evolution, just as they now appear. Whether or not we like it, all of our combined interactions perfectly reflect and express our *being*. Recognizing the life that we witness as the perfect window into the truth of "who we really are" is the next critical step for our evolution. Every person has an equally important role to play in this grand manifestation of our hidden inner life – a "show" that is being *projected* entirely for our benefit. **Eventually, we will each reach the personal threshold where we completely understand that there is only one of us, one experiencer who has been temporarily gifted with this extraordinary opportunity to explore life from as many different perspectives as there are separate living beings.**

WHOLE-ING JUST MAKES LIFE BETTER

Each of us is continuously engaged in the process of whole-ing, whether or not we are aware of it. While whole-ing will unfold naturally and organically by just living our lives, with added awareness we can facilitate this natural opening through a more conscious release of resistance. Using its techniques, we learn how to release our suffering and grow more joyful as we also become more enthusiastic and positive about our day-to-day lives. While we must still "live" our lives to gain the necessary direct experience, with this added awareness, the entire process of life becomes much more enjoyable and far less stressful.

Life on our Earth has always produced an unplanned and unpredictable mixing of cultures and human activity. All of our interactions together have produced this stunning physical expression of human possibilities and activities, but also an expression that often seems confusing or overwhelming. Fortunately, as our collective culture continues to evolve, our awareness also continues to expand, guide us, and illuminate life's amazing unlimited potential. Despite the outward appearance of chaos and danger, our lives and the new emerging paradigm bring only good news. *There is nothing to fear, for all is well; in fact, the deepest truth is far better: the world already is, and always will be, perfect. Exactly as it appears, it has always been and will continue to be, the perfect instrument for the expansion of our collective consciousness. Nothing is wrong with our lives and nothing ever needs to be fixed. Physical life exists expressly for creating, expressing, and experiencing every possible form and activity imaginable.*

THE LATEST QUANTUM THEORIES

A significant amount of recent work in *quantum physics* has involved re-examining the earliest and most established explanations, often referred to as *quantum foundations*, a topic to be discussed later. The inability of contemporary physicists to unify their two most significant theories – *quantum physics* and *relativity* – is the primary theoretical issue that has been confounding physicists for more than 100 years. In the "science" section of the appendix and my earlier books, I discuss this very unnatural and troubling separation of these two spectacularly successful theories. I believe that this problem exists because scientists have been trying to solve a new type of problem while still relying on an old type of thinking, i.e., using a logic that is limited to three dimensions.

The politics within the academic community have, only recently, shifted to allow a growing number of *quantum physicists* to speak more openly and honestly about other interpretations or meanings of their experimental results. For almost three-quarters of a century after the initial discovery of *quantum physics,* a very powerful and well-entrenched academic culture lead by Niels Bohr enforced near-silence and peer banishment for any researchers suggesting alternative meanings for the difficult to reconcile experimental results that consistently emerged from this strange realm of the *quantum.* Because most physicists were afraid (largely for professional reasons) to speak up, generations of physics students, including myself, were taught to ignore "meaning" with the clear directive to just "shut up and calculate." Those that openly resisted did so at great peril, knowingly cementing difficult careers and lives for themselves. Because what was being revealed in the lab defied all rational explanation, and the leading "experts" felt a need to protect their positions and reputations, a code of silence and conformity was strictly enforced. Over the years, several fascinating ideas and interpretations were proposed, but in every case the dominant scientific establishment quickly, and often harshly, marginalized these theories along with the careers of their messengers. Until very recently, new ideas were not safely discussed except within the confines of science fiction or at the frequent alcohol-fueled after-hour discussions between physicists. Fortunately, this newest generation arrives at a *time* of much less resistance, a *time* when many of these ideas have been fermenting in our collective awareness since the beginning of the last century. Through all of this discovery there has been a consistent pattern: the more we learn about this strange new physics, the more likely it seems that one or more of these strange, paradigm-changing ideas might be the best explanation.

Today, in increasingly large numbers, professional physicists and cosmologists are finally speaking out. Noting the large body of enduring evidence that has been carefully documented, they now feel safe to discuss the emerging yet radical meanings of these repetitive but very strange results from over one hundred years of *quantum research*: ***"Our universe is not built of physical 'things' – at its depths, there are no particles, no atoms, no houses, no planets, and no people; instead, in these depths, there are only layers of energetic 'fields' that extend infinitely, and when these 'fields' are excited or 'disturbed,' only then do they produce the particles and everything that appears to us as our physical world."*** In other words, the world that we experience as solid and real is only a temporary *image* or *projection* created by these invisible *energetic fields*. ***Our lives, while appearing very solid and real from our perspective, are only part of a***

secondary image produced by something called a field, that originates and functions in an unseen, yet more fundamental, reality.

Fields, defined as "a physical quantity represented by a number that has been assigned a specific and dynamic value for every point in *spacetime,*" are gradually becoming better understood. The first mathematical description of a *field* was Newton's *gravitational field.* Two hundred years later, Maxwell laid the groundwork for Einstein's most famous work with his description of the *electromagnetic field,* a combined form of a *magnetic field* and an *electrical field.* The "real" electricity we take for granted today is created in generators relying upon these invisible *magnetic fields. Electromagnetic fields* allow the transmission of *visible light, radio waves, x-rays,* and other related forms of *energy* through the vacuum of *space.* Einstein revealed that *gravity* is the result of deformation within a *field* that he named *spacetime,* supplanting Newton's earlier idea of a *gravitational field.*

Today a growing number of *quantum* physicists are proposing that everything witnessed and experienced as our physical universe, including each of us, is only the result of the stimulation of the most fundamental *field* of all, the *quantum field.* **When there is some small shift in this quantum field, a different image is produced, meaning the entire physical manifestation of our universe can quickly and instantly change. So what can produce such a shift in the quantum field? It is gradually becoming clear that, somehow, our intent and participation seem to be important factors.** It seems that we each may be much more involved with the way our universe appears than we could have ever previously imagined. While this explanation may sound like extreme science fiction, more and more this strange new vision appears to be the most accurate description of our reality. It is now seeming likely that, without even realizing it, humans have always possessed this god-like power!

LIFE IS SIMILAR TO A VR EXPERIENCE

Today it has become clear to many that the solid-feeling world of our experience is not the most primary reality. What we directly experience is only the result of a complex sensory illusion, and beneath its appearance lies a deeper "reality" that is far different from any *image* of a creator that we might construct in our minds. This incredible world that we experience, while sensory-rich, real-feeling, and solid appearing, is only a secondary *image* shaped by an energetic disturbance (a signal) that has been communicated at some invisible level of existence.

Instead of a solid, physical, and enduring collection of particles and objects as long-imagined, our universe is, instead, extremely interactive, responsive, malleable, and instantly fluid, not unlike a fantastic, solid-feeling, and tactile virtual-reality experience. If we begin to think of life and our universe as a virtual reality experience, we are moving closer to a clearer understanding of the way our universe works. As Shakespeare long ago reminded us; *"All the world's a stage, and all the men and women merely players."*

I am not saying that our universe *is* a *virtual reality* experience, but I am proposing that the VR model is a very useful tool to help us better understand how existence works. The advent of VR provides a revealing contemporary metaphor that can help us understand our personal experience of the world, and how, at the same *time*, our experience can be completely fluid, changeable, and instantly responsive. With *virtual reality*, we don't need to dig the entire mountain up, scoop by scoop, to move it; instead, we need to change only one *parameter* in its description: its location. Surprisingly, we have been discovering that our "real" lives may work in much the same way.

We are also discovering that not only are we the participants in this ultimate virtual reality experience, but because of the interactive role of the observer, we also seem to function as the programmers. No "outsiders" need to be involved with our programming; we, through our involvement, become the directors of all of life's dramas. *We also are discovering that not only must we participate for there to even be an "experience," but the nature of our participation seems to shape the results.* With this new and powerful knowledge, many will only require a small additional shift of awareness to understand how we can create our own *magical* experience within the very flexible and always-changing *parameters* of this virtual experience we call "real" life.

THINKING IS NOT UNDERSTANDING

The primary objective of this VR-like experience called life is to play, program, and fully explore its *infinite* possibilities so that our awareness can have further experience and grow. Exploring this *space*, we quickly discover that we can change little or nothing with our thinking minds alone. Initially, we may feel disempowered as we discover that no amount of positive or directed thinking or action substantially changes the fundamental nature of life on Earth. Our inability is because the *quantum field* can't be stimulated or changed directly by thought or external action; instead, this *field* is what produces our thoughts and actions. Our programming is set at a deeper level, so changing it must involve something more than just our thinking minds. While our learning and thinking can help prepare to open the mind, it is only the *resonant* change of our *vibrational* core that has the power to change the *quantum field* and, therefore, our lives. Our *core vibrational resonance (CVR)* evolves through direct experience, not our thinking. As many have discovered, our body, thoughts, and egos are not who we are; they too are, only temporary *images* representing one aspect of *being*, in one divided moment of *time*. Again referencing *virtual reality*, the body, brain, and ego together can be compared to a *VR avatar*.

Through the wildly varied experiences of life, we gradually discover that we can't directly change the real nature of the external world with our minds. The basic human condition has remained the same despite many-thousand years of our focused thinking. However, at some point, we begin to discover that through our thoughts and actions, we do have the ability to dramatically change our response or reaction to any event, and this new level of awareness marks an extremely powerful breakthrough. Once understood and integrated, this powerful yet simple idea of changing our attitude will open us to many new types of experiences. Eventually, we discover that through this single change we have completely tweaked our "programming." Attitude adjusts or "tunes" a deeper part of ourselves, a part that is usually unseen and rarely understood: our *core vibrational resonance (CVR)*. This is the normally-hidden, root-level aspect common to each of us, long recognized but rarely seen, that has sometimes been called our *soul*. Our *soul,* or deepest personal essence, communicates and interacts with the entire universe through a form of *inter-communication* that exists and operates at a level far beyond human *language* or *logic*.

Within each of our journeys, we will eventually learn how to let go of our fears and concepts about separation to develop a more direct and fluid connection to this deeper level of *being*. From this different *space* of *being*, everything changes as we happily enter into a more direct experience with the deeper *oneness* of existence. *Intra-communication* is probably a much more accurate description than *inter-communication* because, as some modern physicists are now reflecting, "It is as if there is really only one thing in the entire universe." **Our programmed life is created in the unseen parts of our being, where our thinking only gets in the way.**

THE MYSTERY ONLY BECOMES GREATER

Today, as scientists eagerly explore the deeper nature of our universe, much of what is being revealed is surprising, unexpected, and simply does not make logical sense. This is because existence is always operating at levels and in ways that are mostly invisible to us. *As our incredible science of the last century has flashed its inquisitive light upon "that which we do not know," we have unexpectedly discovered that the "unknown" parts have not become smaller, as one might logically expect. Instead, the more we learn, the more we discover new, unknown, and mysterious aspects that make our universe even more extensive and interesting than previously imagined.* Originating from parts of existence that lie beyond our *conceptual horizon*, the amount of unknown and unexplainable mystery only grows with every new exploration and added piece of *information*. Instead of providing definition and clarification, what is being revealed is raising even more questions, revealing previously unseen but *infinitely* vast new regions of "unknown" yet to be explored. As this immense and growing mystery informs our expanding awareness, it becomes clear that these unknown aspects are *infinitely* vast and harbor more than enough "room" for anything, or every possibility, including many things that we once considered improbable, or even impossible.

For many, including myself, the fact that this vast "unknown" continues to grow and does not diminish as we dive deeper and learn more makes a very strong case for our need to accept and recognize an entirely new paradigm. *Our universe is far different from the physical and "believed to be knowable" expanse that we once imagined. Instead, it is built upon hidden levels, or dimensions of existence, that may always remain invisible to our rational minds, senses, and instruments.*

This new vision for our universe runs completely counter to our "common sense," which is entirely dependent upon our senses, our rational intellect, and our deeply ingrained 500-year-old material view of the world. This new vision requires a change of perspective that defies our logic, our sense of structure, and our long-held understandings about our physical existence. It is a vision that lies so far outside of the bounds of our "normal" conceptual thinking that traditional rational or logical "proofs" shouldn't even be anticipated. This new vision presents a universe birthed in a realm where logic and proofs are meaningless.

THE "PROOF" FOR THIS VISION

Because of these deeper origins, if any reader wonders if this vision is the direct result of the scientific method, the answer must be "no," because, as of now, there are no direct scientific proofs for this vision and, at least for the near future, a comprehensive proof is extremely improbable. (However, there is very solid scientific evidence and even proofs for some of its parts.)

When I studied *quantum physics* as an undergraduate physics major in the 1960s, I became frustrated by having to spend eight or more hours a night manipulating equations with strange or unexamined meanings; equations that were incomprehensible to everyone, even those who developed them. I could not understand, then, that the entire physics community was just as stunned and paralyzed as I was by the potential implications of their discoveries. To feel more engaged, I needed some understanding of what these strange results might mean; I longed to draw some "picture" in my mind that made enough sense to inspire interest for these days of tedious calculations, but a clear "picture" never appeared during my formal studies. After graduating and teaching basic high school physics for two years and then shifting to environmental science, I eventually chose to shift careers and do graduate work in architecture; yet I continued my study of *quantum meaning* for the next five decades until I finally was able to construct that useful "picture" in my mind. My work is far from complete and I still have new questions and ongoing explorations, but the vision that I have assembled comfortably provides me with clear and self-satisfying answers for most, if not all, of my original life questions.

Let me be clear about a "proof" for this vision. I cannot scientifically, or even rhetorically, prove that our universe works exactly as I am describing because no human being can see, understand, or explain this far into the great mystery. Our human brains and sensory systems work well with certain types of input but they have their clear limits; our available rational toolkit simply can't reach this far into these unknown depths. We are bound by the constraints of our three-dimensional realm, which has been built upon physical things, time, space, rational logic, and language. Our greatest *truths* reside beyond these limits, far outside the range of human words or thought. Because this level of *truth* requires communication beyond words, artists are often the ones that can best guide us towards this hidden type of knowing.

The outer limit of our mind's ability to comprehend is defined as our *conceptual horizon*. However, as we expand our vision of our universe, the limits of our *conceptual horizon* will naturally adjust to match our evolving awareness. As we continue to grow through expanding our awareness, our *conceptual horizon* naturally moves out to include more of this "unknown." Later, as our awareness grows, things that seem illogical to our minds today will begin to make perfect "sense." We have already done this many times in human history, the most recent leap of this magnitude occurred five hundred years ago when we collectively moved beyond our flat-Earth limitations.

LIVING OUR LIVES FROM WITHIN THIS VISION

Because we all wake up each day living physical lives, we probably want to make the most of what is being offered right here and right now and, fortunately, this can be facilitated by a better understanding of our universe. While there are no clear absolutes in this great and growing mystery, I am convinced that from our human perspective, the architecture of our universe – the architecture of freedom discussed in all of my books – does a much better job of describing how our universe is constructed and functions than our current but now obsolete model which was built upon the physics of Newton and has been dominating and driving our culture for the last five hundred years. For me, a significant part of the "proof" lies in the dramatic changes that I have witnessed in my own life and the lives of many others who are living by the guidelines that naturally *flow* from this vision. While a scientific and logical "proof" may not be possible, experiential validation is being witnessed everywhere, in both laboratories and individual's lives.

As a young man, I lived life in a constant state of loneliness, agitation, and worry. Highly prone to illness, I was frequently sick, often unhappy, confused, and disturbed by almost everything that I saw and learned about the "outside" world. I even thought this low level of health and happiness was "normal." Today, almost seventy, free of all medication and all physical complaints, I am filled with a permanent sense of joy and a love of life that can't be altered by external events. The difference in my life and outlook is completely due to the integration of this new awareness and, happily, I am far from alone in this discovery. When this vision is recognized and fully embraced, lives quickly change; as the old saying goes, "the proof is in the pudding."

A young friend of mine grew up in an open, creative, and extended family that lived fearless lives embracing large parts of this vision. Even though he has only been treading the Earth for a relatively short *time*, he already has produced many extraordinary chapters in his life. A few years ago he remade himself, again, as the prodigious, talented, and very popular musician, "Shakey Graves." Previously he had been a very successful TV and film actor, and clearly, he will continue to expand and shape new *waves* of expression throughout his entire life. He once said to me, "We are just making all this up anyway, so we might as well create something fun and interesting." He is clear, fearless, compassionate, kind, and immensely powerful, and his large audiences experience a directness of open communication that makes his performance feel much like a collective spiritual encounter. He is a perfect example of how our deep core beliefs and attitudes create and shape our lives. When we learn how to get out of our own way, the unexpected naturally *flows* and the "impossible" happens.

Another powerful personal validation is that I have yet to discover any event, experience, or aspect of my life that can't be easily and organically "understood" or "explained" by evoking this radically different vision for our universe. For many of us, the older paradigm that my generation inherited explained very little about our common, but rarely discussed, "out-of-the-box" experiences such as déjà vu, out-of-body experiences, near-death experiences, precognition or impossible synchronicities. Our older vision never felt satisfactory or complete. It could not because it was built upon human perception and logic, which often leads to the paradoxical dark alleys and dead-ends that leave many of us disoriented and even more confused. This new vision is different; it broadcasts its bright and revealing light everywhere, equally, and for all to see.

To date, everything that I know, have experienced, have been shown, and have observed throughout my life rests easily, naturally, and comfortably within this new model. It contains more than enough "room" for the wide variety of life experiences that I have been seeking, exploring, and engaging with for almost three-quarters of a century. My entire experience of life is fully resonant with this new world-view, and everything about life makes much more "sense" when witnessed through a heart that has opened to this new vision for our universe. What I have discovered through this long exploration has so satisfied my naturally inquisitive mind that I can now speak about it with a clear, bold, and powerful assuredness.

LIFE IS FOR EXPERIENCE

"Experience life in all possible ways – good-bad, bitter-sweet, dark-light, summer-winter. Experience all the dualities. Don't be afraid of experience, because the more experience you have, the more mature you become."
Osho

Even though I must use words to write about my experiences, the deeper answers to life's more persevering questions such as our destination, purpose, or the reason for our *being*, can't be fully realized through the study of words or ideas alone. Such answers only emerge from a deeper kind of "inner knowing," a noetic awareness (*gnosis*) that can only be gained through direct personal experience. The more real, honest, and "true-to-self" an experience is, the deeper and more complete our eventual "knowing."

Because of the way our universe is built, words, by their very nature, will (and must) fall far short. Our universe's architecture allows for *infinite* possibilities, a condition that is required for true freedom, and these possibilities extend far beyond the capacity of our brains, logic, and words. Thinking will never bridge the gap. *At best, concepts and language of any kind can only guide or point us in the general direction of this deeper truth.* This is because *Truth* (with a capital T) originates in a dimensional *space* that sits far beyond the three-dimensional realm that shapes and holds our concepts and words. As the engaged reader will soon discover, our universe is multidimensional, and ultimately non-physical (as we understand this word today), with its critical interacting parts occupying a type of *space* that exists beyond things, words, concepts, and rational proofs.

Words and concepts can only be used to describe things that fall within the bounds of our conceptual abilities. The range of human thought always has an outer limit known as our "conceptual horizon." Because of their inherent limitations, words are inadequate for describing that which lies beyond; language and conceptual thinking become meaningless within these deeper realms. *Words, however, can be useful for describing "that which is not true," and this type of knowledge can help guide us to the point of departure. However, once there, we still must discover this deeper kind of "knowing" through personal experience.*

It is only through our direct experience, often appearing to be extremely difficult and challenging, that we can explore and unravel these hidden depths. This also means that, because of the inherent limitations of our spoken or written words, many of my insights, breakthroughs, and understandings have not been successfully verbalized or recorded. These non-conceptual realizations were uncovered within the depths of my process, and I have yet to find the right approximating or guiding words to describe their contributions to my evolving awareness – they remain ineffable, yet they fully inform and reinforce my deeper "knowing."

IS UNDERSTANDING EVEN NECESSARY?

Ultimately, as many have said, there is no need to ever answer any of these ancient philosophical questions. Knowing the answers to these questions is not life's purpose. Living our lives to their fullest by fearlessly exploring, opening, and expressing our unique individual selves is what most directly supports life's primary purpose: the growth and evolution of our collective being.

While the ideas examined in this book are not critical for living full and meaningful lives, for some they may help facilitate the relaxation of our reflexive, semi-automatic, and largely self-imposed limitations. Internal relaxation helps with heart-opening and that leads to an expanded and freer expression of our full potential, bringing forth more of our important and unique contributions to this remarkable process that we call life. *Each one of us brings a critical piece of the whole, and when we each learn about and embrace our spirits, magic happens: seemingly impossible events and connections begin to unfold fluidly and naturally.*

At some point in this journey, we will each release our dependence on words, concepts, and our need to understand; only then can we fully recognize the awe-inspiring possibilities and freedom that have always been available – these are the great gifts of our being. What we are learning is how to naturally flow within the powerful river that shapes life. Our thinking, while critically important for many other functions, often is what most interferes with our natural ability to flow.

TWO TYPES OF PEOPLE

At a most fundamental level, there are only two polar perspectives that shape our ongoing planetary "tug-of-war" and, because of nature's balancing act, it is likely that they have always been represented in almost equal numbers. *Some believe that we are all separate and competing for dominance and survival, and there are others who believe that we are all fully connected and inter-dependent.* The first group is actively exploring our separation, and the second is actively exploring our connection. Neither half is right or wrong, and both approaches are necessary for shaping and maintaining the structure that appears as our universe: the grand theatre for the play called life. Moving in the direction of more individualization and separation on one side, and towards oneness on the other, these two expressions can also be described as *separation consciousness* and *unity consciousness*. Both are accurate ways of describing our experience in life, and any deeper levels of truth must allow for both to co-exist in a dynamic harmony that not just permits but depends upon this push and pull. Many polarities define our *dualistic* world, but this is one of the two that are most fundamental for shaping our human culture. It explains much of the struggle found within our social systems, political systems, and most other kinds of human interactions. The other major driving human polarity is, of course, our sexual separation.

When examined more closely, we realize that there are two important primary components to this fundamental division: position and momentum. The first variable is just where each individual presently sits on this spectrum regarding their personal beliefs about separation or oneness. However, because life is a process with everything in motion and constantly changing, much more life-shaping is the direction of movement or change. As we each evolve, are we becoming more separated or more interconnected? Is a particular culture evolving to become more isolated and separated, or is it connecting and moving towards more inter-dependence?

In our limited conceptual understanding, it would appear that these two ways of seeing our world are mutually exclusive, but once a deeper understanding of *duality* is fully embodied, all the external conflicts and paradoxes shaped by *dualistic polarities* evaporate to merge into a *flowing oneness*. As we can more fully explore and honor our unique individual nature, we also, simultaneously, discover our *timeless* interconnection through *oneness*. This realization marks an expansion of consciousness where this and all the other apparent paradoxes begin to evaporate as a deeper awareness emerges to unexpectedly make sense within.

ONENESS

Somewhere, deep within, many of us harbor an occasional and possibly even a persistent "knowing" that everyone and everything is much more connected than the divided outward appearance of our lives would indicate. (This is probably the personal experience of many still reading this far into this book.) ***A fundamental truth explored throughout this book is that all of the many expressions of life are actually smaller parts or aspects of "one thing"***

that only appears to us as "many." This apparent division of "oneness" into a multitude can be thought of as a functional mechanism that helps the "oneness" of existence to maximize its "desired" expression of *infinite* physical possibilities and experiences. *The billions of humans and all other living focal centers of consciousness can be understood as the nerve cells of "our collective body," whose job is to sense and collectively experience every part of our extensive environment. Together, mankind serves consciousness by providing eight billion different and unique points of perspective and awareness.*

From our human perspective, it is impossible to fully understand all of the implications of this fundamental *truth* – a *truth* that our latest science, most profound and meaningful experiences, and many ancient and modern spiritual works all agree upon: *"We are all one thing, appearing as many."* The strong human identification with our separate bodies is so ingrained and embodied that it requires a dramatic expansion of everyday consciousness before we can become fully responsive to this hidden level of deep, unifying interconnection. *Because we are always both one and many, to deny either our connectedness or separation, only isolates half of our being; this results in hidden wounds and a profound feeling of incompleteness.* To become whole and healed, we must, instead, build upon this awareness, knowing that we are always, simultaneously, both one and many.

Our journey through life has at least this one primary purpose: individual and collective evolution through the growth and expansion of consciousness. Life is a unique theatre where many different "scenes" become available to be "played" out, granting us the experience that is necessary for a deeper noetic understanding. It is a laboratory where new and different possibilities are first invented and then experimented with so that a deeper form of conscious awareness can emerge to discover and express its full *being*. Through all these explorations and experiments, our playing with separation allows each of us to explore, and eventually open to a full and free expression of our *being*.

The very nature of expanding consciousness is that it will eventually grow to illuminate and include our deep interconnectedness. *Through just the process of everyday living, we eventually discover that our human potential and purpose can only be fulfilled through the practice of inclusion, acceptance, and integration of all aspects of our experience. Life, alone, teaches us that inner growth requires the release of our reflexive and egoic tendency to judge, both "ourselves" and "others."*

Somewhere deep within this amazing journey, we also come to understand that these two – "ourselves" and "others" – are ultimately one and the same.

OPENING

Whole-ing becomes a lifelong process and a lifestyle. It involves learning how to easily moving through endless cycles of allowing, feeling, opening, and releasing, while always nurturing forgiveness and gratitude; eventually, it results in a radical expansion in our understanding of *self*. As we explore *whole-ing*, we learn how to drop definitions, limits, and boundaries –both within and without – as we piece by piece, discover, welcome, and embrace more of that magnificent *being* who we are. **Early in this journey, we encounter the paradox of recognizing both our divided individual selves and our complete interconnectedness. As our awareness grows and we become more whole, all logical difficulties presented by this long-argued paradox easily and completely evaporate.**

The early stages of every individual's "opening" process present numerous challenges so, initially, this process is usually experienced as very difficult, if not impossible, work. Because we are continuously being asked to "feel" into personal shadows that we have unconsciously, but purposefully, been hiding from and avoiding for so long, this particular phase of our transformational work is never easy. The process is also so unique and personal that it is completely natural, at first, to feel alone and isolated, confused, discouraged, and even afraid. **It is quite common and very normal for this stage of the process to appear like a lonely and seemingly endless struggle to reach an often confusing and extremely obscure goal.** Fortunately, we live in a *time* when we can be assured through the work of many others who have followed this path that this profoundly challenging feeling is completely expected and destined to pass. There are no personal or individual maps for exploring this deep inner territory that, previously, was completely unseen. Every new adventure or experience will always last exactly the amount of *time* required for each of us to make the necessary adjustments to prepare for the next step of our journey – not one second more or less. As the healing process unfolds, it can't be rushed; it is timed perfectly for each aspect of our *being*. *Time*, as we will come to understand, is a very flexible and mutable element of our experience and universe.

At some point, the very nature of our opening process begins to shift, especially after we start to recognize and embody another critical and fully-related truth: **"*this physical body is not who we really are.*"** When this radical concept is first introduced, the body/ego/mind will immediately and harshly reject it, because this idea is always seen by *ego* as a very direct threat to its authority. ***Fearing for its very existence, each of our frightened egos engages a mighty struggle to resist this deeper awakening; the ego correctly recognizes that this signals a loss of its power.*** In truth, we do not ever wish to destroy the *ego* because it has a critical and important purpose: the maintenance, operation, and protection of our separated body. In actuality, the real struggle is only about the *ego's* demotion to its proper place in our *being*, that of a loyal servant. As *ego* adjusts to being reassigned, we discover that the once difficult process of life begins to gradually become easier, more rewarding, and more joyful. Guided by this new deeper awareness, everyday life begins to unfold in more interesting, revealing, and often unexpected, directions. What was once extremely subtle or unseen can now rise to reveal new, significant, and completely unexpected meaning.

After this realization, life's ongoing processes (life, death, conflict, crisis, emotions) do not disappear, but the way we respond to them changes radically; our ever-expanding understanding of the bigger picture completely transforms all of our programmed concepts about our lives of separation. ***At this important stage in our process, every experience can be observed with more internal illumination, which now washes and colors what once was only experienced as darkness and shadows in a much more luminescent and positive light.*** Somewhere within this process there also begins the additional awareness that these internal changes are not just fleeting – they are here to stay and will continue to fully infuse every aspect of our lives. Once integrated at this level, our new awareness can never be lost because now it resonates with every part of our expanded *being*.

For many of us, the most difficult and challenging step in the entire whole-ing process is the tremendous hurdle of fully recognizing that all parts of existence are critical and important contributors to the balanced whole. What this awareness looks like "on the ground" is that everything we now judge and push away must be entirely reevaluated. This includes absolutely everything, all that we consider to be good, bad or even evil – every dark turn in our personal and collective history, must now be integrated and brought into the harmony of wholeness.

The "letting go of all judgment" is a common liberation theme that has often been proclaimed by saints and mystics from all corners of our planet. However, this step is even more challenging, profound, and life-shifting than it seems when first contemplated. While this step is essential for our liberation, it is also the most challenging threshold for almost everyone, often more difficult than the reassignment of *ego*. Because constant judgment is our reflexive cultural "go-to" default-mode, for full integration, the "letting go of all judgment" requires a long and dedicated period of focused practice. Seeing all that we have reflexively judged and continue to judge through the illuminating light of love and understanding runs completely counter to all of our old concepts and ideas involving good and evil. *Embracing the blasphemous-sounding idea that everything that we consider to be evil must be welcomed and not pushed away is easily the most challenging step of the entire whole-ing process.* (Note: Possibly surfacing from the collective guilt of my Germanic heritage, the focal crisis of my most difficult and life-changing cultural "let go" was my strong and unambivalent judgments related to Hitler, the holocaust, and the complicity of an entire nation. I described this difficult and extremely challenging process in my earlier books.)

FREEDOM IS AVAILABLE HERE, RIGHT NOW

We live our lives in a framework in which our experiences are always defined by *time*, *space*, and separate physical things; and at the same *time*, because of the last 100 years of science, we now know that *time*, *space,* and physical *particles* are not, as we once thought, fixed, rigid, or absolute. *Not only is our world not rigid, fixed, and definable, but many contemporary quantum physicists are now saying that physical things like atoms and people do not independently exist in our universe. Recently they have been speaking more freely about how their experiments are revealing that there may be no things at all, at least on the kind of continuing basis that we usually associate with material objects. Instead, what exists at a more fundamental level are nested layers of energetic fields that constantly respond to a specific type of signal or disturbance. These excited energy fields then somehow shape and produce, in each and every moment, the very particles that define our physical existence. Their research seems to be indicating that our entire physical universe may be nothing but a temporary "image" that is being projected by these energetic fields. This means that everything that we see, touch, and experience – our entire physical universe – appears to us only as a secondary result arising from the stimulation of these hidden energetic fields. While currently unclear, it is my belief, and the belief of many others, including some of these physicists, that this signal or disturbance may involve consciousness itself.*

It is now universally recognized by all physicists and cosmologists that our universe is radically different from what has been our understanding for the last five hundred years. What does this mean for us? *What might change about our lives if all individuals understood that we are not separate and alone, but instead, we are deeply connected to everyone and everything else in our universe? What would change if we understood that each of us has the powerful ability to alter our personal experience at any moment by just shifting our awareness? What if both birth and death were fully recognized as just two handy markers that denote nothing more than our entry or graduation to new chapters in the continuing story of our eternal being?*

The profound scientific discoveries of the last one hundred years inject a multitude of fresh insights that demand a completely new vision for our universe; one that defines an immensely expanded potential for our lives. Each passing day, our universe continues to reveal more of itself from its treasure chest of unseen life-changing secrets. Even through our first "early morning" and hazy glimpses, this new vision is stunning. Today, with the immense and growing body of experimental scientific evidence, we are beginning a profound reassessment of our long-standing beliefs about life, our purpose, and our understanding of who, or what, we are. As this new understanding begins to inform our lives, the *flow* and quality of our inner lives will then gradually shift towards the miraculous.

While many of our cultural assumptions are based on now-obsolete science and have been shown to be inaccurate or even wrong, they still actively define our thinking. This means that obsolete and even wrong thinking continues to shape all of our daily activities. Our spiritual traditions and many of our more "out-of-the-box" personal experiences have always signaled the presence of connections, interactions, *spaces*, and entire worlds that lie beyond our existing and dominant, but very limited, cultural vision. Our newest science, our personal experiences, and our most important spiritual traditions all mesh to demonstrate that we have influences, powers, and deep intimate connections that extend far beyond those defined by our current belief system. *We live in a universe that we are only beginning to recognize: a completely flowing, interconnected and responsive system – a form that is impossible to comprehend or explain using only our inherited system of rational and time-ordered thought, concepts, and words.* We are quickly discovering that, like the prisoners of Plato's famous cave (discussed fully in Part Five), our limited understanding of reality is based entirely upon our interpretation of the dancing shadows that we see *projected* on the wall of our relatively limited conceptual cave.

Because the scientific exploration of this fundamental *truth* about our existence is in its infancy, the collective integration of this new awareness will, by necessity, be gradual. These *truths* transcend any of our thought and words, so this deepening of our awareness requires direct experience and this, of course, requires *time*. As this new awareness becomes embodied, individual by individual, it changes everything that we once thought we understood about life.

After this new awareness touches each of us, it becomes impossible to think of ourselves as only isolated individuals. As we integrate this new recognition of our "oneness" into our *being*, we begin to more directly experience our deeper connection to all other living things. This shift unfolds slowly, but one day we realize that all these "others" are nothing less than critical aspects of our "oneness," and therefore they are also parts of ourselves, but very misunderstood parts. From this new awareness, harming other creatures or damaging the Earth, solely for the short-term purpose of self-enrichment, will not even arise as a reasonable consideration. With this shift of awareness, we are no longer subject to the reflexive impulses of power, greed, or fear that once pounced from their hiding places, shadowed deep within our "darkness."

For those who are ready, their "reality" will actually appear to shift. Violence, aggression, and intentional harm become obsolete responses as, the focus of effort becomes, instead, exploring the wonderful possibilities *flowing* from our interconnectedness. Behaviors that might harm others or create conflict and division are no longer seen as evil, but simply as "confusion" caused by fear-based misidentification with our separation. Instead of fueling more conflict through rejection and isolation, we will choose to treat these confused "others" just like young, misguided children who need to be embraced, nurtured, loved and healed – children that are to be brought back into loving wholeness. Rather than rejecting them and just perpetuating more human suffering, these confused and isolated individuals are recognized and treated as that which they truly are: lost but important parts of ourselves.

Within each of our journeys, we will also be granted endless opportunities (the exact number and kind of opportunities that we each may require) to embrace this new vision and completely change our way of *being*. We won't "miss the boat," and we can't "mess this up" because *"time* is on our side."

No Need to Wait for "Others"

"We don't experience the world as it is, we experience it as we are."
Anais Nin

What may be the most unexpected and radical aspect of this new awareness is the eventual realization that this loving and peaceful world that we seek already exists. In addition, to experience this magical and loving world ourselves, we need not wait for others to change their ways. Entire worlds exactly like the one I just described already exist (refer to the chapters on "infinity" and "Many Worlds theory") *and we are always free to "enter" and enjoy them at the precise moment that our personal, but completely fluid, "core vibrational resonance" (CVR) allows. Not only does this magical transformation of our world require no others, but this type of profound personal change can only be achieved from a deeper change within oneself.* From the deep perspective of "oneness," no others are ever involved because no one else even exists; there are no "others" to even wait for. *What we discover is our deeper nature where there is no difference between the inner and the outer; in this realm where these are one and the same, if you change yourself you have changed the entire world.* The appearance of an "outer" world is only the direct reflection of what is churning deep within each of our inner worlds. This is an extremely radical idea that is not easily seen, but once it becomes clear, it changes absolutely everything.

Freedom Means All Paths Are Always Available

Ironically, this new "freedom to choose" from all possible already existing worlds must also mean that the "old world" paradigm of separate selves that are endlessly competing (ultimately against oneself) for limited resources, also will continue to exist. In our magical expansive multiverse with its architecture of complete freedom, all worlds and all possible views will continue to co-exist simultaneously. This means that the endless cyclic tread-milling on the old "karmic wheel" of suffering continues to be freely available as a "choice" for every being.

The continuation of human suffering might be viewed by many as an ungodly evil, but it is not. It exists, instead, as a necessary mechanism for a natural process that unfolds in *time* and leads to a deeper experiential type of understanding and growth for all *beings*. All possible "worlds," and all ways of *being*, an *infinite* number of them, must exist side by side so that any and every "world" can be freely "entered" and experienced through this profound manifestation process guided and governed by each individual's *soul* or "*core vibrational resonance (CVR)*."

It is this, our deeper resonance, and not our thoughts that determine and shape our outer world. Our thinking is still critically important because our thoughts will adjust our actions and our actions shape the growth of our CVR. However, the manifestation of our "outer" world is determined entirely by our unconscious belief systems being expressed through this deeper type of vibration. Shaped by our hidden shadows, our CVR is also completely "fluid" and free to change at any moment.

We are all free to expand the range of our CVR by addressing the hidden darkness or shadows that can be found within every one of us, and then gradually illuminating this darkness with conscious awareness through a healing process that has been called "enlightenment." The less that remains "shadowed" and hidden deep within our unexamined darkness, the more joy and freedom we can and will experience. The illumination or "enlightenment" of our shadows is absolutely necessary for personal freedom. Freedom means that we have become more open and available to experience and abide within the natural flow of the "river of life." To achieve personal freedom, we must first heal ourselves through a process of illumination, inclusion, and integration, a process which I have named "whole-ing." Freedom can't and won't exist where there are still vast areas of unexplored darkness creating separation and division.

AWARENESS UNFOLDS WITH PERFECT TIMING

Of course, there will always be many individuals who are not yet ready to embrace, or even explore, this new and radical level of inner change. This is never problematic because, in all natural growth, such as with newborns, it is simply a matter of *time*; the process is ongoing and each step unfolds with perfect timing. *What this means for all of us is that our world will appear and continue to operate in the exact way that best prepares each of us for the next stage in the expansion of our awareness. It may not look like the world we "want," but it will be exactly the world that we "need" for our continuing evolution.*

Wars may continue to alter our lives, and greed, judgment, and prejudice will have repeating opportunities to create new drama for the facilitation of our experiential knowing. Poverty and epidemics will continue to plague regions and nations, and not without irony, this same chaos creates a multitude of new opportunities and experiences for those who choose to devote their lives, either through compassionate action or the political process, to the mitigation of these persistent human "problems." Through this process, all social activists ultimately discover that while their efforts do facilitate human interconnection and temporarily relax some immediate level of human suffering, *the deeper human "problem" that creates this kind of suffering over and over, can never be "fixed" by external actions of any kind.*

In our fully interactive and responsive universe where there is no such thing as an "accident" or "mistake," our human "problems" must then exist for a deeper reason: they exist to illuminate our current state of awareness and facilitate experiences that help us to open to our deeper nature. As long as shadows exist within individual human *beings*, there will continue to be these types of external "problems."

This also means that those who are practicing compassion by attempting to alleviate human or animal suffering, whenever and wherever it appears, will always have many opportunities and places to engage. This is because when one "problem" is improved through these temporary external means, new issues will always rise, continuously, not unlike the old carnival game of "whack-a-mole." *The range of possibilities for human expression is endless because all possibilities must manifest somewhere to maintain balance and completeness in our dualistic universe. All these different types of expression, however they might be judged, are equally important to the evolution and wholeness of our universe.*

For each of us, no matter how many acts of kindness are offered, witnessed, or experienced, the outside world of human suffering will not appear to transform until the shadow world inside each of us is illuminated. As always, the outside world is only a reflection of the inner: "as within, without." While this is an ancient truth, with our recent scientific discoveries, it has never been clearer.

OLD PROBLEMS REQUIRE NEW SOLUTIONS

At first brush, this perspective will likely seem naïve, callous, uncompassionate, uncaring, cold, or even cruel. At its rooted core, it is the complete opposite. For all of the 100,000 or more years that humans have walked this Earth, we have been repeatedly encountering the same types of inter-personal "problems." While the technologies, scale, visibility, and the number of people involved might have changed, the basic human "problems" of greed, fear, hate, lust, prejudice, judgment, and isolation remain. No amount of external "fixing" (politics, laws, treaties, religion) has ever changed this aspect of human nature in a fundamental or enduring way. Today, through media and easy travel, we all can witness the deep and desperate human suffering that now ravages more souls than ever before. Despite our access to new devices, medicines, methods of sharing, philosophies, and massive worldwide humanitarian organizations, these old types of "problems" persist, only now they seem to be much more complex, multi-layered, far-reaching, and even greater in their scope. It has become clear that our constant reflex to correct our problems through external means is not working. Our traditional externalized attempts at "fixing" human suffering can be compared to treating cancer with aspirin, and only after it has already spread and ravaged. *We are attempting to manipulate the surface expression while completely failing to recognize its far deeper source.*

Our most profound problems are often our best opportunities. Through these, life is repeatedly signaling us to find a different approach. Again and again, we are being asked to go within and re-examine the deeper nature of our "problems." This is the journey where we eventually realize that we have been completely misunderstanding our relationship to the "outside" world; we discover that all we once considered to be "external," actually spring from within.

The entire "world" that we experience as "outside" of ourselves, is actually created from within and only expressed as the physical so that we may witness the *core vibrational resonance of our being.* Our CVR is always in flux, and with each shift, we are brought directly to the next perfect experience for our evolutionary development. Our next experience will always be determined by the precise way that we are *resonating,* at that moment, in the core of our being. *Our personal experience of the world is always the perfect reflection of our most fundamental vibrational state of being. As we have always been reminded by spiritual visionaries, "The outside world is just the mirror of what is occurring inside."*

The most profound and important insight to be gained by this awareness is that we each have the ability to change the way that we resonate and thus completely alter our experience of the world. No one else can do this for us; it is and must be a completely individual and internal process. Others can inspire us, guide us, or act as role models, but all the real work of change must be done within. There are as many paths as there are voyagers, but all paths will eventually lead us to the same home – to a deep awareness of our connected oneness with everything in existence.

POSITIVE THINKING

Let me repeat myself to be very clear. This is not just another reference to "positive thinking" as some might assume at first glance. **Personal freedom will not and cannot be obtained through "thinking" of any type, positive or otherwise.** It is only through the *vibrational expansion* of our deeper *being* that we can begin to explore and experience the true freedom that is our birthright.

However, thinking is a critical function for our *being*; it one of the primary and most important energetic expressions of our human form. As Gandhi said, "Your beliefs become your thoughts. Your thoughts become your words. Your words become your actions. Your actions become your habits. Your habits become your values. Your values become your destiny." Words and thoughts are important *energy* forms, making them an integrated part of any energetic process. However, words and thoughts alone will not lead to freedom; they represent only one small part of the feedback process. Popular techniques such as "positive thinking" and "manifestation" are instructive and contributing practices because they help us better understand our awesome ability to create – even entire worlds. Unfortunately, as taught today, these techniques often only focus upon personal gratification of short-term egoic desires. While a *spiritual materialism* that only explores the most superficial aspects of our awesome potential can be very useful for attracting new students, its misinterpretation and misunderstanding will also lead to significant confusion and disappointing setbacks. Through many popular New-Age practices such as *manifestation, positive thinking,* and *ascension technology*, we discover that we actually do have the power to create our ego's desires in the short term, only to find ourselves unsatisfied and no happier in the end. Ultimately, these explorations help us to recognize that clear knowledge of what is "best" for us requires a much better understanding of our relationship to the universe. **The path is never as we plan, for our egos are very myopic and instinctively avoid the very types of difficulties that often lead to real breakthroughs.** Joseph Campbell, the great educator, reminds us, "If the path you are on seems very clear, you are probably on someone else's path." While self-improvement workshops can be helpful for many things, including short-term fixes, after many of these, one day we realize that **our constant search for external "fixes" only postpones the one real path for illumination: the internal journey of self-discovery.**

At this point, I must also be clear that this path is also very different from the popular practice of learning how to "vibrate at a higher level." In a now very popular belief system, higher vibration is seen as good, while lower vibration is implied to be bad. However, this widely taught concept, where "higher" means "better," only creates more separation, division, and judgment. *Whole-ing and the growth of consciousness must always be inclusive and expansive; we can't grow by rejecting aspects of our being that we judge to be less favorable, and then somehow shifting to those we consider to be better or "higher." We can only adjust our "core vibrational resonance" by including and embracing more of "that which we already are." This must include everything in existence, all that we might judge as "higher" or "lower." Instead of rejecting lower vibrational states we only can grow by expanding our range to include all vibrational states.*

The *whole-ing* process described throughout this book rejects nothing and leaves nothing behind; it is entirely based upon the expansion of our *being* and the inclusion of absolutely everything that life presents to us. Because all parts of existence contribute to the healthy whole, the "higher" and the "lower" contribute equally. We can discover wholeness and deep healing only through the inclusion and embrace of "all that we are." This means everything: all that we previously divided and labeled, including all of the good and all of the bad.

Our lives are like a great symphony. Instead of our "instrument" shifting its frequency to vibrate at a "higher" level – equivalent to playing a higher note – life's growth and expansion must instead be recognized to be more like adding additional instruments to the orchestra, providing more vibration at all harmonic levels and helping to create a larger symphony that has a much fuller sound. And yes, the often maligned banjo, cowbell, and accordion must also be welcome in this expanded orchestra.

Due to the entrenched beliefs inherited from our long-existing culture, for many, this new vision may seem very strange, improbable, or even impossible. However unlikely it may seem, this new vision is much better supported by our science, experience, and spiritual traditions than the old, fixed, rational, and deterministic world view that shaped the last 500 years of our history. *Today, we have the opportunity to embrace a new way of being that was previously only imagined in dreams. In every country, all around our planet, individuals are rapidly discovering that we already have, and have always had, the ability and power within to completely change the outer expression of our world. This occurs the instant we are fully ready for this transformational change. We depend on no others to manifest this change in our world; it requires only a shifting of our core*

beliefs through a sincere exploration of the buried and deeply hidden shadows that can always be found within.

SCIENCE AND OUR PARADIGM SHIFTS

The two parallel scientific gems of twentieth-century science, *quantum physics* and *relativity,* are far from new untested theories. After more than 100 years of development, experimentation, and integration into our culture, these two theories have been proven again and again to be extremely accurate predictors for new discoveries and they have both led to a multitude of incredible new technologies. They are fully recognized as our best and most accurate descriptions of the universe, and together they have completely changed our understanding of natural phenomena. *Our modern world's entire technological infrastructure is built, and fully dependent, upon the accuracy and practical applications of these two well-tested theories.*

Although seemingly impossible and conceptually very different from each other, both of these new theories have fully established their validity by repeatedly and consistently proving themselves to be astonishingly reliable and accurate. Together, they have led directly to the production of a dizzying array of very practical, real-world devices that we use and completely depend on every single day. From this "common-sense-defying" physics, our engineers have discovered and developed the following: refined transistors; laser; amplifiers; neon, fluorescent and LED lights; *microwaves*; MRIs; CDs and DVDs; computers; atom bombs and atomic energy. Using this revolutionary science, we have predicted new *elements* and *particles*, launched *space* probes, developed accurate GPS systems, constructed the Internet, and gained a much more detailed understanding of our expanding physical universe. As of 2015, it was estimated that more than one-third of our national Gross Domestic Product (GDP) was based upon or built from products developed through *quantum theory* alone. The technology is advancing so rapidly that this estimate may represent a gross understatement by the time this book is read. Today, devices built from these technologies have been incorporated into almost everything we use: from our cars and houses to our phones and coffeemakers. Once unimaginable, products developed from these two theories are now as much a part of America as apple pie. *Quantum physics and relativity have become a critical and integrated part of our lives. Time after time, both of these strange theories have proven themselves to be as real as anything else is in our world!*

Along with these practical technological devices and developments, the last one hundred years of physics have also uncovered principles, interpretations, and possibilities about our existence that simply do not make sense when viewed from within our existing, but now very aged, *paradigm*. **Despite more than 100 years of new revelations and discoveries, our world is still operating from its old worldview, one based upon a far less accurate scientific understanding that is now more than 500 years old and clearly obsolete.**

Major cultural changes take *time*. Historically, mankind has always required a great deal of *time* to collectively gain new experience and shift its most deeply held beliefs. Today, we are just beginning to witness the emergence of this new, and life-changing, global paradigm. Eventually, the magnitude of this shift will be even more extreme than the changes introduced by the last jump initiated over five hundred years ago. Then, it took a long *time* for most people to accept that we were not living on a "flat Earth" that was the center of the creation, but instead, we are just passing visitors on this small spinning sphere that is a minuscule part of an enormous, cyclic, and dynamic universe. **We must remind ourselves that five hundred years ago, before Newton, Galileo, Kepler, and Copernicus, the <u>only</u> existing worldview was this "flat-Earth" paradigm.** That limiting worldview did not disappear quickly or easily; it required many deeply committed scientists, explorers, rulers, and religious leaders to give their full efforts – and often their lives – before that strong cultural "flat-Earth" shackle could be finally broken.

There are still a few "flat-Earthers" proselytizing today, but ironically, their focal belief that we are situated at the center of the entire universe has suddenly gained profound new meaning. Because of the latest discoveries about the *holographic* nature of our universe and the illusionary nature of *time* and *space*, we are now more easily able to accept the strange-seeming idea that every one of us is a unique access point to the entire universe. J. Krishnamurti reminded us almost one hundred years ago that "You are the World"; and again, fifty years ago, the great Buddhist Chögyam Trungpa declared that, "Opening to oneself fully is opening to the world." Today, physicists are now muttering similar and equally strange-sounding phrases.

There was then, as there is now, an established, organized resistance to dramatic societal change. Resistance to change is completely natural and often healthy for living systems because it helps to maintain and support a steady-state for their existence. Large cultural shifts require enormous societal redirection and restructuring, so this level of dramatic change will always be disruptive of the existing social order and strongly resisted by the established power structures. This is an understandable response because corporations, institutions, and powerful individuals are deeply vested in the existing status quo and don't want to risk their privileged positions or great wealth.

Some amount of built-in resistance is a good thing for the maintenance of a culture because it functions as a buffer, helping to keep society stable by reducing or *damping* radical fluctuations and instabilities. This resistance can and should be understood as a normal, organic part of the process of change. Functioning like the shock absorbers of a car, a certain amount of this built-in resistance can ultimately make the ride smoother for all of us. *It is important to always keep in mind that rapid change, either cultural or personal, can be destabilizing, destructive, and counterproductive. We always need to remind ourselves that "time is on our side."*

Six hundred years ago, the "flat-Earth" paradigm, with the Sun obeying a god's daily command to rise on one side and set on the other, was just a part of everyday life; it was impossible for most people to even begin to imagine a different model. In 1543, waiting until he knew that he was about to die, Copernicus published his description of a radically different solar system in which the Earth was no longer the center—physically or even philosophically. With this dramatic shift of perspective, humanity began the slow process of giving up its self-imagined concept of being the most central and important element of the entire universe.

Of course, initially, the general public had no context from which to visualize this degree of change, and because this new worldview directly threatened the Christian church's absolute authority and hold on power, this new science was, of course, declared to be blasphemous. In the century that followed, many brilliant and courageous scientists were persecuted by the church for introducing or spreading these "dangerous" and heretical notions. The then-universal belief, that our Earth existed as the physical center of the universe, could not, and would not, die easily as Galileo and many others tirelessly and courageously developed and presented additional clear, physical evidence for Copernicus' new vision of our solar system.

These cosmological pioneers and the new science that they brought to light continued to suffer terribly at the hands of the church and state, but eventually, a few monarchs and rulers were convinced to support limited explorations, primarily because of the potential economic benefits associated with the plundering of a much more expansive planet. Eventually, once gold, silver, spices, and other riches began to return, adventurers like Columbus and Magellan were granted outright state and church permission along with financial support. *Thus began the full-scale exploration of what was then a "new" and very strange territory: one that previously did not, and could not, exist within their older world-view.*

Those first bold explorers proved to the disbelievers of science that our planet was round, and not flat, by physically sailing "over the edge" and then successfully returning, with previously unimagined treasures. For the masses, these voyages to strange new lands initiated a life-changing paradigm shift, which gradually became increasingly real as more and more riches and wondrous tales of new lands flowed back to their world. It took generations, but eventually, all the world's people realized and accepted the new and world-changing idea that our planet was only a small part of a much greater system.

Later in this period of rapid change, Kepler, and then Newton, refined the physical "laws" of motion for planets and other objects. Their contributions did a remarkably accurate job of describing why large physical objects, such as apples and planets, behave as they do; but even with all this new scientific and experiential evidence, the full global integration of this new and very strange paradigm still took many additional generations. Gradually, as the world adjusted to this new vision, humanity's understanding of the universe, and even life itself, was forever changed. *It took hundreds of years, but what was once unimaginable eventually became completely understood and therefore quite normal. Today, this once-unimaginable paradigm is so firmly established that it is almost impossible to think that we ever could have embraced the flat-earth vision. Now that older, but once certain, flat-Earth world-view even seems laughable.*

Today, after the discoveries of quantum physics and relativity, this once-revolutionary world view has, itself, been demonstrated to be quite old and obsolete – it is now clearly showing its age. Yet, despite over one hundred years of research highlighting its problems, this almost six-hundred-year-old deterministic vision for our universe remains our dominant cultural paradigm. Even though our current world-view is clearly understood to be obsolete, it still defines and shapes almost everything about our lives on Earth. Over the years, Copernicus' original vision has gradually expanded its physical range to include the rest of our local galaxy, and billions of other galaxies beyond, but it is still fundamentally grounded in the old deterministic view: one that sees our primary reality as a machine-like, three-dimensional physical expression, that is also fully knowable and has defined limits.

Once again, in our long history of change, we find ourselves facing the same type of large-scale societal challenge because a full century of experiments and discoveries from relativity and quantum physics have repeatedly shown our current model to be inaccurate, and sometimes even wrong. Our science, philosophy, real-world experience, and social issues have once again pushed a cultural paradigm beyond its breaking point. Today, we find ourselves at the beginning of another radical societal shift as we begin to rethink our existing, but now obsolete, worldview. *We are, once again, at the first stages of learning how to live within a new, completely different and strange seeming paradigm—this new "Architecture of Freedom."*

What Science Is Now Saying About Our Universe

The following is my short-list of the most unexpected and paradigm-changing ideas that have been illuminated by the last one hundred years of science along with some older, more established, scientific ideas that now embody new meaning. *This list is just a brief introduction; all of these ideas and others will be discussed later in greater detail. In the Appendix, I include a more thorough discussion of the science that illuminates the origins of many of these bold or impossible-seeming statements.* A more thorough discussion of the science behind these ideas also can be found in my earlier books – *The Architecture of Freedom* and *A Path to Personal Freedom* – along with many books by other authors, some of which are listed in the resource section of this book. What follows are thirteen of the most important take-aways. Because they are all so interconnected, they can be read, thought about, and digested in any order:

1. Our entire perception of life is shaped by the type, quality, and quantity of information registered by our human neurological system, a very limited system that restricts and defines the types of information that we can sense and process. Everything that we perceive and experience begins with an energetic stimulus to our neurosensory system, which then must be processed by our brain. Through it, we can only experience a small portion of the "known" physical universe, and we are now beginning to realize that this "known" physical universe is only a small part of existence.

We all recognize that our human biological systems have their limits: for example, we cannot sense *radio waves* or hear, see, taste, feel, and smell things that dogs and other animals can sense. We have developed special instruments that can detect or measure some of these otherwise invisible types of *energy* such as *radio waves*, *microwaves*, *infrared*, high or low pitched sound, along with invisible *matter* like *molecules, subatomic particles*, and distant stars. However, even with all that we have uncovered, there still exists an unknown but enormous amount that remains completely hidden from both our senses and our instruments. Because we lack the receptors, neurological systems, instruments, and knowledge to process many types of *energy* signals, there is a lot that we don't sense or understand just within the limited bounds of our physical, three-dimensional universe.

Beyond the recognized physical universe, we see many hints of enormous parts that we don't understand, and we completely lack methods and instruments for peering into the even more vast parts of unknown that we have yet to discover, explore, or name. And beyond all of this, an untold amount of mystery will always exist because the great majority of "all that is" resides beyond our ability to even recognize its existence, lying in regions or dimensions that reside far beyond our *conceptual horizon*. **The very finite nature of our physicality will always limit the range and scope of our physical experience within the known parts of the universe; and, on top of this, we are almost completely blind to that greater part which likely exists beyond the physical. We have no way to even begin to sense what we don't know about that which we don't know we don't know.**

2. Nature is always trying to maximize diversity. The natural physical world never sits still. Strange and unusual organisms evolve to fill every niche as the natural world always propels itself to try and test every possibility. In our *dualistic* physical universe, *contrast* is not just a desired quality, it is completely necessary. What meaning could hot have without something colder for comparison, or big without something smaller? These polarities create an enormous amount of diversity and nature's "desire for diversity" is an important driver of our own need to explore, inhabit, and reconfigure every environment and discover every possibility. *Contrast* and diversity give shape and meaning to our vast and dynamic playing field, creating an *infinite* range of opportunities that excites and propels us into exploration, learning, and ultimately growth.

3. Time does not exist in the way we commonly perceive and understand. The "linear, one-directional marching *time*" that we think that we understand has been shown through numerous scientific experiments to be quite variable; it adjusts to velocity, relative position, relative motion, and *gravity*. *Time* also does not exist independently; it is folded into a four-dimensional *field* that Einstein named, *spacetime*. Even with this new scientific knowledge, our current understanding of *time* does not begin to explain common personal experiences of unusual "time-based" phenomena like precognition and déjà vu. Some scientists are now thinking that our sense of passing *time* seems to be regulated by the *frequency* and quality of our interactions. *Time* seems to be related to our mind's method of organizing and restricting *information* so that it can be effectively processed by a limited local processor: our brain. *Time* may just be the biggest misunderstood joker in the game of life. Whatever it is, *time* is certainly not the absolute, relentlessly marching, one-directional, defining master that we usually imagine.

4. Science has only been able to see or measure, at most, about two to four percent of the "known" physical parts of our visible three-dimensional universe, and beyond this, we are completely "in the dark" about the vast amount of what is still "unknown" or "unseen." Studying the movement of heavenly bodies, cosmologists have determined that we are unable to see, explain, or directly detect at least ninety-six percent of the forms of *energy* that are stirring or interacting within just the observable part of our three-dimensional universe. Ironically, it seems that the more we learn about our cosmos, the more mystery we uncover because, even as we learn "more," those parts that "we know we do not understand" expand at a yet faster rate.

It seems clear to me that the most fundamental problem with our attempt to "understand" is that we are trying to explore the "*timeless infinite*" using the constraints, logic, and language of our rational "*time-bound*" realm. Of course, building upon what is already "understood" is how all human knowledge, including good rational science, must progress, but trying to understand things from only this limited conceptual perspective can also make simple multidimensional ideas appear to be extremely complex (see discussion of the Victorian era novel *Flatland* in Part Four).

5. Since the time of Einstein's theories, we have clearly understood that matter is equivalent to energy. This means that everything we experience as physical is only energy being expressed in a different form, and this includes our bodies. Our bodies are just another expression of *energy*. Einstein also described *time*, and therefore much of our reality, as only " a very persistent illusion." Our internal and enduring sense of the world as solid, material, and marching through *time* is an internal perception and not an external absolute; it is formed, and then perpetuated, by our brains and central nervous systems. Experimental results from *quantum physics* indicate that not only are our beliefs and intentions somehow involved, but they may even be the most important elements for the creation of the world that we experience.

Ultimately this, taken with other discoveries, also means that our powerful individual sense about the ultimate realness of our physicality, as well as all our beliefs about our separate *selves*, are also parts of this same "persistent illusion." For human life, this also happens to be a very convenient and important "illusion" because it allows for, and supports, our entire physical existence and experience within this realm. For all of us, this experience of life on Earth is simultaneously both "real" and "illusionary." Being "illusionary" does not mean that our physical lives are not real; from our perspective, they are quite real. Instead, our thinking becomes confused when we believe that this physical reality is the ultimate, or only, reality. Our confusion comes from our misunderstanding of the nature of this persistent illusion – it originates from our internal sense of existence being most fully tied to these temporary physical bodies.

6. The familiar physical world of three dimensions is only a very small piece of a far greater multidimensional expanse of existence. We typically think of ourselves as "three-dimensional" *beings*, but for more than one hundred years we have known, from even before Einstein's work, that there are at least four integrated *dimensions*, a realm that we have even given a name: *spacetime*. Even knowing this for several generations, humans still perceive, function, and think in only three *dimensions*, with a misunderstood added layer of one-directional *time*. Today, we only partially understand and certainly haven't integrated, most of Einstein's now more than 100-year-old discovery.

While we know that our universe is built from more than the three *dimensions* that we can most easily understand and experience, some of our most recent physics indicate that we may inhabit a *multiverse* that has at least eleven *dimensions*. In any case, our physical experience is restricted to a very small and limited portion of a much greater existence, and even within this very limited and restricted region of only three *dimensions*, we still understand relatively little.

7. Though our physical world feels very "real" and solid to us, some physicists now believe that our physical universe is, instead, only a

projected image that comes from a remarkable hidden geometry. This also means that all of the familiar objects of our lives: birds, bees, flowers, trees, rivers, fish, *atoms*, planets, stars, and even our bodies, are only the secondary effects of this hidden and deeper reality. If true, our entire world and lives can be better understood as being only an *image,* similar to a movie that has been *projected* onto a three-dimensional screen.

Only recently have a substantial number of scientists felt free to publicly express their musings about this and these other very strange-sounding conclusions. In the past, the entrenched scientific establishment marginalized or ruined the academic careers of those who expressed interpretations that ran counter to the mainstream. Today, the weight of scientific evidence for these once "irrational" interpretations has finally reached a tipping point. More and more, modern *quantum* theorists are talking and writing about their gradual realization and reckoning with the emerging idea that there are no actual "things" in our universe and, instead, there are only *fields* of *information* that become 'disturbed' and somehow manifest, on the fly, the *particles* which shape the "things" that we recognize as our world. Many scientists now believe that our entire physical world, including our human bodies, only appears as the secondary result of energetic ripples or *waves* disturbing a particular hidden *field*: the *quantum field*.

Because of this, much of life's mystery can be understood as the result of misinterpreted "artifacts" or "shadows" cast (*projected*) from this deeper dimensional geometry: a geometry that is entirely hidden from us. For example, for hundreds of years, we thought of *gravity* as a *force*, but Einstein demonstrated that it is, instead, the indirect way that we experience the curvature of four-dimensional *spacetime*: a *field*. Our common sense tells us nothing about *fields*, and yet we now know that they exist and shape our entire experience. Electric generators and radio antennae were two early practical applications of *field theory,* and we continue to discover, harness, and employ *fields* throughout our modern technological world. Many physicists now believe that the *quantum field* forms and re-forms our entire universe, over and over, making it the structural foundation of all the other *fields*. With it, we are also getting our first scientific glimpses into something that sounds eerily similar to our past descriptions of gods?

8. Every part of our universe (actually multiverse) is in constant and instantaneous communication with every other part. This is one of *quantum physics'* most profound observations. Surprisingly, this relationship remains true even if there are trillions of miles, billions of years, or multiple dimensions separating these "parts." This instantaneous intercommunication means that the speed of light is not even a limiting factor. In three-dimensional terms, it is as if every part of the universe is always in direct contact with every other part. Physicists describe our universe as completely *enfolded* and *entangled*. Many prominent physicists and philosophers believe that this extraordinary degree of interconnectivity means that there is actually only one thing in the entire universe. Our entire universe is so fully and instantly interactive that it behaves like a single organism.

9. There are no accidents in the universe. Since the universe is so thoroughly interconnected, it is also fully interactive; and, at some deep but unseen level, fully *causal*. Even though *quantum physics* has shown that at the sub-atomic level, our universe appears to be probabilistic, the deep inter-communication and interconnectedness mean that nothing is ever random or accidental. Everything we experience is always the result of *information* or *energy* that is being exchanged at unseen levels of existence. Our physical experience is never actually random or accidental because it is always the result of the structure and workings of a deeper universe that is always acting in ways that lie far beyond our ability to detect or understand.

10. Whenever we participate in an experience, we influence its outcome. *Quantum physics* has repeatedly demonstrated that before our interaction or involvement, an *infinite* number of outcomes are possible, but once we engage, a single outcome becomes "solidified" and appears to us as "real." How this happens and what this means no one knows, but from our perspective, it is becoming quite clear that our expectations somehow influence the outcomes that we eventually experience as "real."

Our most recent science also indicates that there are not even "real-seeming things" unless first there is some kind of interaction; "things" only appear as a direct result of interaction or inter-communication. Solid matter arises spontaneously from a *quantized energy field* (the *quantum field*) whenever it becomes disturbed by some type of interaction. It is this *energy field* and not matter that is the most real and fundamental thing in our universe. In one sense, we are creating our universe as we go, on the fly, through an internal process that originates and operates from a hidden place, a place that lies far deeper than thought can reach. Our solid-seeming world seems to be very responsive to our intentions and may not even exist without our conscious participation.

11. We know that the majority of cells contained within and on our bodies are not ours, or even human. We are not, as we have long assumed, individual and sovereign entities. Each of us already exists, instead, as a fully interactive colony. What we have been discovering is that every healthy individual can more accurately be described as a bustling, interactive colony, supporting and completely relying on millions of viruses, bacteria, mycelium, protozoa, and other small, helper organisms. When these helper organisms are killed by antibiotics, pesticides, or toxins, our health suffers. We depend on the other members of this beneficial colony for our very existence.

Going in the opposite direction: out, we encounter other critical relationships. We breathe air, stand on the ground, and rely on food and light for our continuing survival. We can't exist without the earth, air, water, and sun, so how are these *elements* not also a part of our *being*? We exist as an enormous vibrant colony of life-extending far beyond the once-imagined bounds of our bodies.

Once we fully understand that we are this varied mix of different interconnected *beings*, a new question emerges. Where, in this soup of interactive organisms, does the "I," that we identify with, reside? At this fundamental biological level, our long-standing concept of existing as separate sovereign individuals requires a complete re-examination. Even at this most basic level of biology, we must question our deeply ingrained concept of existing as a single, individual, and separated entity.

12. Vibration is the universal language of existence. Many of the top minds in physics are now saying that the most basic thing in the universe is *information*, much of which is then shared and communicated through various forms of *vibration*. Vibration, as we understand it, is measured in cycles-per-second, making it *"time-based."* This implies that "outside of *time*" there must be some related form of interconnective and energetic communication that only "later" appears, to us, as *time* -measured *vibration*. What this might be, no one knows. Miniscule vibrating *strings* (*string theory*) or a vibrating *field* made from *quantum* bits of *spacetime* (*quantum gravity theory*) are the two best current guesses. *Universal Mind* and *Cosmic Consciousness,* are terms that philosophers and writers have used which are very relevant to this discussion.

13. It appears very likely that our universe is infinite. The latest cosmological research strongly supports earlier evidence that our universe (*multiverse*) is *infinite*. If any of the substantial numbers of leading theories that demonstrate an *infinite* universe proves to be correct, then everything we think we understand about life instantly changes.

The following will seem like a "big leap" for many, making it a very difficult idea to easily grasp; but an *infinite* and *timeless* universe also means that somewhere in its endless expanse there must be an *infinite* number of *parallel worlds* that are exactly like Earth, populated with humans that are identical to each of us. Today, as thoroughly discussed later, this once-crazy idea is starting to be seen as a probable fact. While the implications of this are completely impossible for our conditioned minds to assimilate, this vision is backed by very solid science and math. According to many of our greatest philosophers and scientists, this matrix of *infinite* possibilities is the most likely form for our universe.

The meaning of this realization for our lives, as all my books explain, is beyond even incredible or fantastic. For example, since we now understand that all parts of the universe are instantly interconnected, this means that we are always fully connected to the other "identical selves" that are also a part of this *infinite* expanse. The more we learn about our cosmos, the more this extremely strange, impossible-seeming, and mind-boggling conclusion appears to be a part of the deepest truth.

After reading these summaries, some readers will react incredulously, wondering about the origin of these "crazy-sounding" statements, while others may only need or desire more elaboration before continuing. I encourage anyone with these doubts or concerns to skip directly to appendix one which is a history of the science that leads to these conclusions. That more extensive discussion will help to establish the solid ground upon which I build the case for many of the strange-sounding ideas discussed throughout this book. I placed this scientific history in the appendix because this topic is so well covered in my earlier books. Also, even though it includes very little actual science, its inclusion early in the book might discourage some readers. While it is certainly not necessary to understand what scientists are seeing and saying to benefit from this book, it should help quell a great deal of very natural and expected disbelief; and that is extremely important. What all contemporary scientists are seeing and many are saying is truly beyond even "astounding."

INTRODUCTION TO THE TWELVE PRINCIPLES

From these takeaways and other scientific realizations, our long history of spiritual traditions, and our unexplained personal experiences, clear principles emerge that help us to better understand this strange-seeming architecture of our universe. *From our limited conceptual perspective, these principles describe the way our existence appears to us.* They are only offered as helpful concepts and visualizations for constructing a new model, one that can help to guide us as we gradually learn how to embody and embrace more of "that which we are."

I have defined twelve principles that describe, from the perspective of our limited human understanding, the way our existence appears to be shaped. These principles, and, more importantly, the vision of the universe they naturally evoke, form the common backbone of all of my books. *The principles have been sifted from the common territory where science, spirituality, and our unusual (but universal) personal experiences intersect.* The division into twelve principles is completely arbitrary; I could have described it in only one principal, and I could have just as easily described fifty. All of these principles describe the same thing: the deeper-level "oneness" of our universe. *Since this "oneness" exists outside of the very time and space that define the boundaries of our physical existence, its deepest secrets are not describable through verbal, rational, or conceptual means.* Because of this, the understanding and eventual integration of this new vision will be extremely challenging for most readers; it requires a kind of "learning" that is unfamiliar. Its division into these twelve principles was simply a functional choice to facilitate the digestion and eventual "understanding" of what is, in truth, a single revolutionary idea. It has only been divided so that it can be imbibed in smaller bites for better digestion. Though impossible to understand through our familiar concepts of *time, space,* and language, when viewed together as an integrated whole, these twelve ideas present a more helpful and accurate description of how existence works than does our current, but now very obsolete, deterministic paradigm.

The gears that move existence operate within realms that lie far beyond words and concepts, so any book (or any other attempt at a verbal or logical description) must be, by its very nature, limited, incomplete, or inaccurate. Because the deepest Truth lies far beyond our ability to perceive, understand, and reason, it is unlikely that there will ever be a logical, rhetorical, or scientific proof for many of these principles. However, there still is empirical "proof"; it is what we discover through our direct experience and the much deeper "knowing" that organically emerges from our personal growth.

As discussed throughout this book, all paths eventually guide us to this same deeper "knowing" so many philosophies or methods work equally well – some are ancient, some have yet to be discovered. However, if we live lives guided by this set of twelve "lesser understandable truths," this alone will completely change how we each experience our lives, and that change alone will deepen our "knowing." Following the presentation of these ideas as twelve principles, I will express them again, differently, as a set of guidelines for living our lives in greater freedom. For some who try this method, the resultant "shift within" will be proof enough. Living lives founded upon these twelve "little truths" will also prepare us for further growth and adventures that are completely impossible to describe or even imagine today.

These twelve principles all point to the single greater truth: we are one thing only appearing as many. Other traditions divide this same single truth differently, but the conclusion is always the same: the recognition and embrace of the eternal and fundamental "oneness" of all existence. Since the numbering of these principles is completely arbitrary and since they each represent different views of the same single idea, all of these principles will seem to overlap in different ways. All of the ideas discussed throughout this book also directly evolve from these principles.

As previously mentioned, for those readers who are having a difficult time embracing these conclusions, or for those who want to understand more about the *quantum physics* and *relativity* behind these principles, the science section, presented in a form that should be easy for a lay-person to follow, is included in the appendix.

THE TWELVE PRINCIPLES

What follows are the twelve principles that can help us to better understand our universe and its architecture of freedom.

1. WE ARE INFINITE BEINGS: Thinking that we are only limited to these physical bodies is mankind's greatest and most confusing illusion. The truth is that we are already unlimited, eternal, and infinite. Even though we may be completely unaware of this, our influence and actions always extend far beyond these bodies, through both time and space. We are fully connected and communicating with everything, everywhere, and throughout time.

2. WE ARE MULTIDIMENSIONAL: There exists, now and always, an infinite and fully interconnected structure, not unlike a woven web, built from the information for every possible outcome of every situation, choice, or thing that has ever been, or will be, possible. It is within this infinite fabric of creation that we exist, act, and play—a perfectly designed structure that contains every possibility for everything that did happen or could have happened, ever. It exists, along with each of us, now and forever; everything possible, and all that we think of as "real," already and always exists "outside" of time. This incredible Web is built upon more physical dimensions than our familiar three directional coordinates, but we are not able to directly "see" or experience most of this interwoven Web. It is invisible to us because it lies beyond the limits of our physical senses and our limited ability to conceptualize: it is not, in any familiar sense, physical. This mostly invisible "Web of Infinite Possibilities" is the multidimensional and infinite fabric that forms the very structure of existence: the field where we live and express ourselves in all of life's various forms.

3. WE ARE VIBRATIONALLY BASED: Physical form (bodies, things, words, concepts, ideas, and beliefs) is of secondary importance within this Web. Everything in existence begins with information that is then communicated to and between us through vibration. The entire Web is alive with this vibration. Just where each individual awareness appears and expresses itself in this dynamic Web depends on the current resonant qualities of each soul, and its expression will also change moment by moment. All physical form, as we recognize it, only appears temporarily as a secondary result of the way information is expressed and shared through vibration. In our three-dimensional realm, vibration is always described using time; frequency is the amount of time it takes to repeat one cycle. The frequency of the vibration can be nanoseconds, eons, or anything between or beyond. In our realm of time, nature appears to be cyclic and eventually repeats itself.

4. WE EACH ARE SMALLER ASPECTS, OR PERSPECTIVES, OF A SINGLE VIBRANT LIVING BEING: Everything in existence is alive with vibrational life and is intimately connected with everything else. It is this energetic life vibration that temporary shapes the appearance of all the divided forms that we encounter, yet we are also so closely connected with everything that, at the deepest levels, we are only one thing. We all interact and work together as a single entity, whether or not this is our awareness. Our appearing to function as separate beings is helpful to facilitate the creation of a multitude of exciting new possibilities and opportunities for growth. As mentioned, when we allow ourselves to be open to more of this unbounded expansiveness, that part of us that we experience as the "I" will expand to intermix and co-join with "others," but the core awareness of the "I" will remain, no matter how we evolve. That "I" part of our being does not vanish—instead, it grows and expands and, as it does this, it still will identify with the old, familiar self. This process is expansive and inclusive; nothing real is ever lost. Through this evolutionary expansion, we lose nothing except our limiting ideas and concepts. This journey transcends all temporary physical form, including our bodies. As we evolve, we expand, grow, and deepen our interconnection with all "others." Our direct resonant connection to each other and source is often most clearly experienced as "Love." Love is not something that is given, it is, instead, a place of being—a place that we can always choose to enter and dwell within. Love is that place within where we can best experience our deep interconnection with everyone and everything.

5. FORM IS AN ILLUSION: All the physical forms that we encounter in our daily physical realm are only shadows or dreamscapes, even though they feel and seem very real. Our universe is built upon many more dimensions than the three to four that we understand – it likely contains at least eleven dimensions, but we only can participate in that smaller part of existence that can be "seen" from our limited, three-dimensional perspective. All form is created through the appearance of separation or division within the greater "oneness." Form unfolds so that our "oneness" can be explored and experienced through the special perspectives of three-dimensional duality. From our perspective, form always appears containing aspects from pairs of dualistic opposites: good or bad, right or wrong, hot or cold, long or short, dark or light: the deeper oneness has been sub-divided this way to create all form. All physical forms, no matter how they appear, are only the illusionary shadows cast from this oneness. Also, the world of form that we understand is only one small aspect of existence; both beyond and within it there is so much more. Our real and solid-seeming world is far from what it seems. Since our bodies are founded in form, they are also shadows and "not who we really are"; we, too, are so much more. For everyone, this part of the illusion: form, creates the greatest amount of confusion.

6.   TIME IS NOT LINEAR OR ABSOLUTE: Time, as we know and understand it, is another part of the illusion. Our sense of time passing allows our three-dimensional minds to process information in discrete chunks and a linear fashion: time is how we keep our type of conceptual "thinking" organized. All experience happens "outside of time" in the "now," but our brain's machinery can't process this way. However, the "now" is fully accessible to us, and learning how to access it facilitates our conscious connection to the part of our being that always exists beyond thought and time. The moment we think about anything or try to explain what it means, the "now" is lost: it instantly becomes the past or the future.

7. ALL POSSIBILITIES ALREADY EXIST – RIGHT NOW: From within our realm of time, it could be said that all potential possibilities for form and life are already being expressed in an infinite number of other pre-existing "worlds." Some of these worlds are identical to ours, while others are very similar; however, most of these infinite pre-existing worlds are very different from our "world." These "worlds" are positioned in this fully interactive and multidimensional Web so that all possible "worlds" are directly adjacent and "parallel" to each other, allowing instant intercommunication and access. This type of mixture is called enfoldment. From our three-dimensional perspective, typically we consciously observe only one of these "worlds" at any moment of time; however, as we learn to step "outside" of time, we will become able to broaden this awareness. Eventually, we might even be able to learn how to travel freely between these "worlds," enjoying a type of freedom unimaginable today.

8.   EACH OF US ALREADY HOLDS THE KEY TO CHANGING OUR EXPERIENCE: We can (and do) move continuously and automatically between these parallel "worlds." Every one of us is constantly shifting our viewpoint, or position, within this fully expansive Web. What we imagine to be our single "world" is actually many different parallel "worlds" that are continuously shifting, changing, and shuffling. That which we understand as the "I" is always expanding, contracting, and migrating between these different "worlds." In any given moment, the "I" will always manifest in the specific part of the Web that most resonates with our individual, but always changing, core vibration (core vibrational resonance or CVR) – an always evolving quality of every individual which is sometimes thought of as our "soul." We each have the ability through a personal transformation to consciously adjust, or tune, our CVR and when we do, we resonate and appear within new locations or other "worlds" within this Web. In this way, we are constantly adjusting and changing our experience by moving throughout an infinite collection of universes. Since these are truly different "worlds," these shifts can, and do, completely change the way our outer world appears to us. This incredible journey is almost always hidden from us. Because of the way our brains function by employing cognitive dissonance, smoothing, and interpolation, most of the time we don't even notice these shifts unless we are particularly open or they are extremely dramatic.

9. LIFE IS ALREADY PERFECT: Since our universe is dynamic, self-reflective, and completely interactive, it is always in perfect balance and this balance can be witnessed everywhere and in everything. Due to this complete interconnectedness beyond our world of time, at every moment, everyone is always in perfect harmony with the entire universe, and nothing is ever wrong or out of place. All events and interrelationship are therefore causal – there are no accidents. It does not matter how our lives appear; they are always the perfect expression for, and of, our being at that moment. The universe is holographic: meaning that every part, no matter how small, always reflects the whole, and the whole reflects all the parts. For each of us, this also means that the whole can be most easily accessed and most thoroughly explored by simply going within. Throughout our history sages have said, "We do not experience life as it is, we experience it as we are." What could be more perfect?

10. DEATH DOES NOT EXIST: Since time is not an absolute and form is an illusion, our concepts about the absolute nature of the birth and death of our physical body are also part of this illusion. A deeper understanding reveals birth and death to be just one mechanism for shifting positions and journeying through this expansive Web. Physical death is a limited three-dimensional concept; the inevitable outcome of mapping a marching time in only one direction. Our reflexive fears of death continuously inhibit us from fully experiencing the extraordinary potential of life. Since form and time are not absolutes, our typical concerns about the inevitable and pending death of our present forms reflect only the depth of our misunderstanding; they are completely misplaced, being nothing but artifacts within the illusion of marching time. Our thinking that we are limited to these bodies, believing that when these bodies die we also die, is the most inhibiting illusion shaped by our sense of time

11. OURS IS AN EVOLUTIONARY JOURNEY: *Our physical lives are only part of a greater process. Always an integral part of existence, we grow, experience, and expand awareness so that we may "know and become one with everything in existence." Each evolutionary step represents an expansion of consciousness as we gradually sense, feel, integrate, and merge with that which lies beyond. Each new level of our expanded beingness will always include everything that came before. We never lose our sense of self; the self just gradually, but constantly, expands to include more of our potential being on its way towards embracing, "knowing," and becoming its greater Self. This is our common journey and it is the real purpose of life.*

12. WE ARE HERE FOR THE EXPERIENCE: *We are here in our physical lives to encounter, embrace, and know every corner in our special part of this amazing Web. Our "mission" is to explore and participate in its every nook and cranny, no matter how wonderful or terrible it may seem. One reason for the appearance of many separate individuals is simply to facilitate this total exploration of existence. Through this diaspora of broadening experience and the embrace and incorporation of everything that we encounter, we expand and grow together to know more of our greater Self. As our awareness expands, concepts that were once judged and separated into "good and bad" become seen and understood through a much different light. Everything, once divided, then becomes an integrated part of a beautiful flowing wholeness. We will know true freedom only when we no longer experience ourselves as separate physical individuals, living within a limited amount of time, expecting to one day die and disappear. These long-held beliefs are all part of the primary shadow illusion of our dimensional realm. This process of "knowing and becoming" through our opening to all of existence is the ultimate purpose of our strange and awesome journey that we call life.*

LIFESTYLE-CHANGE RECOMMENDATIONS

What follows next is my list of recommended and practical lifestyle changes that directly and organically flow from these twelve principles. *These lifestyle*

changes, if practiced fully, quickly and dramatically shift our personal experience of life. To participate in this shift, there is no need to understand the new physics or even express an interest in science. Believing any of this book's conclusions, mine or those of others presented here, is not a prerequisite for this experiment. ***If a reader sincerely seeks to change his or her life, then they only need to begin to integrate these lifestyle changes.***

If the universe acts the way that I describe, then this understanding alone will naturally lead to a different way of thinking about, and walking through, life. As an experiment, just assume for a limited and set *amount of time, that* this proposed vision of a multidimensional, fully interconnected, *infinite*, and *timeless multiverse* is an accurate representation. For a short trial period, envision and live your life from this new perspective and just observe what happens. See if your life and the "outside" world you encounter seem to shift towards something that feels much more peaceful, joyful, real, and empowering: something that maybe feels more like "you." An important overall principle for guiding this experiment is "attitude changes everything." This is a very powerful idea because it is clear to most of us that while we may have little or no ability to control outside events, we do have total control over our attitude and subsequent reaction. I believe that developing an "attitude of gratitude" is the single most powerful change a person can make in their lives. As many sages throughout history have uttered, "Life does not happen to you, it happens for you." This single recognition can make all the difference.

Deep internal change is never a fast or easy process, so real healing (*whole-ing*) requires extreme patience, perseverance, and experience. This is a life-long process, so eventually must just become a way of walking through life. However, there are still some positive shifts that can be experienced very quickly, and these may provide the inspiration and motivation necessary for entering and persevering the deeper work. When these recommendations are adopted, the first noticed changes are usually very subtle. We may only sense a small shift in the ambiance or tonality around our lives, a very delicate change of flavor. We might find ourselves enjoying a meal more than before, or colors and sounds might be a bit more vibrant. With this first tiny, but positive, feedback, we might find ourselves encouraged and emboldened for a longer or bolder excursion into deeper waters; and once there, we might notice that other things have also significantly changed. Maybe a difficult relationship suddenly improves, or we unexpectedly discover that an old problem is no longer a visible influence in our life.

As already stated, to experience the type of freedom that I describe, there is no need to understand science. There is also no need even to believe that the vision of the "universe" presented in this book is more "real." The only necessary step is to begin to integrate these lifestyle changes into our lives, and then these changes alone will gradually and incrementally modify the *core vibrational resonance* of our *being,* and this alone changes the way our world appears to each of us.

Some of these proven lifestyle changes are:

- **Live our lives without fear**, especially our common fear of physical death. Be courageous and fearless in our adventure of self-exploration.

- **Learn how to "quiet" our minds.** This is an extremely broad, yet important, step for creating deep change, and techniques for learning how to do this are discussed later. Silent meditation retreats are a great tool for learning about the constant "chatter" that our brain (ego) generates on its own. Quieting the mind means learning how to release the dominant mind-chatter so you can distinguish between the messages that are forming in your heart and those that originate from the ego or mind.

- **Live our lives as if we all are separate, interactive parts of one being.** Treat all others as we wish to be treated and do not intentionally harm others. We may still make "mistakes" but we will grow from these. This is, of course, the "Golden Rule."

- **Live your life as if everything that happens has been designed for you and is working to assist you.** It is! Everything that unfolds in your life is specifically formed to maximize your experience of life and facilitate your transformation.

- **Remember that Love is always, and only, available from within.** Quit searching outside of ourselves for this Love. Love does not originate from the outside and it is never dependent on another person. Love is a place inside: a place that we can visit any time we are open and ready for the experience. Love is a particularly wonderful and special *energetic space* to share with others; it is the first doorway to our experience of "oneness." Always present, we miss the richness and depth of the internal experience of Love only because of our programmed choices and habits.

- **Live as if we are creating our universe anew with every thought and feeling.** Speak and act with clarity and awareness. Remember that every thought or word is vibrational and has great yet unseen communicative power. Our words instantly communicate and connect with everything in the universe. We are never alone and everything we do or think has ripples that constantly "stir" the cosmos.

- **Learn how to feel inside ourselves more deeply.** Learn to find and feel into all our empty, dark, and frightening spaces—individual and collective. At first, this idea may appear confusing; we may not even understand what this might mean. Even when better understood, this inner journey may still appear frightening; we all have many things that we want to, or would like to, forget. The truth is that this "darkness" is just asking, even begging, for illumination, expression, and inclusion. This practice eventually becomes a lifestyle helping us to find and illuminate our hidden darkness, allowing our true spirit to shine brightly.

- **Learn also to feel "others" more deeply.** Deeper within, all others are only different expressions of the same self: You. Eventually, you will experience all others, even those with whom you have "issues," as beautiful expansions of yourself. Developing empathy towards all others is an early step in a deeper type of "self-awareness."

- **Practice forgiveness.** Forgive ourselves through learning to forgive all "others." Ultimately, there is no difference. Examine every hint of blame. At deeper levels, all judgment and blame are forms of self-loathing. Forgiveness, non-judgment, and freedom are closely related, they all describe the same way of walking through this world.

- **Always remember that we are so much more than just our physical "body and mind."** Do not become lost in the illusion of the small self, the single body called "me." Meditate, breathe, and feel further within. From there, learn how to feel those parts of *being* that lie beyond and deeper within.

- **Live as if there is no such thing as a secret.** Everything is so completely interconnected; there is no place to hide a secret—all thoughts and actions are always expressed somewhere within existence. Everything hidden and unspoken still creates *waves of vibration* that communicate instantly throughout the entire universe. Learn to be impeccably honest with yourself and all "others." This is how we can work with, instead of against, the universe; life *flows* much more fluidly when this truth is recognized.

- **Avoid waiting for, or expecting, "others" to change.** Remember that no other person needs to change for your world to change. *In fact, you can only make this change by changing yourself— only you can change your world! The help you are looking for is within, not without.*

- **Live as if there is no such thing as a mistake.** Instead, there are only new and different opportunities to deepen and enrich our experience and awareness. Every type of experience presents a chance for learning and an opportunity for growth. Your choices may shift your path within *time*, but all paths eventually arrive at the same place: a deeper knowing of oneself. There is nothing to ever regret, for all experiences combine to make your awareness greater. *There are no wrong paths*; some are possibly less "efficient" when viewed through a perspective that involves the "lens of *time*." Outside of *time*, in our real home, it makes absolutely no difference what path you enter; all of life's possibilities contribute to our expanding awareness.

- **Always remember that everything changes instantly with our participation.** Our focus, intent, and attention completely influence the unfolding of events. Do not pre-judge a potential experience based on past experiences. Every experience is unique because, each time, even if circumstances appear similar, our participation and attitude will influence and change the unfolding of events.

- **Remind ourselves often that everything we fear today will completely change its appearance once it has our full love and acceptance.** Once there is a full acceptance of our shadows and fears, integration begins and it is only through this step of completion or *whole-ing* that the *transmutation* or alchemy can occur. Things that seem scary or horrible to our minds and bodies completely change their nature once fully integrated. *Divided, both scalding and frozen water do not support life but mixed and integrated they produce the warm water that is essential for life. When we accept and embrace our dislikes or fears, they unexpectedly change quality upon integration.*

- **Practice finding the "positive" aspects that are inherent in every situation, person, event, and experience.** All things in life can always be viewed from many different perspectives. Remember that our physical world requires *contrast*; therefore, all things must contain what we consider to be "positive" and "negative" aspects. While we must never hide from the "negative" aspects, we must also not focus entirely on them. From our perspective, every situation will have both "positive" and "negative" appearing qualities to discover and explore: they are equally important for balance in our *dualistic* world. Always remember that our world of *duality* is built from these opposites; they must exist together to create the necessary balance for our *dualistic* universe and life itself. When "negatives" are embraced as an equally important part of the whole, they mysteriously *transmute* to become powerful allies. Recognize, understand, honor, and embrace the necessary balance in all things.

- *Stop rushing through life.* Because *time* is not what we once thought it was, there is also no real reason to hurry; we are already "here, now" and there is nowhere else to go. Slow down and savor each step of the journey. Examine the moments between the moments. Find and experience the *magical space* that is always available between the activities of our constant "doing."

- *Learn to be more patient.* Things will always take exactly the amount of human *time* that is required. We are always provided the correct and perfect amount of *time* for everything necessary. Allow life to *flow* naturally. Learn how to "be a passerby."

- *Give loving attention to whatever is right in front of us at every moment.* Live fully in each "present moment." Life automatically presents each of us with what we require for our perfectly paced growth. Every moment is presented as an opportunity to open up and become more of who we are. To have more clarity about what we should be doing at any moment, we require nothing more than our simple observation of what life is offering. Dive fully into each one of our present-moment gifts from the universe.

- *Always remind ourselves that everything is always perfect, no matter how it may appear to us at any one moment.* The universe is all-knowing and self-correcting. In our constrained realm of *time*, there may be the temporary appearance of imbalance, but this is only a human judgment of a hidden and always perfect process.

- *Learn what "attachments" are, and how to let them go.* Life is an experience that only unfolds in the *flow* of the present moment. All of our *attachments* to people, things, concepts, intentions, and experiences are related to wanting to control our environment and path. We tend to try to control our lives either through "manifesting" our desires (living in the future) or recreating experiences (living in the past). Within this journey called life, we learn that we actually have the power, as promised, to create many things. We are, after all, awesome *beings*. However, over *time*, we also discover that the results we manifest and make physical do not satisfy us the way we had imagined. This is because all types of *attachments* disturb the *magic* and the *flow* of the present-moment experience.

- ***Get out of our own way by learning what it means to live an "agenda-free" life.*** Develop a trust of life that allows for a more natural *flow*. Stop sabotaging the natural *flow* of our lives with our preconceived ideas and personal agenda. Once again, embracing what is right in front of us in every moment allows us to *flow* perfectly with the organic changes of life.

And…

- ***Develop an enduring "attitude of gratitude" for everything in life – even the difficult parts. Always be grateful for this gift of physical life, no matter how it may appear in any given moment.*** The physical expression of life is an amazing opportunity and gift. ***Everything that appears in life is involved in bringing us to each precious "now": the magical moment of unlimited possibilities.***

Through these reminders and practices, deep internal changes are initiated, as we discover how to love "others" and ourselves in new and often surprising ways. These guidelines are very helpful for exploring this phase of the *infinite* process of "becoming"; the part of the opening process that lies within the small but very critical part of existence that we think of as physical life. As with the twelve principles listed earlier, all of these practices overlap because they all describe a single thing: the natural and loving way of living that becomes obvious once we recognize the "oneness" in everything. ***At first, during the early stages of practice, this different way of "walking through the world" is not easy; it creates many new challenges for our conditioned minds. However, once this new approach is integrated and the natural inner shifts begin to unfold, everything about our lives will start to change in wonderful and completely unexpected ways. Today, there are many individuals who are living outwardly normal-appearing, yet very joyful, lives (myself included) who represent the "living proof" that these practices don't just work; they, in fact, work beautifully.***

Commit to these lifestyle changes for a short, but fixed amount of *time*. For clear insights, try a full six months; there is nothing to lose by "going for it."

There is no potential downside to this experiment because if you try and it doesn't work for you, the worst that might happen is that your world will not appear to change.

These recommendations, when integrated into daily life, support the deep, individual, and continuous process of opening or whole-ing. We all must be prepared for this process to be very difficult at times; there will be moments when everything will seem to be working against us. However, at other times, life will flow so easily and naturally that it will be perfectly clear that the universe is "rigged in our favor," and we will then find ourselves wondering why this ever seemed difficult.

Many writers, speakers, and workshop leaders have encouraged similar lifestyle changes through related approaches or technologies. Some of these people and methods are listed in the Resource section of this book. Try any of these techniques; there is no perfect or right practice for everyone. Any practices based upon the "oneness" worldview will point to the same place: a life steeped in a deeper truth, a life of freedom. *The greatest truths lie outside the limits of our ability to sense, conceptualize, or even imagine, so if we desire to experience this deeper level of truth, we must always be willing to explore beyond what we once understood to be "real."*

THE DEEPER DIVE

Eventually, inspired by small but significant changes, we may fully commit with a deep dive into the *whole-ing* process. Once we are in these deeper waters, our critical but invisible core belief systems can be fully witnessed, recognized, and allowed to expand. Like all changes in physical life, inner growth is a process that must unfold over *time*, so a trusting kind of patience is required. *It is also helpful to understand and remember that when we participate in this type of expansive growth, we are being asked to leave nothing behind except our limiting judgments.*

Ultimately, the attitudinal changes that make the most difference cannot be accomplished by any exercise or simply "changing" thoughts; instead, we must honestly examine, recognize, and fully "meet" our deepest beliefs: conscious, subconscious, and those buried and enfolded deep within the physical body itself. The roots of our existing belief systems are almost always completely invisible to our *ego*-controlled rational minds. Both physical and energetic, they can be discovered buried and hidden deep within muscles, organs, and even the bones of our bodies to then be powerfully reinforced again and again by our culture and its dominant paradigm. *Our core beliefs are rooted so deeply that, over time, they literally become physical – integrated into our bodies. To move beyond our obsolete, but deeply integrated beliefs, they must be fully seen, recognized, fully illuminated and met at every level. This change is completely contrary to our old "logical" belief system that automatically tries to exclude, change, dismiss, and ignore our inner darkness. We must, instead, fully welcome it into our being by honoring, appreciating, and embracing our darkness exactly as it appears. By recognizing our darkness as nothing more, or less, than those judged, rejected, separated, and "orphaned" parts of our once-whole being, we recognize that if we push any part of our whole being away, we are just continuing the old and tired human pattern of separation. Instead of dividing ourselves and creating even more separation, we begin our whole-ing by welcoming everything that we discover within.*

This part of the process, the exploration, integration, and embrace of our "darkness," may seem very strange, backward, unnatural, or completely wrong to many people. Why explore that which we have already rejected, dismissed, or escaped? There are structural reasons that can explain why this extremely challenging step is so critical, but those reasons all seem to evaporate once we begin to experience the surprising gifts we receive when we reconnect with our internal wholeness.

To be effective and satisfying, this healing process must create change at the deepest levels of our *being*, and because this involves a physical process, it takes *time*, focus, trust, and persistence. As our core beliefs gradually adjust, our hidden *resonant* vibrational patterns will naturally and organically shift. *Through this process, we gradually come to understand that our lives are, in fact, vibration-based and that absolutely everything we perceive and experience as reality begins, communicates, and manifests through the quantity, quality, and flow of this deeper-level vibrational information. Once we understand and integrate this critical step in the process of whole-ing, the begins to happen.*

We must understand that living this way is a process – a lifelong one. In our world, one defined by *time*, we will experience breakthroughs, plateaus, and even temporary regressions, and through it all, we must not expect an endpoint; there will be no single point of arrival, graduation, or *enlightenment*. That which we seek will instead appear to us, again and again, as a never-ending process of gradual *enlightenment*. If we have specific objectives, goals, or don't fully accept and allow for this to be an ongoing process, we will quickly become discouraged by the perpetually elusive nature of these old hopes that we harbor. From our perspective, locked in *time*, we naturally crave resolution—we perpetually seek conclusions and final results. Living is a process that does not work this way; it is truly a never-ending story. At the same time, when viewed from a deeper level, one that sits "outside of *time*," all of this "happens" in an instant. ***At a still even deeper level, nothing ever actually "happens," for there is only the "oneness" of being and everything just "is."***

PART TWO – THE HEALING JOURNEY

"These pains are your messengers. Listen to them."
Rumi

"What we don't heal within ourselves we seek
as a missing part in someone else."
Lyna Rose

THE WHOLE-ING BEGINS

Whole-ing: *The illumination, acceptance, and integration of all our divided and misunderstood parts, so that we can become a more whole and complete being and profoundly experience the sacred gift of life.*

We belong to a deep "oneness" that encompasses everything in existence. Because this "oneness" describes everything, it must include all that we judge to be "good" along with all that we consider to be "bad." Whole-ing involves learning how to live our lives free of separation and judgment so that we can discover, accept, and embrace a greater part of what is being offered right now to help us become more complete *beings*. This lifelong process finds and clears our hidden shadows and resistances, allowing us to become more open vessels for facilitating the *flow* of love and light. With this shift, the river of life can finally flow through undammed to guide and move us effortlessly within its free-flowing and joyful waters.

Whole-ing is based on new scientific interpretations that support the ancient principle that we are intimately connected to and therefore a part of everything in the universe, just as everything is also a part of us. We usually experience life through our cultural perspective of separate things and *beings*, but, at levels hidden from our senses, we are always fully interconnected to everything in existence. Because we function as part of this much larger entity that is "oneness," we are never only "this" or "that." We always exist as an integrated aspect of "all that is," yet at our current level of awareness, this stunning interconnectedness is almost completely invisible.

Because we are fully connected to absolutely "everything," this must include all that we judge to be good or evil, along with all that is judged right or wrong, as well as wonderful or terrifying. We have been conditioned to fear or reject half of life – all that is judged by us to be negative. When instead we learn how to open to life's full expanse, then unexpected and completely impossible-seeming magic can and will unfold. The *very* things that we habitually fear and push away will internally *transmute* to re-appear in a completely different light or form. Instead of actualizing our worst fears, they unexpectedly transform to become our most precious allies, adding to our lives in ways that we could have never previously imagined.

Growth is rarely an easy or straightforward process. As young children, we all had "growing pains." We banged our heads and skinned our knees because we did not yet fully occupy our bodies. Similarly, we are now discovering and learning to occupy more of our potential being, a being that is not just limited to these separated bodies. It helps our process to remember that any frustrations and difficulties that we experience are only "growing pains."

Whole-ing involves discovering how to move beyond our own personal and cultural resistance to reclaim our missing pieces and occupy our "full and entire being," so that we can better engage and enjoy the full flow of life. With whole-ing, we become much more available for all of the amazing, but often missed, opportunities that life is continuously presenting.

When we are fully open and available, our lives unfold through a series of unexpected but perfectly orchestrated encounters, synchronicities, and experiences, which we, alone, never could have imagined or constructed. When we are open and authentic in this way, the universe always seems to create and provide the perfect plan without ever relying on our conscious involvement. Through more experimentation, we then discover that if our conscious minds later engage and try to direct this natural process, the magic quickly collapses. Authentic living is an incredible process, but experiencing it fully requires a profound shift in perspective, one which presents an especially difficult challenge for the Western-trained mind.

The process of whole-ing is relatively easy to describe, but it requires extensive and broad experience with life to fully understand, appreciate, and integrate. Once intellectually understood, beginning the actual healing process might take only an instant, or it still could require lifetimes. Regardless of how quickly the *whole-ing* process is entered, its integration lasts forever; it is a never-ending process, an entirely new way of living each day, and one that is completely life-changing. Once we are able to commit and fully engage, the once-difficult process of living on Earth shifts to become more joyful, loving, completely sustainable, and largely free from personal suffering. *At its simplest and most distilled, whole-ing means completion: a re-integration of abandoned and rejected parts of our being that have been cast away. The old familiar feeling of void or emptiness within is then replaced with a permanent, full-filling, unwavering, and ecstatic joy.*

WHY WHOLE-ING?

"The gold is in the dark"
Carl Jung

Our lives are wonderful experiments for exploring our separation and interconnection. Most of us have glimpses that communicate that there is much more to our universe and life than most of our everyday activities and thoughts reflect. We sense that we are living lives that barely touch our potential for engagement, joy, and love. Most of us tip-toe through life tethered to a pervasive type of deep angst that leaves us feeling empty, incomplete, and driven to the endless external search for that intangible "something" that will fix our multitude of problems. Artists of all kinds have captured the feeling of both the angst and the imagined liberation, but rarely, if ever, are we offered practical and useful pathways or solutions that can be applied to our actual lives. We fill our *time* with activities; most represent futile attempts to distract ourselves from this ongoing angst that draws from our primal fear of what might emerge from the felt darkness within. Hoping to close this gaping wound, we strive to achieve more and consume more as we attempt to separate and define ourselves through careers, achievements, riches, or fame. Fearing our inevitable death as each passing year invites it closer, we may fall into a deep depression or build what we hope to be a more permanent form: a "legacy" that we pray will outlive these fragile bodies. Not only does this approach not work in the long term, but upon our inevitable failure to satisfy ourselves through such attempts, we often drop into even deeper despair.

Our culture teaches that we create happiness and immortality through external means. This pervasive belief is applied to almost everything we do; our entire culture is built upon the roots of this dysfunction. By the *time* we are adults, we are fully conditioned to compete with each other through fear and a culturally programmed need to "win." Instead of discovering ourselves through the joy of play, as we did when we were innocent children, we create even more separation and isolation by striving to define and distinguish ourselves as better, special, or more important. Then, this misguided programming expands to influence the agendas and policies of the organized groups we form in our neighborhoods, churches, and schools. This striving continues to grow, evolve, and mutate to become the insatiable hunger for complete dominance that infects and dominates many corporations and nations. These larger entities aggressively compete with each other for control of a market or region, employing any type of "warfare" that will increase their wealth and power. To this end, this level of confused and empowered expression of ego has demonstrated that it will sacrifice the welfare of individuals, cultures, and even our planet's health in its attempts to meet its corporate or national goals.

At the national and corporate levels, this push to expand, control, and dominate leads to wars, the destruction of our environment, poverty, hunger, disease, and immense human suffering. Basic stability and sustainability are rarely corporate goals because, in most of today's corporate boardrooms, stable profits and slower growth are seen as a serious problem to be immediately corrected; instead, a very unsustainable rapid expansion is valued the most. Our standard current business model places corporate profits and stock value above stability, the health and welfare of people, and the planet's environment. Within this unbalanced type of system, everything is assigned a value based solely on its ability to be bought and sold. In our corporate boardrooms and governments, whenever the welfare of the general population is considered, it is often only because their model requiring constant expansion also needs a steady stream of new consumers to purchase its products or build its armies. We live with a dominant economic paradigm that requires constant and rapid growth; a reflexive and unsustainable strategy that completely ignores the finite and fragile nature of our physical planet.

However, this depressing and cynical depiction of our lives only addresses one small part of our much greater, more balanced, and more complete existence. Nature is always working to balance opposing forces, and this natural drive to balance ultimately drives our entire physical universe. Because the universe will always seek equilibrium, when our lives appear to be out of balance, all of the countering forces also exist and are operating, even though our awareness of them might be obscured by the limits of our physical senses and our mind's interpretation of passing time. The forces and ways of being that create equilibrium are always operating, but are just hidden beyond our everyday awareness. In this physical expression defined by *time*, mankind is currently expressing a great amount of *separation*, but this only represents one position of the always swinging arm of life's pendulum. It has been propelled to this one extreme at this one particular moment in *time*, but because of our particular limited relationship to *time*, we only see a small part of the much bigger picture. When any pendulum swings in one direction, that action will always be followed by an equal swing the opposite way: a natural correction by nature to re-establish balance. In truth, this swinging back and forth between *separation* and *unity consciousness* is a very useful mechanism, for it allows us *time* and repeated opportunities for deeper exploration. **These swings of life's pendulum are exquisitely designed to provide a rich tapestry of life experiences that help to free us from our habitual judgments: particularly our need to judge things and actions either good or bad.** These radical swings, constantly altering our perspective, help us to reconsider what is possible within our physical realm. Life's always changing nature provides multiple opportunities for fulfilling mankind's primary purpose of exploring every corner and niche of *time* and *space*.

In this realm of separated *time* and *space*, life continuously appears to us through cyclic or repeating patterns. Throughout history, humans had to focus most of their *time*, *energy*, and resource on their basic human survival needs. As today's lifestyles and growing population place more demand on the world's resources, just meeting this basic need is very difficult or even impossible for many people. Fortunately, today, there are also many places on Earth where our basic needs are more easily met and where people have the *time* and resources to focus some portion of their *energy* elsewhere. While many individuals use much of this newly freed *time* to satisfy the relentless demands of their *egos*, more people than ever are choosing to focus on the further exploration of the body, mind, and our relationship to the rest of existence. Today, a large part of the world's population is experiencing and enjoying this expansive phase of the eternal pendulum's arc.

Only after thoroughly exploring many different types of external possibilities, will we be ready for a process as demanding as whole-ing. Only then can we fully recognize that we are still left with a profound sense of incompleteness: a shadowy void inside that is tied to life's meaning and purpose. Reflexively, we might seek to fill this void through external means (sex, drugs, success, money, and fame), but then, and only through their active exploration, we eventually discover these to be only distractions destined to leave us feeling even more unsatisfied and wondering what we are doing wrong.

In the deepest of truths, nothing is ever actually "wrong," with any of our explorations because life is for gaining exactly this type of personal experience. Here, deep into exploring our separation, we have purposefully, but only temporarily, lost sight of our interdependence. This phase of our pendulum swing, where we are still actively striving to achieve happiness through external actions, is a healthy and normal phase of *contraction* that serves to activate and energize our next major opening: the *expansion* of *being*. The deeper driving force behind this back-and-forth journey within *separation* is our deep need to uncover, "know," and integrate every aspect of our true nature. From a greater perspective, this is but a short phase of the much greater cyclic wave that constantly reverberates through *time* and *space*. One by one, we each come to realize that the outer is only, and always, a reflection of the inner, so all these global issues are only the external and secondary manifestations of our inner shadows.

From an even broader perspective, all is not just well, it is in fact, perfect. Just like when our children enter their terrible twos, we are calmed knowing "this too shall pass." For us to move beyond any *contractive* phase, and acquire its important insights, *time* is all that is necessary. No matter how hopeless it may look on the surface of this temporary and illusional physical manifestation, in the critical depths, all is always unfolding in perfect balance and harmony.

THE NEED TO CONTROL DESTROYS OUR FLOW

As we move into this new way of being, one of the biggest challenges for the Western mind is recognizing and adjusting for the cultural conditioning that drives our relentless need to be "in control" of outcomes. We completely lack the cultural context for understanding the great positive benefit of allowing life itself to move and guide us. The feeling of completion and wholeness that we all seek requires a true, broad, and deep availability that must include an openness to all possibilities, especially those we reflexively try to avoid. This idea will initially seem completely counterintuitive, but our lives are more "in control" when we entirely let go of our habitual need to control outcomes. Only after fully releasing our cultural need to control, are we able to more fully experience this wonderful gift of life that is being presented to us: a flawless choreographed flow that sequentially unfolds in ways that no one could have ever planned or predicted. Each moment it reveals is created, communicated, and coordinated at invisible levels to compile a living and engaging drama that is perfect for our individual and collective evolution.

The process of learning how to "flow" with life is not unlike learning how to drive a car down a winding country road. When we first learn to drive, we tend to hold the wheel too tightly as we misjudge curves, and often oversteer. Eventually, we learn how to relax and roll with the curves, and then the ride changes to become easy and flowing. Almost all of us have learned how to "let go" when driving while remaining alert, present, and ready to react if necessary; here we have already mastered flow. The most wonderful and enjoyable way to travel down the "river of life" is much the same.

For the "river of life" to flow freely, powerfully, and fully move us, we must not hold resistance of any kind: hidden or visible. Just like boulders in a river, all resistances, even those which lie unseen and unrecognized below the surface, interfere with the flow of the river. In the most extreme cases, they will form a dam that can completely stop the river flow. The ride through these impediments and resistances can be exciting and challenging, or it can be frightening and dangerous, but as long as the pathway still has boulders of resistance, it will not be smooth and flowing.

Despite all external appearances, options are always available to everyone. There are certainly many times when a wild river ride is the very thing that we desire; ultimately we are here to explore all possibilities. However, when we eventually tire of the endless obstacles, frustrations, and dangers and are ready for a different type of experience, we can then begin the process of clearing the rocks and debris. To change the nature of our ride, we must first learn how to "find, excavate, and remove" our accumulated resistance; this is the first part of the healing process or whole-ing. There is no "right way" or "wrong way," to engage life, for the raging river and the flowing river can both be enjoyed and have their important place in our lives; the key is recognizing that it is our hidden and visible shadows that determine which type of ride we experience. We can't simply choose with our minds or thoughts; the nature of our ride is only determined through deeper-level vibrational shifts within our entire being.

THREE TYPES OF RESISTANCES

We interact with our world through a phenomenal and extremely complex system: the physical body. Our body senses its environment through its powerful neurological system and its enlarged nerve center: the brain. To generate our thoughts, experiences, and feelings, the body utilizes many interwoven systems: many are unknown, some we don't understand, and some are only partially understood. In addition, our bodies rely upon types of communication and signal-processing systems that either we don't understand or of which we are completely unaware. *Within, throughout and between bodies, information signals are constantly flowing, dancing, and connecting in every direction; but within all of us, the most nuanced, interesting, and illuminating pathways are often blocked or slowed by our many resistances – both hidden and visible.*

Within this complex, we build and embody three primary types of resistance: mental, emotional, and physical. These resistances were not formed accidentally; at one *time* each had a purpose, useful for providing the support that was once important to our bodies and egos. All three types of resistance often began through functional and healthy expressions required for survival: mechanisms that were built for maximizing chances of physical survival and designed to enhance quick responses to dangers in our environment. However, for various reasons these same mechanisms can also work against us, become misdirected and confused, and habitually continue to alter our responses, long after their initial period of usefulness. For example, the "fight or flight" reaction can become subverted and gradually "normalized" within, and when not recognized and addressed, this once-functional pathway becomes the semi-permanent impediment named PTSD.

Our habitual resistances continue to interfere with our experience and our body's health, until at some point, after hiding and being ignored for far too long, we discover that they have fully manifested as chronic physical and psychological conditions. While once serving us, they have outlived their original purpose and now direct our bodies through the shadows of our unconscious, and from this unilluminated darkness, they influence our entire lives. The only long-term solution is to learn how to detect and clear these unwanted habits and resistances much earlier in the process before they have the *time* to manifest so powerfully.

DRUGS AND MEDICAL INTERVENTION

Modern psychiatry and psychology incorporate many useful tools to help address physical tension and emotional suffering. Unfortunately, the most recent medical trend has been for doctors to prescribe more psychoactive drugs instead of using their older and much more fundamental tool: psychotherapy. For severe and chronic cases the pharmacological approach can work well for emergency intervention, but it has also become clear that, over *time*, the continued use of these powerful drugs results in many undesired, and often dire, side effects. Once the life-threatening phase of the emergency has passed, psychological counseling becomes their best available tool for the next part of the healing process. To facilitate deep, real, and permanent healing, this standard medical approach also needs to integrate a number of the newest holistic techniques that address the entire body and greater energetic *being*.

HOLISTIC THERAPIES

One of the most effective methods that I am aware of for addressing the deep-level tension at its source involves the integration of several forms of bodywork. We might initially resort to bodywork for temporary pain and stress relief; this also creates extra *time*. Bodywork from a well-trained and skilled therapist not only facilitates tension release, but it sets the stage for gaining new insights into the deeper *attachment* issues that initially created this tension. A skilled therapist not only locates and works on the physical areas of tension and damage but also can suggest correlations, insights, and possible causes. At the very least, once we understand just where our bodies are holding physical tension, we can begin to pay more specific attention to these areas; and eventually, they will reveal internal clues about their energetic source. ***Through skilled bodywork, we can begin the process of learning about the deeper causes of our tension, while also receiving the needed, short-term relief.***

Along with insightful bodywork, it is always beneficial to include physical therapy and alternative, body-centered techniques such as ecstatic dance or yoga. There are also numerous new forms of psychological therapy such as *Emotional Freedom Technique (EFT™)* that can add to and greatly improve the effectiveness of psychological counseling. Another technique that I developed and named *Assisted Whole-ing Experience (AWE™)*, helps to build conscious awareness of our energetic bodies, their interactions, and their profound ability to interconnect.

Throughout the 1950s and 1960s, psychotherapists had great success with *psychedelic-assisted therapy,* but largely for political reasons, research into their effectiveness was completely halted in the 1970s. Several privately funded organizations such as the Multidisciplinary Association for Psychedelic Studies (MAPS) have been working ardently to make this once-promising therapy legal again. As discussed later, Michael Pollan's book, *How to*

Change Your Mind, does a wonderful job covering this fascinating but generationally forgotten, story.

If approached with an open mind, all of these, and dozens of other similar or related techniques, can work together, synergistically, to help us heal our minds and bodies. This adventure is a lifelong process of discovery; through it, we will continue to uncover more about our bodies and our even greater expansive *being.* The primary goal is always to enhance the free movement of *energy* and *information* (*flow*) so that we can live and express to our maximum potential. Once we are prepared for, and open to, this type of growth, we begin to use these tools more effectively, learning to identify and then better understand and release our long-hidden *attachments* to people, concepts, and things.

PREPARATION FOR WHOLE-ING

Some of our internal *resistance* was acquired just through living, and, because we have not yet learned how to sustain full awareness and illumination, new *resistances* continue to be secretly built within, and on top of, these existing blocked (shadowed) areas. However, we must always keep in mind that much of our *resistance* is not personal at all; it was inherited at birth. **Whole-ing involves the uncovering, illumination, and re-integration of these many layers of misunderstood, hidden, rejected, and isolated parts of our greater selves.** Discovery, opening, and integration are one way to break this cycle and initiate profound personal change and expansion. It is my personal belief that this type of growth is also an important part of each individual's deeper life purpose. I have become convinced that this experience of physical life is expressly for the expansion of our collective awareness, a process that must include the opening and clearing of each of our personal vessels.

Without even changing behavior, and simply by living our lives, we are already engaged in this important and eternal process; together we are continuously creating the full range of necessary experiences. **However, a more conscious approach and focus, such as with whole-ing, makes what was once a long difficult process more visible, direct, fluid, stress-free, and even enjoyable.** It also provides direction for moving us from this initial preparation phase of gaining experience, into the actual healing. Once engaged, this healing process, by itself, provides exactly the right amount of additional motivation and internal guidance for the rest of our journey into greater wholeness.

As we will discuss, there are three major phases within *whole-ing*, yet there is a fourth pre-stage that is always the most *time*-consuming and critical part of the entire process. It consists of the life preparation necessary to even become ready for the first phase of the actual *whole-ing* process. This preparation is the most difficult and *time*-consuming part of the entire process and it can't be rushed. Fortunately, life always creates exactly the right amount of *time* for each of us to reach our full readiness. Each *soul* must first encounter a broad range of life experiences to establish the motivation, clarity, and balance that is necessary to begin and then sustain the rest of this deep, often difficult, but always life-changing process.

Only through our "everyday" lives can we accrue the broad range of experience necessary. We continue our drama until that moment when we find ourselves desiring a very different type of experience. Then, instead of more of the same, such as more things, a better partner, or a bigger house, we develop and begin to grow a deep and driving longing for something that initially seems quite ineffable: a persistent desire for an entirely different way of walking through this world.

Only broad experience, with all of its difficult and challenging moments, along with those that are wonderful and smoothly flowing, will stimulate that powerful and enduring voice inside that passionately shouts, "There must be a better way." There are no short-cuts to this preliminary phase of preparation; it takes a full dose of life experience before one is completely prepared for diving into this life-changing process. The next step, the actual initiation of the *whole-ing* process, can only be successful once an individual has enough broad experience to become dissatisfied with the status quo, motivated to change old patterns, and deeply committed to real internal growth. This preliminary step required to enter the process can never be rushed, and each individual must reach this critical threshold before attempting more. *Time* is the wonderful gift that has been granted to us just for this preparation; it is always on our side.

ILLUMINATION OF OUR SHADOW WORLD

Each of the three major steps of the actual whole-ing process is integrated with techniques for uncovering and discovering our resistances. While some of our resistances may be visible and glaring, they are usually hiding deep within the shadows of our unconscious mind and body. All three phases of the *whole-ing* engage a common critical sub-process for the discovery, uncovering, and illumination of our personal and collective shadow world that lies hidden below many layers of egoic defense. This sub-process is a constant throughout *whole-ing* and with practice and *time*, this discovery and excavation sub-process become progressively easier, more precise, and much clearer. **This subprocess is not unlike combing out badly tangled hair; it is impossible to do all at once and very difficult at first, but with each pass, it becomes easier, smoother, and more rewarding.**

As discussed, we discover our shadows in many ways; different methods will work for different people at different stages of the process. *Psychoanalysis* and *psychological counseling* can be used to uncover beliefs that may be rooted in our shadow world. There are hundreds of forms of bodywork, some mentioned earlier, that along with many other new integrative therapies can offer surprising insights. As mentioned previously, *psychedelic assisted therapy* also has demonstrated enormous potential for exposing hidden beliefs and fears, along with demonstrating great promise as a professional tool for helping to rewire our reflexive reactions. While currently illegal in the United States, it now appears to be a "breakthrough therapy" for treating PTSD. Since it is also far less expensive and produces faster results than other currently available therapies, the legality of using *psychedelics* for professional therapy is now actively being reconsidered by our government.

While professional help can be very important and even critical early on, particularly if PTSD is involved, our greatest teachings and most informing experiences are always discovered internally, through events that arise by just living our everyday lives. Our relationships, physical issues, diseases, traumas, habits, problems, and old patterns all serve to illuminate our buried resistances. Once we learn how to listen more carefully, we discover that we have, and always had, this remarkable, yet very personalized, inner guru.

OPENING TO THE SHADOWS WITHIN

All aspects of *duality* contribute to shaping our always fluid landscape. As described throughout this book, our world is built from *dualities,* and by its very nature, this allows for any and every possibility. Through this mechanism, mankind did its job and co-created an enormous amount of diversity. Nature naturally seeks a broad range of expression for survival, so this wild diversity found throughout our world provides an extremely healthy and positive setting for physical life.

One problematic result of this extreme diversity for people is that there is little possibility of clear agreement, because even simple, non-emotional physical concepts such as "warm" and "cold" will be understood differently by those living in the Arctic and those living in equatorial deserts. Because humans are also "blessed" with a strong tendency towards forming opinions and beliefs, for every opinion or belief that someone has determined to be "correct," there will always be another person, somewhere in some situation, who will hold a very different view. All of these divergent parts are important contributors to the whole and, together, help make us who we are. An early step within each of our journeys is our opening to, and recognition of, these important contributions from all of the diverse parts of our greater *being*.

The process of becoming a more open and fully integrated human requires that we embrace all parts of our selves. This means not just an examination of our dual nature and all of the hidden shadows within, but also meeting them fully by recognizing and fully honoring their important contributions to our combined greater *being*. **However, the process must always begin with the sometimes challenging step of just recognizing the existence and importance of inner darkness. This recognition initiates the process of whole-ing, introducing the first sparkle of light that eventually brings these once rejected and hidden parts of our being into the full light of consciousness.**

WHOLE-*ING* IS LARGELY A SELF-GUIDED PROCESS

Whole-ing is, ultimately, an internal process that must be driven and guided from within. Others can coach, advise, and support, especially in the early stages, but the real work involves clearing the energetic pathways deep within our soul. Here we discover the ineffable landscape of hidden feelings, images, and sensations, through which only an illuminated and quieted self can find the clear path.

Throughout this entire process, there must be a continuous checking within, a feedback process that allows each of our unique energetic systems to open safely. As we encounter glitches of any type – physical, emotional, or mental – we must then slow down to better feel the newly discovered resistance. As we allow and receive whatever arises and explore our inner landscape, we must stretch, breathe, meditate, and seek to relax all aspects of our *being* so that we can feel more, not less of our inner selves. Just as we have been trained to do in most forms of yoga, we never force any aspect; all expansion and progress are built upon a very gradual and natural relaxation and release. We must slow down to breathe and relax into these newly discovered sensations and feelings until they become so comfortable and familiar that they easily release on their own. Only then do we begin to recognize these long lost parts of our *being* as the misunderstood orphans that they truly are.

THE THREE PHASES OF *WHOLE-ING*

The first major phase within the actual whole-ing process is learning how to curb and eventually stop the accumulation of new stress and resistances. A wound can't heal until the infection or "bleeding" is stopped. The first step must be to stop creating, building, and compounding new energetic blockages. Our typical and habitual Western response is that once we recognize that a "problem" exists, we tend to become impatient and want to "fix" it quickly. Our cultural programming usually makes this first step the most difficult. Moving through and beyond our initial desire for a superficial "quick-fix" requires an enormous amount of focus, motivation, clarity, and enduring trust in the entire process. Only experience and *time* can build this trust.

If there is hidden PTSD (*post-traumatic stress disorder*) present, it will rise and demand attention early in the process, because a brain rewired by PTSD will not support the open and clear *space* required for this level of healing. It seems ironic, but many, even most, who reach the threshold for entering the full *whole-ing* process do so only because they have already been struggling with PTSD. While PTSD is always a very troubling psychological condition, it is also a great motivator. Not only does it loudly announce a wound, but it often stimulates the desire for change that directly leads to a more profound level of healing.

Once the *flow* of new resistance is slowed, the second phase of *whole-ing* becomes effective. *The second phase is to "clear" the old embodied resistances that have already found a home within. Because these resistances have accumulated over the many years of living, they have become integrated parts of our being: not only do we unknowingly live our lives through these hidden shadows, but they also become structurally integrated with our physical bodies.* As we learn how to find and release these areas that have become blocked, our health, *energy*, and awareness improve as a multitude of previously unseen possibilities begins to emerge spontaneously. Beginning with this stage, the entire *whole-ing* process lightens and it can even begin to become fun.

Once the body and mind have learned how to release resistance and become more flowing and fluid, then integration, the last and final phase of whole-ing, becomes the focus. This part of the process allows each of us to move beyond our past limitations and boundaries to become more of "that which we are." In this phase, we learn how to drop our attachment to our personal identities (body, ego, mind) to heal, and make whole, our entire being. Ultimately, because the universe is *infinite* and we are a part of everything, as everything is also a part of us, there is no visible separation or ending to any of these phases. The entire *whole-ing* process just becomes a natural and fully integrated way of walking through life: a wonderful way of living and *being* that has always been available to everyone.

PHASE ONE – STEMMING THE TIDE

This first phase of *whole-ing* involves slowing, then halting, the ongoing buildup of new resistances, and this must include the treatment of any PTSD that exists. As discussed, it is often through the recognition and treatment of the life-altering effects of PTSD that a person is first led to this healing process. *Whole-ing* is not possible without first addressing any existing PTSD, and most of us have some degree of this deeply formed and completely automatic resistance. Buried within, these common but disruptive experiences must become more conscious before healing can proceed.

POST-TRAUMATIC STRESS DISORDER

> "It is through the cracks that the light gets in."
> *Leonard Cohen*

PTSD (*post-traumatic stress disorder*) is a new name for an ancient brain "disorder" that is now widespread. *Through exposure to extreme or repetitive trauma in critically stressful and often difficult to interpret or understand situations, our brain will rewire itself to bypass the slower, more analytical, and logical steps of its normal processing. Nature does this to provide a temporary mechanism that speeds reaction time for critical life-or-death situations.* Our brains automatically shortcut those parts of our thinking process that are not critical for physical survival to initiate quicker reflexive solutions. While intended to be only temporary, the deeper or more repetitious the trauma, the more permanent and solidified this short-circuited wiring can become; over *time* the pattern can, and will, become fully "burnt-in." In the short term, this re-wiring is a very useful adaptive feature because it increases survival chances for the body, yet the continued or extraordinarily deep arousal of these accelerated and heightened reflexes can be very problematic if this pattern becomes the active default mode for our "normal" lives.

Post-traumatic stress disorder (PTSD) causes actual physical changes to the brain's normal functioning and these changes also mean that we lose the ability to consciously observe and monitor our reactions. When significant PTSD is involved, healing requires a different and very careful initial approach. When the body and nervous system have been overextended by trauma, this neurological damage (trauma-induced rewiring) must be addressed first before the self-healing phase of *whole-ing* can be initiated. The PTSD "short-circuited" nervous system, makes it impossible to slow down the building of new resistances because innocuous everyday events become misinterpreted at the pre-conscious level; the normal stimulation and response pattern is no longer conscious, ordered, and direct. Once this deeply traumatized, the human nervous system jumps ahead of the normal healthy-brain step of being processed and recognized by the *medial prefrontal cortex,* where these everyday life-triggering events would be understood for what they actually are – harmless. Deep within our brains, but outside of our conscious awareness, the altered PTSD wiring is repeatedly shorting-circuiting and stimulating a response at the primal level of instinct, causing a hyper-arousal that automatically, cyclically, and continuously builds and compounds new levels of hidden resistance.

Deep within the more primitive parts of PTSD victims' brains, the *fight or flight reflex* has become automatic and habitual as the *hippocampus* shrinks and the *amygdala* swells. The *hippocampus,* the part of the brain that is most responsible for our memories, loses fluids and shrinks, making it harder for sufferers to retrieve memories and understand events in ordered *time*; past events are then experienced as if happening in present *time.* The *Amygdala,* which then enlarges and becomes a much more dominant player in the responses, is directly responsible for most of the symptoms we associate with PTSD: traumatic memories, negative mood, extreme startle response, and habitual avoidance of anything that might stimulate the trauma. Those that are afflicted can suddenly find themselves without normal conscious control of their minds and bodies; they can be quickly thrown into a rage or brought to tears over casual, everyday non-traumatic events. Victims may inexplicably experience insomnia, heart palpitations, memory loss, uncontrollable shaking, nightmares, and poor concentration. The constant hypervigilance creates elevated levels of stress hormones, and this makes it very difficult for the body to relax, control, and regulate itself.

PTSD has been observed for thousands of years, often described within chronicles of warfare as Shell Shock, Combat Fatigue, or Hysterical Neurosis. While PTSD was described over two thousand years ago by Homer in *The*

Odyssey, the ancient tale of a returning soldier, it is only recently that it has been officially recognized as a real and significant problem by the US government. Today, though officially recognized, it is often not openly addressed for a wide variety of political, financial, and personal reasons. Until a few years ago PTSD's important societal repercussions were suppressed or ignored, largely because of political and social pressure related to the recruitment, maintenance, and financing of our armies. Over the past forty years, the United States has seen a very dramatic rise of PTSD in large part because of, but not limited to, our military adventures. Soldiers who seek help because they were injured in this way are still routinely ostracized and treated as failures by their peers before being discharged and left to cope mostly by themselves.

While our culture's new level of recognition of PTSD contributes to its current visibility, the high-pressure environment created by the industrialization of a heavily populated planet and the rapid blur of modern life combine to produce many new avenues for trauma. After a traumatic event, especially when experiences were so extreme that they couldn't be interpreted or understood by our conscious minds, the trauma then becomes buried and hidden deep in the shadows of the unconscious, where the brain's wiring will not (and cannot) automatically re-adjust and correct itself.

The depth and effects of PTSD vary greatly among individuals, events, and situations, but it is almost guaranteed that, at some point in our lives, all of us will experience events that cause some re-wiring. In many, this trauma will trigger automatic responses that don't serve us in our day-to-day lives. If we have been having unexpected, or overly dramatic, reactions, such as unconsciously blurting-out or getting angry, teary, or frightened for no clear reason, then these responses indicate that some degree of PTSD may be present.

PTSD THERAPY

Because PTSD was not officially recognized until recently, research into therapies is still in its infancy; fortunately, PTSD is reversible, at least in many cases. The enlarged, over-sensitive *amygdala* can be helped to shrink and quiet down; the *hippocampus* can be trained to store and retrieve memories again, and the *sympathetic nervous system* can be coaxed away from its automatic and reactive mode. There are some very promising new therapeutic methods and "tools" available for this special level of healing, including the use of certain *psychedelics*. These methods will be discussed later in this section.

When addressing PTSD, professional help is always advised because it offers experience, insights, direction, and effective tools. Once we can consciously identify and remember the event and are fully aware of how it relates to our present-day reactions, the nature of the problem shifts to become more within our power to address. Through a more focused and conscious awareness, we can gradually un-knot the problematic wiring of PTSD. ***Professional intervention and therapy are required as long as the events are still confusing, there is still memory loss, or there is still the potential for reactions that might result in abuse or harm.***

SLOW OR STOP NEW RESISTANCES FIRST

After untangling any PTSD, we can begin the first phase of *whole-ing,* which is to effectively address the habitual and daily accumulation of new resistance. Every day in our lives we reinforce old habits that invisibly work to interfere with our freedom, but for the most part, we are completely unaware of doing this. When we are new-born, our minds are open to freely receive *information* from all parts of our environment and, in this open state, we learn and grow rapidly. As we grow, our brains start to make assumptions based on past experiences; this occurs naturally in the service of efficiency. We categorize and file our past experiences, and later we pull these up from our mental archives to make quicker decisions that save us *time, energy,* and occasionally even our physical lives. While the faster response *time* is often helpful, through this same process, our minds also gradually close to many types of unusual or unexpected experiences and *information.*

What we are witnessing is the operation of the brain's Default Mode Network (DMN). This mechanism is the tool that permits us to quickly repeat learned reactions so that we can skip steps and respond more quickly. We could not play music fluidly, participate in sports, or drive a car safely without this mechanism doing its job. However, because the entire system relies on the brain making assumptions about what we are experiencing, later, when we do encounter something entirely new, the DMN may automatically choose from its data bank of previously cataloged experiences to make an inaccurate or completely wrong determination. ***As we mature and gain experiences, new data that does not fit our crowded experiential programming may not even register in our conscious brains; we will often miss it entirely.*** This extremely interesting mental phenomenon, named *cognitive dissonance* by psychologists, will be discussed later at length.

We are constantly missing new data and repeating previous assumptions or responses without ever being aware that we are doing so, but this is completely normal and even healthy behavior because our brains are designed to do exactly this. ***Through this process, we are also constantly reinforcing older actions, decisions, and beliefs.*** Our judgments, fearful reactions, and isolating decisions of the past, both conscious and unconscious, are repeated, reinforced, and deepened. As we age we tend to tunnel deeper and deeper into our habitual patterns and all of this creates a kind of darkness within that divides through building hidden *shadowed* areas where the light of awareness does not easily penetrate.

As long as this process remains unexamined, it remains entirely automatic. Within the depths of our unawareness, our subconscious is always hard at work building new *resistances,* while at the same *time* strengthening these hidden cobwebs of old darkness. When we remain unaware, our *shadows* continue to grow by adding new layers of habitual *resistance,* so there is no *time,* or *space,* to even begin our examination of any older blockages that have long been interfering with the free *flow* of our *energy.* If we have not first addressed this automatic pattern of layering new *resistance* over the old, when we uncover older existing blockages and attempt to do the work for clearing them, we only get confused and frustrated as new issues compound faster than any clearing we can achieve. Due to layering, interweaving, and mingling of new and old, the clearing process can become so clouded and convoluted that we find it impossible to even discover where the old darkness resides. Operating in this confused and overwhelmed state, most of our attempts at clearing and opening will be completely futile. Before we can even hope to clear old blockages, we first must stop, or at least greatly slow, the accumulation of new resistance.

Learning how to stop, or even slow, the compounding upon our existing "*shadows*" is usually the most difficult phase of the entire *whole-ing* process and it must be almost entirely self-guided. During these early stages, the rewards and benefits are often difficult to see or understand, while the traps and temptations are many, so staying with this step of *whole-ing* requires an enormous amount of personal commitment and deep trust in the overall process. ***This is why we must approach whole-ing fully prepared with the right motivation and deep life experience.***

Whenever we attempt to explore our *shadows,* progress is always being made, yet progress is rarely visible, measurable, or rewarding during this first phase. This step always requires *time* because the inner discovery techniques and methods that we develop in this step will gradually be refined, and improved to become the essential tools we use throughout the rest of the *whole-ing* process, as well as the rest of our lives. It is completely normal for this step to take many years, possibly the majority of a lifetime, but these will also be wonderful and enjoyable years because as this process unfolds, it progressively becomes easier and much more fun. Much of this section is devoted to this first critical phase since the rest of the *whole-ing* process largely involves the deepening and refinement of the new tools and awareness that is built within this first phase.

PHASE TWO – CLEARING THE OLD

"How can I begin anything new with all of yesterday in me?" *Leonard Cohen*
"Only the hand that erases can write the true thing."
Meister Eckhart

Once we have successfully recognized, addressed, and abated the development of new *resistance,* the path to *whole-ing* simplifies greatly. With new resistance slowed, we can now focus on clearing the old and embodied tensions that continuously interfere with our fluidity, functioning, and health. At this important milestone in our process, we can recognize that even though the buildup of new *resistances* has ended, there are still physical and emotional knots and kinks that continue to operate and interfere with our lives. We find these hidden within the three primary expressions of our physical existence: our minds, bodies, and emotions. Deeply embedded, these older and more embodied resistances are fully supported and reinforced by our culture and unexamined personal habits. They have become ingrained and integrated through the years, some beginning even before our present body began its physical journey; we arrive in our physical form with resistances already embodied deep within, inheritances from both our culture and our evolutionary genetics.

The experiences of our lifetime will add new layers, but the general patterns and their deeper roots are nothing new; they have appeared, again and again, throughout human history. Over our long human history, the *collective consciousness* has assembled a vast library of human experience. In our depths, we have access to all of these deep-level patterns common to all humans and living *beings*, including the primal survival fear that has driven so much of our suffering, thought, emotion, and activity. At this deep level, the resistances discovered are no longer personal; they are even beyond just being species-wide; they are deep archetypal fears of life itself. Here in our *timeless* depths, through a non-verbal internal process that taps into our long history of *holographic* embodiment, we can excavate and witness the entire energetic record of mankind's evolutionary and cultural history.

Once the building of new resistances has been addressed and slowed, the *illumination* of our embodied resistances becomes an easier and much more comfortable process. Not only has the process become much more fluid and joyful but, by this point, it also has become irreversible. Again, reaching this important and happy milestone is the most difficult part of the entire process because to arrive at this point a great amount of dedication, clarity, perseverance, and most importantly *time* are required. **Having first worked to eliminate the constant building of new resistances, we have not only averted the potential flood of confusing new issues but we have also created a safe and more comfortable space, where all discovery is welcome, supported, and able to be freely explored. Simultaneously, we also have gained many valuable self-help tools to assist the rest of this process and our journey beyond.**

In this second phase, the process changes dramatically and becomes more manageable because we are no longer attempting to swim upstream against a raging river in the middle of a rainstorm. Slowing the rushing tide of new resistance makes it possible to relax and explore within at whatever pace we find most enjoyable. By this phase we have started to see clear benefits and the more stable healing environment allows us to feel completely safe, so we trust the process even more. Now the old challenges of life can be seen as new opportunities, as a deep and resonant undertone of joy also becomes an integral part of our everyday experience.

As we continue to open, over *time*, we begin to experience the availability of greater amounts of personal clarity and *energy*. Instead of pushing against our resistances and exhausting our *energy*, the river of life appears to widen and flow more gently, and with that shift, our ride also becomes easier. This is exactly what happens when we remove our finger (the resistance) from the end of a garden hose: more water flows, but there is now less pressure, no misdirected spray, and far less work. This new level of available *energy* and enthusiasm is reminiscent of how we once felt as children.

Eventually, we rediscover how to just "float" and let this powerful but gentle river guide us more fully. At this point we begin to recognize that the "river of life" is actually doing the "heavy work" for us; life does become easier and we begin to feel a "lightness of *being*." Everything about living seems to change as we gradually realize that our once-difficult life has quietly become naturally easy and even fun. Instead of constantly fighting the head-on current, we have learned how to float upon it, hitching a free ride on the river of life. We have relearned how to play in our joyful freedom, but now, unlike children, we are permanently open and wise through experience.

PHASE THREE – OUR BECOMING

Once the second phase has begun, our awareness begins to expand at a rapid rate. Life now feels much lighter, more gentle, and *flowing*. What we once thought impossible, rapidly becomes very normal and every-day. Unimagined ways of *being* that were never a part of our old awareness become wonderful new possibilities. We even drop the entire concept of "impossible" as the miraculous becomes our daily norm. These are all changes that occur automatically. *These changes mark the beginning of the third phase, the opening of new doors to the rarely seen world that Christ and other mystics describe. This transition brings the realization of our life purpose and requires no help or guidance from the outside; residing beyond words and logic, it is, and must be, entirely communicated from within. This phase also marks a new level of recognizing our self through the witnessing of our direct connection to everything in existence, along with a fully embodied knowledge that the outer world we experience is only a reflection of the inner.* At this point, we have built a solid new platform from which the *soul* can step beyond the old, but entirely self-defined, limits that we previously assigned to our physical existence. It also marks the start of our ability to enjoy the fully conscious experience of our eternal and collective *being*. *In phase three, we experience our wholeness almost continuously and the few small gaps pass quickly and rarely disturb our lives. We recognize that this was always our birthright, but consciousness desired that we first fully explore and know this for ourselves, through our direct experiences in the separated realm of time and space.*

WHAT IS "HEALTH"

The answer to the question "What is health?" depends on whether we are asking about just the physical body or the health of our entire *being*. While *whole-ing* is focused upon the health of the entire *being*, it also has a direct, and important relationship to the physical health of our bodies.

Each of us may have a slightly different interpretation of the word "healthy." For most, "healthy" usually means having no, or few, medical "problems," having a strong and muscular body, and possibly being pain-free. We may also notice the additional quality of "sparkle," "glow," or powerful spirit that seems to surround or radiate from some people.

While all these qualities can sometimes be observed together in "healthy" people, a person may also be *whole,* fully healed, and spiritually very healthy even though their physical body is actively failing. Using this definition, a "healthy" person is one who is communicating with and freely resonating through the many invisible layers of their full *being.* **While the spiritually healthy person, living in freedom, will fully employ the river of life to direct and move them, at some point in time, for even the most physically healthy, this movement must include the release of their temporary physical body. Throughout this entire transition process, one that we commonly think of as "death," an aware individual will remain spiritually healthy.**

Sometimes nature's requirement for complete healing may be the "dropping" of a particular body form, one that once served but has become a limiting factor for the growth of *being.* If we truly understand that this physical life is only an experience, one that our *timeless* spirit creates for a special adventure in *time* and *space,* then our understanding of "healthy" will be different from that of our predominant culture. If we fully know that we are so much more than just "these bodies," then we can more easily understand and abide in the deeper wisdom that recognizes purpose in the transitional nature of these temporary forms. **Our lives become free once we fully realize that our being is only having this specific experience of a body in time and space, and when this experience is complete, we will be automatically guided to the next perfect form for continuing the expansion of our awareness.**

If we are living lives confused by separation and isolation, judging and comparing ourselves to others, and not examining the roots of this separation or the personal darkness within, then regardless of external appearances, our bodies will eventually reflect the inevitable suffering created by this way of living. It is impossible to live in an unexamined state of separation for an extended length of *time* and remain physically healthy. On the other hand, while we can expect the river of life to present unexpected, unplanned, and sometimes difficult adventures, some involving difficult physical challenges, spiritual growth, even when demanding, generally supports better physical health.

Even when a person's body may not appear to be whole or fit, nature still finds unique ways to fully utilize its form, for true health transcends these temporary bodies. *Many examples demonstrate that it is largely the limitations and problems that each of us encounters throughout our lives that drive, and carefully shape, each of our unique and perfect adventures. Examples abound and one needs to look no further than the example of the recently transitioned physicist Stephen Hawking. Confined to a wheelchair with minimal and rapidly diminishing communicative abilities, his life allowed the world to witness an extraordinary example of how life can express itself powerfully, despite or, as Hawking himself noted, largely because of a dramatically restricted body.* While his technical contributions were enormous, his demonstration of the power of spirit is even more significant. While spirit and body relate, his example demonstrates that a powerful expression of human spirit does not require what most of us consider to be a healthy body. The timing of his life was such that the entire world could witness and follow this remarkable miracle, one that dramatically demonstrated the power of spirit to express itself fully, even if its form is given only the tiniest window for this expression.

Once we have fully integrated the inner discovery of our "oneness," it becomes clear that resisting anything or anyone is tantamount to rejecting some important aspect of our greater self; this awareness must also include any resistance about perceived physical "limitations." Through the exploration of our separation, we eventually reach a point in our journey where we can be truly thankful for all of our gifts, including those of our limitations: physical, cultural, economic, and social. The deeper understandings that there are no accidents or wrong paths in life, and that all of our varied experiences add to our personal and collective wisdom, only become strengthened through the embrace and exploration of our limitations. By fully living our lives as they are presented to us, we deepen our personal and collective "knowing," and through this, our awareness grows as we continue to expand the bounds of our *being.* Life, just as it is presented, is always showing us exactly how to get out of our own way.

Cognitive Dissonance

Over millennia, as we collectively evolved, our culture gradually formed and then solidified consensus beliefs about how our world and the greater universe works. This set of assumptions, understandings, values, concepts, and practices shapes our cultural paradigm, which is then instilled into each of us during our cultural upbringing; it becomes our "guidebook" and largely determines what we think is possible in life. That which we call "common sense" is only a sub-set of this worldview and is completely shaped through our cultural programming.

Today, the vast majority will not even think to question our general cultural agreement about the "realness" of our physical existence; only a very small minority of disappearing subcultures embrace a significantly different view. Each of us has many personal experiences that fit and support this collective and deeply entrenched paradigm, and this constant reinforcement makes the illusion of a fixed and solid universe appear even more real. Ultimately, because we are extremely powerful creators who continuously re-manifest this same "agreed-upon" version of the physical world from our collective beliefs, this "agreement" also functions as a self-fulfilling prophecy.

Each of us also has a much less-understood catalog of personal experiences that do not easily fit this collective vision, and these we manage in a wide variety of ways. These "out-of-the-box" experiences often involve things appearing unexpectedly or out of ordered *time*, such as with déjà vu, synchronicities, precognition, or ESP. While these unexpected situations are unfolding constantly, in our day to day lives we seldom even notice or register those parts of any experience that fall outside of our dominant cultural agreement. This happens because we all have a natural filtering system within our brains that hides and obscures certain types of *information* received from our environment. *Cognitive dissonance* is the name psychologists use to describe this, our brain's unique way of filtering out certain *information*.

Most of the *time*, our brain only can process new *information* if it fits with what we already understand, believe, and therefore expect: our previously understood worldview or paradigm. This means that things or experiences that fall outside of our "agreed upon" worldview are often not even recognized, seen, cognized, or processed. **We must always build our newer concepts from the bricks and mortar of older concepts: those that our mind already understands.** Our brain has evolved this way primarily for efficiency and survival. This automatic filtering or restriction of the amount and types of conscious input is largely to help our brain work quickly and efficiently in our three-dimensional universe and to prevent it from becoming constantly overwhelmed. *Cognitive dissonance* is the direct result of this "conscious awareness filter. Through this largely automatic process, we cut out most of our strange or unexpected experiences; they are not even "seen" by our conscious awareness and certainly not remembered. Due to *cognitive* *dissonance*, we are usually not even aware of these many "out-of-the-box" experiences that continuously unfold every day.

Occasionally we may notice some of the more unusual aspects but because we have no real way to understand or relate to these, we usually remain quiet and just try to ignore them. If what we accidentally notice disturbs our sense of well-being, for our self-protection, our neurological system might then go further and move to "forget" the incident by burying it within our subconscious. If these strange incidents seem to be happening frequently, a common way to avoid these disturbing signals is to keep our minds very busy and engaged in unrelated activities, such as socializing, gaming, working, or playing.

On certain occasions, we might recognize these "out-of-the-box" occurrences as the "paradigm busters" that they are but still choose to ignore them because of fear or other social or personal concerns. Knowing that there is no acceptable context for processing these events within our current culture, we usually find ourselves too embarrassed or uncomfortable to even share these experiences with others. We sometimes attend churches or join spiritual groups in our attempts to normalize, discover meaning, or provide a language for these unexpected experiences. However, not all of our unusual encounters are lost in "the prison of our paradigm." Some will break through the veil to be expressed as visual art, books, poetry, song, movies, insights, spiritual experiences, and direct personal growth.

MY PERSONAL INVITATION TO *WHOLE-ING*

Our universe is structured to create wide-ranging diversity, vibrant experiences, a multitude of possibilities, and an *infinite* number of opportunities for realization. Within this astounding, experientially rich, and fecund playland there are as many unique paths to explore as there are conscious *beings*; this is not a coincidence, for diversity is one of the most fundamental driving forces in the cosmos. While many paths will bring us to *whole-ing*, all eventually lead to the same destination: the recognition, knowing, and the eventual integration of the deep interconnectedness that is the foundation of our "oneness."

WHERE SCIENCE, SPIRITUALITY AND EXPERIENCE INTERSECT

My description of our universe speaks to all aspects of our lives, including those parts that often appear mysterious and are likely to remain unprovable through rhetoric, logic, science, or mathematics. ***While my primary entry point to this unconventional vision was through the sciences, my work***

and conclusions rely equally on what I have gained from direct experience and my lifelong study of, and experience with, many of our long-standing spiritual traditions. All three human endeavors: science, direct experience, and cultural wisdom were needed to build a foundation that was strong enough for me to trust.

It was the science, in particular my questions about the meaning of the *quantum physics* and *relativity* that I was studying, which initiated and then largely guided my lifelong journey down what at first, and often thereafter, seemed to be a very lonely path. **Today, I never feel alone. Happily, I have discovered that I am but one part of a broad, diverse, and growing group of individuals who are living and sharing their discovery of this deep and powerful knowing.** I now live a life free from many layers of the imagined constraints and bounds that once interfered with my ability to see and enjoy this deeper truth. Today, every aspect of my life is experienced through this expanded vision, and this has changed everything. Every day is now met with joyful anticipation and every challenge is recognized as an opportunity, although not always in the first few breaths. A continuous level of peace and joy fills me, no matter how my outward circumstances shift and change, as what was once a very difficult struggle has evolved into a joyful and always interesting *flow*.

THE BODY'S SPECIAL WISDOM

A critical breakthrough in my early explorations was learning to listen to my body's quiet wisdom. As a child, I had many allergies, respiratory problems, reoccurring infections, tight muscles and tendons, and I was often "sick." Because these physical problems kept appearing in my life, and western medicine offered little relief, I was eventually forced to pay extra attention to my own body. After an immense struggle, I eventually learned how to recognize and accept my physical issues and illnesses as my guides. I discovered that once I learned how to listen within, my body provided wise, but often very unexpected, answers. Over *time*, I learned to decipher what my body was trying to tell me about my habits and the way I had been living my life. One *time* in my twenties, before I learned how to listen nonverbally, it even communicated a critical message which I heard in clear and distinct English, demonstrating that this guidance can sometimes even arrive with words attached. Because life is always a process, this same type of clear verbal communication was needed, once again, much later, when I was sixty. However, for me, such clear verbal directness is rare, because, throughout my life, the most common form of communication with deeper spirit has been through the forms of sensation, impulse, insight, and the release of tension.

In retrospect, I can now see that my physical issues were a great gift; they acted as my very personal "canary," and through them, I discovered the deeper value of yoga, massage, athletics, movement, and dance. Then, through these body-oriented techniques, I gradually explored my hidden and expressed tensions and discomforts that were the initial messengers signaling habitual resistances and holding patterns. Eventually, I became thankful for my body's extra "sensitivity" because I began to recognize how it helped guide me (kicking and screaming of course) to more lasting and healthy solutions, and ultimately to my most profound realizations.

BECOMING FAMILIAR WITH THE EMOTIONAL BODY

Learning to recognize and accurately interpret these often subtle signs and signals is a critical part of *whole-ing*. I gradually discovered that my body's aches, pains, illnesses, and tensions were far from accidental or random. They were, instead, deep-level communications asking me to pay more attention, and, once I learned how to listen to this inner voice, a series of astonishing journeys, insights, and changes quickly followed. I also discovered that my body was never separate from my thoughts or emotions. All three were fully intertwined, but they all communicated to me primarily through my body.

Gradually, guided by this physical form of *self-inquiry*, and through several dramatic and paradigm-shifting experiences, I came to understand that our bodies, thoughts, and emotions are all tightly interwoven, and the physical world we experience is not as solid and immutable as I once thoroughly believed. By paying attention to my experiences, especially the unusual ones, I became aware of the existence and importance of our underlying energetic bodies. However, I still felt that I could not freely share this knowledge because I feared that most people would think it was a little crazy.

HAVING THE EXPERIENCE OF A BODY

As decades passed, a few contemporary physicists began to speak more freely about the energetic quantum field being the primary reality and our physical world being just a secondary manifestation. With this important shift, I began to feel more at ease, knowing that some of my earlier internal realizations were also being recognized by mainstream scientists. Only then did I become comfortable enough to openly discuss and share my earlier realizations of the profound connection between my personal life, our spiritual traditions, and this exciting new physics.

I have always had great trust and respect for the processes of our natural world; nature tends to not waste energy and as we learn more about our world, we always find deeper reasons for its methods. As I gained more experience, it gradually became clear to me that we must be having this experience of separation employing bodies, space, and time for some greater purpose. Eventually, I began to understand that this purpose must be to explore, play, and come to know more of our infinite being. I now understand that the way our embodiment appears to us is the most direct form of communication possible: a type of message that we are just learning how to properly read. *The physical manifestation of our lives provides each of us with a remarkably clear road map, communicating exactly where we are in both our individual and our collective journeys. If we learn how to read our personal roadmap correctly, without our usual preconceptions, judgments, expectations, or fears, it will always guide us directly to our next perfect step in the never-ending process of discovering our full being.* If we resist its clear message, it still guides us, but we encounter more difficulties along the way and suffer deeper and longer.

Over *time*, I gradually was able to also integrate the life-changing idea that we are not tied to these physical bodies which are, instead, just handy but temporary devices that we are utilizing for this unique and illuminating part of our journey. *My path was built upon discovering that my body was a great guru; while it raised many questions, it also always provided the most meaningful answers.*

EXPLORING THE SHADOWS OF MY TRAUMATIC PAST

It was only many years later, while exploring the sexual trauma of my youth, that I was introduced to the works of Bessel van der Kolk, M.D. and Peter A. Levine, Ph.D., two of the pioneering researchers of PTSD. (Recommended readings by both are listed in the resource section.) I had studied Freud and Jung much earlier, but only here, through the direct exploration of my PTSD and the personal trauma of my youth, did I begin the part of my journey where I came to understand the significance and power of our shadow world.

THIS IS NOT A RELIGIOUS IDEA

While some of my conclusions may sound similar to the beliefs of several existing spiritual and religious paths, as I discussed earlier, my journey was not initiated or focused there. I arrived at my understanding from a very different direction; my path began with the merging of two personal interests:

the understanding of my unusual personal experiences and my fascination with the meaning of the most recent science, in particular *quantum physics*. My mind was filled with what seemed like unrelated scientific and philosophical ideas, while many of what later became life-changing revelations were first being signaled through my physical body. It was only much later, after decades of gestation, that I recognized that I had been exploring the territory where the scientific, experiential, and spiritual paths all converged. Today, thanks largely to my very demanding and personal "early warning system," I am as healthy as any person I know. *I have also come to realize that disease, chronic illness, and even accidents only follow later when these early signs and signals are repeatedly ignored. Life seems to be "designed" so that we eventually will get the "message" and evolve. The only thing we can adjust is the timing.*

PERSONAL PARADIGM-BUSTING EXPERIENCES

Through this lifelong process, I have discovered so many interesting things about my *being*, along with satisfying answers to most of my many early questions about life and meaning. My experiences have consistently brought me to new paths for exploration, which then, in turn, led to the discovery of even deeper truths. I feel very fortunate to somehow have been able to recognize a significant number of these dramatic "out-of-the-box" experiences early in my own life. These experiences helped open my mind to other possibilities for our existence and dramatically altered the way I understand *time, space,* and our physical reality. In my earlier books I discuss many of my most significant "paradigm-busting" experiences—some from my own life and some from good friends, family, and others; what follows are just a few of the most pivotal and thought-provoking.

When I was seventeen, I had an experience with *time* slowing and stopping so radically that I had the all the *time* I needed to react and save my life in a high-speed auto accident. After that experience I was left with a deep understanding that *time* was not as what I once thought; it was somehow changeable, variable, and able to be controlled by the mind in special circumstances. My next conscious encounter with this level of the unexplainable was when a good friend inexplicably had such severe pain that it required her emergency hospitalization, while unbeknownst to her, her twin sister was dying after an auto accident eight thousand miles away. This event radically shattered my old worldview and helped to initiate my interest in our interconnectedness and types of unexplainable intercommunication. A few years later, a mysterious but invisible voice spoke to me in clear English, offering fresh and needed personal advice in a very loving and fatherly way. That advice was that my life would improve once I practiced searching for the good and positive that are inherent in all situations. I followed that advice closely and over *time* my life shifted in a wonderful way. The next strange personal event was a spontaneous and irrefutable case of my conscious awareness leaving my body to view a previously unseen landscape from a vista point twenty miles away. When I later drove to the spot and confirmed that this was the scene that I had clearly but mysteriously viewed earlier, my understanding of awareness being rigidly tied to our bodies was forever changed.

Over the next twenty years, an extended series of impossible-to-arrange synchronicities and encounters continued to remind me that the unfolding of life was far from just mechanical or random. Several *pre-cognition* and *déjà vu* experiences added to my growing sense that our collective idea of a one-directional marching *time* was not so absolute. During this same period, I experienced several physical problems that did not respond to normal medical treatment but responded well to sometimes strange-sounding alternative methods. When I turned sixty, a difficult internal opening resulted in an emotional shift that instantly healed a severe medical problem. This very dramatic personal event is what finally convinced me of the awe-some transformational power of *forgiveness,* and the degree to which our hidden and not understood emotional and *energetic* bodies shape our physical lives. Here are the full versions of these life-changing experiences that completely altered my old worldview.

TIME SLOWS IN AN AUTO ACCIDENT

I place this particular story first because it was the crucial life event that marked the actual beginning of my conscious exploration of the edges of our awareness. It was probably not the first *time* that something like this had happened to me, but it was the first *time* that I consciously felt the strong need to reevaluate my inherited worldview.

I was seventeen years old and driving my parents' car to a friend's house that was about twenty miles away when this unexpected experience completely changed the way that I understand and relate to *time*. My family lived in a rural area, one where the long stretches of empty, two-lane highway invited young male drivers to push the limits of speed. Feeling bulletproof, as I often did in those youthful testosterone days, I was driving over 100 miles per hour on a mostly straight and usually empty road. On a long straightaway, another car inexplicably pulled out right in front of me. Rather than slow or brake, I casually swerved into the clear oncoming lane. Completely unaware of my car or my maneuver, the older couple in the other vehicle suddenly turned left directly in front of me; they never saw me—a car traveling over 100 MPH was probably unexpected and outside of their *viewport*. With no place else to go, I reflexively swerved off the road into a very rutty orchard filled with rows of large closely spaced trees.

At over 100 miles per hour, all of this happened very fast; but unexpectedly, at that same moment, I found that suddenly I had "all the time in the world." My sense of movement and *time* was instantly altered; my awareness became a slow-motion slideshow revealing one clear *image* at a *time*. Each view held as long as I needed for study and reaction, and then it automatically shifted to the next frozen scene. Each sequential scene was a perfectly arranged, still picture of exactly what *information* I needed to see and understand for the next critical maneuver, and I was always given the perfect amount of *time* to look at everything. I was aware of every direction, every corner of the car, and every tree; and I do not think that I ever turned my head. From my perspective, *time* actually stopped and then progressed "one frame at a *time*" in slow motion and somehow also with a full three-hundred-and-sixty-degree view; I had all the *time* and *information* that I needed to decide which direction I should steer and how to go about each maneuver. I slowly and deliberately executed the movements required to negotiate every single still frame before checking again, and I repeated this process with each slowly advancing frame. It was as if I had a remote control and my slideshow was clicking ahead six inches at a *time*, pausing just long enough for me to observe, react, and then assess the next six inches of necessary movement. I cleared dozens of trees by inches and then, a few hundred very bumpy yards later, I brought the car to a halt in a cloud of dust. Once fully stopped I began to return to normal consciousness, and I could finally study and analyze my overall physical situation. I noticed for the first *time* that my heart was pounding uncontrollably, and after a long pause to catch my breath and slow my heart, I assessed that not only was I alive and well, but I had not even visibly damaged my parents' car (although I am quite sure that its suspension was never the same).

A sudden awareness dominated my entire *being*. "What in the world had just happened?" I turned around and there I was, now sitting in the middle of a potato field – a thicket of big trees behind me and a clear set of tire ruts marking the radical maneuvers that I had just completed at over 100 mph. "This really happened!" At that moment, my life was forever changed. I had encountered something that my training and upbringing could not explain, and was suddenly thrust into a completely different perception of *time* and reality!

Later, I was talking about *time* perception with my sister and she recounted a similar story. She was involved in a multi-car accident on the freeway, and for her *time* also slowed to almost a stop. She had *time* to observe the rapidly changing situation and knew exactly when and how to duck to avoid flying objects and glass.

Many athletes have similar stories of *time* slowing, and being in a "zone" while playing their sport. Motocross racers will talk about the *time* they had to make conscious decisions while flying through the air, mid-accident. Just recently I crashed my bike, and as I was flying through the air, my first thought was, "This is going to be bad so I might as well relax and go with it." At that thought *time* slowed and I was able to break my fall and roll perfectly to the applause of a crowd of bystanders. In my musical career, I have had similar experiences. When playing a fast and complex musical phrase, I occasionally have all the *time* necessary for feeling, phrasing, and executing the quickly passing phrase. Similar *time*-altering experiences occur in many situations throughout all of our lives. Our personal sense about the passing of *time* is extremely malleable and not nearly as fixed or rigid as our timepieces, schedules, and culture would have us believe.

Close Connection Between Twins

Almost fifty years ago I was living in Wisconsin and my girlfriend's best friend was a twin. Her twin sister, whom I had never met, was living over four-thousand miles away in London. One night my girlfriend received a call from her friend, who was in terrible pain and needed a ride to the hospital. All night long, the hospital staff conducted tests, searching for the cause of her excruciating pain, yet all the tests proved negative, leaving all the doctors thoroughly baffled. In the morning, the pain suddenly stopped and she was sent home, only to receive a call within the hour informing her that her twin sister had been in a terrible auto accident in London, survived for several hours, but ultimately died. The timing of her sister's accident precisely correlated to her pain the night before!

This type of special intimate connection between twins is well documented. In our culture, even though we recognize these unusually close relationships, we can offer no clear explanation for the existence of these particularly strong connections. We are now discovering that through the deeply interconnected and *non-local* structure of our universe, we are all intimately connected; but it seems that the level of connection between some people (and especially twins) lies much closer to the surface. This is the first *time* that I remember noticing the existence of these strong but invisible connections that cannot be explained from within our old paradigm.

HEARING THE VOICE OF "GOD"

A most unexpected encounter occurred on the deck of a sailboat in the middle of the Pacific Ocean, while I was living in Fiji and working with the Peace Corps in the late 1970s. One beautiful evening I decided to visit a friend's sailboat moored in the marina. Lying on the deck, while taking in the full moon, the stars, and the night sky, I noticed that the spinning wind-meter (anemometer) at the top of the mast was nearly the same size and shape as the round, full moon, which was also right overhead. My playful fascination with shapes and form took hold, and only a small shift in my position was necessary to place the circular anemometer directly in front of the full moon; it was a perfect size match – the spinning disk exactly covered the entire moon. Within a few seconds, the illuminated spinning wheel unexpectedly created a powerful strobe effect and I suddenly became very disoriented. Fully immobilized, I felt as if I were being spun through a tube of spinning light; my entire body felt like it was twisting and undulating as it flew towards the moon. After a few long, disconcerting moments, the spinning stopped; I steadied, and then I heard a deep booming voice, that I, having been raised in a Christian culture, instantly "understood" to be "the voice of God." Male, deep, warm, and laughing tenderly, this voice felt extremely loving and caring as it commanded my full attention. I lost all sense of my body and everything else around me as it carefully and clearly spoke the words, "Tim, you always worry about everything. Just remember that there are many different ways of looking at anything. Everything has positive and negative aspects or interpretations. Look for and find the positive that exists in every situation. Focus on these positive aspects and your entire life will change." Suddenly it was over; the voice was gone, and I was, once again, lying there on the deck with the clear night sky, moon, and stars above. While this voice was as real, resonant, and vocal as any actual physical voice, I later confirmed that no one else on board heard anything.

I have no clear explanation for what happened that moonlit night, except that the spinning wheel centered on the moon somehow facilitated a change in my mental state, not unlike hypnosis. The incident made such a deep impression that I immediately began searching for ways to change how I viewed and approached every part of life. It took me many years and many additional reinforcing insights to fully understand this message and its meaning, but today this awareness is a foundation stone of my life philosophy. *The message was distilled, directed, and worded so that I could hear it with my mindset that night, but it was essentially relaying one of the most fundamental principles of this book: "Everything within life can be viewed from different perspectives and if we adjust our attitude, we will change our entire view of the world." Every possibility exists, so the world that we perceive is always, and only, a reflection of our most deeply held beliefs.*

MY DRAMATIC OUT-OF-BODY EXPERIENCE

We normally assume that our awareness is tightly connected to our physical body, but this does not have to be the case. In special circumstances, our center of awareness can move away from our physical body—sometimes far away. We might suddenly witness our awareness focused above our body, behind it, off in the next room, or moving about freely somewhere far away and unrecognizable to us. Many of us can remember having this type of experience at least once in our lives, most often when very young or while involved in some physical crisis.

According to doctors, patients, and many others, this "Out of Body Experience" (OBE) is a common phenomenon, especially when great stress or traumatic injuries are involved. Sometimes, patients involved in accidents or emergency surgery discover that their conscious presence was fully aware, watching, hearing, and understanding the EMS workers or doctors who were working on their otherwise unconscious bodies. They relate how they suddenly found themselves watching from the tree above their wrecked car, the corner of the operating room, or from some other remote place as others attended to their damaged bodies. In this remote but very conscious state of awareness, some also had life-changing spiritual experiences, including instantaneous reviews of their entire lives. Occasionally, they can even recall having made the thoroughly analyzed and conscious decision to return to their original, but now damaged, body.

One variation of this occurrence is so common it has been named the *Near-Death Experience* or NDE. Many stories have been collected and recorded by patients, doctors, researchers, and psychologists; scores of books examine this interesting phenomenon in great detail. Of course, these types of memories can only be told by those who have experienced this traumatic separation from their bodies, and later were able to return to their bodies and remember the experience. We hear little or nothing from those who made the other decision to not return.

Quite spontaneously, and in a very dramatic way, I also had an "out of body" experience. This experience of my awareness leaving my physical body was also so emotionally disruptive and frightening that it left me feeling quite unsettled for several years. At the *time*, I was in my late twenties, newly married, and living in Lake Placid, New York. While fully enjoying the circumstances of my personal life, I was also profoundly disturbed by what I then perceived to be the general condition of the "outside" world. The widespread poverty, wars, human injustices, environmental problems, and our inherent cultural shortsightedness all disturbed me to my core. At this *time* in my life, I had been meditating daily for almost ten years, and it was rather automatic for me to leave behind any thoughts and worries about the day-to-day world while practicing. Our apartment featured a large picture window that directly faced the beautiful snowcapped peak of Whiteface Mountain, which sat majestically some twenty miles away. Every morning I would meditate for half of an hour, using this mountaintop as a visual *mantra* (*yantra*) to help focus my attention and clear my mind.

On this one particular day, instead of dropping into my usual relaxed meditative state, I felt overcome by a sudden and very powerful bodily sensation, a sonic "whooshing" that vibrated my entire body (not unlike that hypnotic state on the sailboat a couple of years earlier) and I quickly found myself rapidly moving through a bright tunnel-like *space*, as if flying. In the next instant, I was suddenly seeing and experiencing the world from the very top of Whiteface Mountain, looking directly back at the town of Lake Placid, where my actual body was presumedly sitting. I had no awareness of a physical body or any sense of being cold, but my conscious presence and visual sense were fully centered right there on that mountaintop. After a few moments, I had the clear and strong "knowing" that I had an important and critical decision to make: I could continue this journey into this exciting new *space*, or return to my body. I also somehow "knew" in that instant that, while this state of *being* was very tempting to explore, if I continued further, I would lose the will and ability to return to my new wife and the life that I had been living. With that thought, I panicked, lost my impersonal meditative focus, and my consciousness was instantly pulled back through a darker and more rapidly spinning tunnel. I then found myself again sitting in my living room, but now deeply shaken and completely covered with sweat. A couple of weeks later I drove up that mountain and confirmed that the view, looking back to town, was exactly as I had witnessed.

I know of many others who also have, for various reasons, found their conscious awareness temporarily shifted to someplace other than within their own body. When my brother was a teenager, he often found himself observing his own back while walking down the road. Another friend, as a pre-teen, repeatedly floated outside her house and was able to watch her sleeping body through a window; and my wife once watched herself in conversation with friends, from above looking down at the top of their heads. While the center of our awareness is normally associated with our physical bodies, it also seems that, under special conditions, it can also roam free of this physical boundary.

After that experience, I stopped my deep-meditation practice for almost a decade, afraid that I might have a similar experience and not have the ability, or more accurately, the desire, to return. Upon reflection, it is now clear that during that period of my life I was fundamentally unhappy because of my perception of the troubling and problematic nature of the external world and our human existence. Back then, my connection to my body and this life was not always joyful, and some large part of my awareness wanted to escape my troubles. I also somehow understood that it was not my *time* to make this transition, as I still had much more to do and learn while in this body. The path I chose produced two beautiful children, an exciting career, many great adventures, dear friends, a sense of humor about the human condition, and a strong appreciation for all of life and its incredible mystery.

I am now certain that our body and our conscious awareness are two different and potentially independent things; they are interrelated and interconnected, but still separate. **We are not just our bodies; we have an existence outside of our bodies. This is one of the most important realizations on the road to true freedom because, once it becomes clear, this understanding serves to liberate us in many wonderful and unexpected ways.**

My Unexpected Healing

While he was still living, my relationship with my father was distant and strained. A depression-era, badly wounded and traumatized, WWII veteran, he was also, as common for his generation, an absentee dad. Extremely productive and successful, he was always busy with his projects either at the office or home, and I have few memories of *time* with my father except when he was directing me to perform tasks around the house. Like many in his generation, he was never comfortable expressing his feelings. His inherited, and then battle-hardened, belief system included the idea that showing affection or love would not be "manly" and would only soften his boys. He, therefore, never told either of his sons that he loved them, but instead displayed his love through strict discipline and financial support. Only achievements and positive results were rewarded; and if I ever came home with a ninety-eight grade on a test, I knew that I had to face a stern, "What happened to the other two percent?" Because I grew up in this environment, all of this seemed perfectly normal and I fully modeled his way of walking through life.

This was the starting point for my journey: it was all that I knew. It took me many decades to learn how to feel within myself, and then many more years to understand and let go of my resentment for his restrained expression of affection and love. A difficult physical crisis in my sixtieth year became the messenger that helped me open, realize, integrate, and finally heal my relationship with my father.

I now know that between the ages of ten to twelve, I had been sexually abused at the hands of a once-trusted neighbor, but my awareness of this abuse only began during my junior year at college when I received a large mailed envelope from my father. Inside the manila envelope was a recent issue of the weekly news magazine, "Newsweek," and on its cover was a picture of an unfamiliar man in a powerboat. Adding to my confusion, the front-page story was about "uncovering a large child pornography ring in Florida," and across the cover, in large, black-marker lettering, my father scribbled, "HA, HA, NOW I KNOW WHY YOU ARE SO WEIRD!"

This communication was completely confusing because I had never lived in Florida, I did not recognize the person involved, and nothing about this story was the least bit familiar. I called my father to ask what this was about and he replied, with a somewhat nervous laugh, "Don't you remember? That man lived four houses down from us on Long Island. For several years you used to go out on his boat almost every summer weekend."

This was a complete shock to me; I had no memory of such a person or these frequent boat trips during that period when I was only ten to twelve years old. I responded, "Dad, this is serious; don't joke." His response still reverberates in the uncomfortable depths of my *being*: "Oh, he wasn't a 'homo' back then (just seven years earlier) or I would have known it."

Today such a response sounds unbelievably ignorant and out of touch on multiple levels, but many things were treated very differently in 1970, a *time* before the church scandals when homosexuality and child sexual abuse were often equated and completely misunderstood by many very intelligent and well-meaning people, including my father. While I was completely stunned, I was also very busy with my life and I somehow "chose" to just ignore and bury this dramatic new revelation. It sat quietly within the depths of my memories, secretly affecting my *being*, until twenty-five years later when, without warning, it aggressively leaped back into my awareness.

During those twenty-five years, much had changed in my life; I married, completed a graduate degree in architecture, had children, and began a very successful and satisfying architectural career. I also found myself driven to explore alternative techniques and philosophies such as meditation, tantra, massage, acupuncture, yoga, and many others, as I searched to address the deep shadowy angst and chronic physical tension that always seemed to be pressing upon my awareness. Without, I led the perfect life; but within, it was clear that something was not right: I had several chronic medical issues; I was always tense with continuous joint and muscular discomfort; I had large impenetrable gaps in my memory, and I would often inexplicably feel like bursting into tears, yet never could.

Another seemingly isolated but very challenging part of my life was also beginning to integrate. Earlier, from the *time* that I was seventeen, I found myself in and out of hospitals, struggling with a chronic and severe infection of my prostate gland that appeared to have no clear cause or source. After years of antibiotics and lesser surgical procedures, doctors were recommending that I have my prostate removed, a brutal procedure that offered no possible solace for a young man desiring a normal marriage and sex life. Fortunately, a young doctor who had just finished his service in Vietnam, where he often witnessed first-hand positive results from acupuncture, recommended that I try it before consenting to this disfiguring surgery. This was a different world in 1974; at that *time* his fear of losing his medical license was realistic, so he shared this advice secretly and made me promise that I would not mention his recommendation to anyone.

This one piece of advice completely changed my health, focus, and life direction. After nine years of painful and disruptive hospital visits involving many uncomfortable and invasive tests, weeks of hospital stay, being fully "scoped" many times, cauterized twice and fully anesthetized three times, one single visit to an acupuncturist brought an end to my chronic, painful and disruptive medical problem. As I explained my condition to the eighty-year-old Chinese acupuncturist, he blurted, "Oh, very easy," a phrase that instantly stood in stark contrast to my many years of hearing doctors express confusion about my very "difficult and complicated" problem. This complete reversal of the professional healer's attitude and the dramatic healing that followed caught my full attention. Not only did this treatment initiate a major change in my health, but it also led to the emotional and intellectual opening, which kick-started my lifelong interest in acupuncture and other alternative methods of healing. The full version of this story is told in my earlier books.

At this point in my life, I still did not associate this medical problem with the fishing-boat trips of my youth. Many years later when I was forty-seven, I was visiting a friend who also happened to be a very intuitive and quite talented massage therapist and "energy worker." She was widely recognized for her very powerful and accurate intuition, an unusual talent that I had witnessed many times as she helped others through difficult issues. We were just chatting over tea as we often did, but at some point in our conversation, she stopped and said, *"Tim, this may seem very weird, but I have a strong sense about this. Would you please go over to that corner and get down on your hands and knees and tuck your head into that corner where the walls and floor meet."*

Perplexed, but curious and trusting her and her unique way of experiencing the world, I carefully followed her directions. She moved up behind me, got on her knees and put one hand on my hip, another on my shoulder, and pushed forward forcibly from behind, pinning my head down into the corner. Suddenly, a completely unexpected and torrential flood of fearful emotional sensations, quick visual vignettes, and powerful physical tremors filled my *being*. I could not believe what was happening; it was as if a spillway to some dammed up, but completely hidden, part of me had suddenly burst wide open. ***I recognized these surprising and painful physical, visual, and emotional "memories" as my own, but it was as if I was experiencing them for the first time.***

That event marked the dramatic beginning of a long process of conscious recovery as I gradually opened to these hidden and painful memories. Over *time*, the memories and feelings slowly became more visual, remembered, and clear as an enormous amount of chronic tension was gradually witnessed, acknowledged, addressed, and finally released. I eventually reached a level where the process, while still not easy, was fully steeped in acceptance, forgiveness, and love.

It is now clear that between the ages of ten and twelve I was a victim of childhood sexual abuse, some intentionally humiliating and including other young friends, at the hands of this once-trusted neighbor. These events involved more than just abuse because this man was also a person with whom I had bonded, who was generously offering his *time* and companionship, filling a void that was shaped by my perceived lack of relationship with my father.

Looking back through my life, there were many obvious clues and hints that I, my parents, my teachers, and other neighbors had completely missed. These included the unexplained infection of my prostate, flashpoint angry behavior, not being willing to talk about my fishing trips, acting out in school, and most dramatically, the complete loss of memories. This loss included not just the memories about these multiple boating trips with this person, but a nearly total loss of all my memories about anything that happened during that period of my life. ***Only much later would I recognize my response and memory loss as classic symptoms of PTSD.***

The healing process that was initiated then reshaped my life in so many ways. Other than the enormous relief and insight gained from eventually addressing my buried trauma, I also learned at a profound and direct personal level just how effective our conscious brains are at controlling the perception and memory of events. ***We remember and retain only what we are capable of processing – conceptually, physically, intellectually, and emotionally.*** If something falls outside of our limits, we automatically engage what psychologists call *cognitive dissonance* (discussed earlier), and this event can then become repressed and filed in our body's equivalent of a "dead-letter drawer." This is one way our bodies, minds, and egos conspire to quell the dissonance within. My personal experience illustrates clearly how our minds and senses will filter or entirely miss *information*. ***It also taught me that these memories are still stored within our bodies, even if it seems that we have no direct conscious access to them. Also illustrated is one more example of a person displaying a type of talent that cannot be technically explained by our Western science.*** In our culture, intuitive skills and insights that can't be explained scientifically or rationally are undervalued, ignored, and often ridiculed. We lack useful cultural examples or archetypes for processing disruptive experiences that fall outside of our *conceptual horizon*, and even those that are closer in but don't fit within our worldview.

THE MISSING PIECE

Once my "Pandora's Box" of raw emotional and physical memories was opened, almost every day brought new and unfamiliar feelings and sensations. I could not always identify the emotion and I often had no specific memories associated with my many unusual sensations and feelings; the process was lonely, very difficult, and usually confusing. I increased my practice of yoga and meditation, signed up for numerous workshops, movement therapies, and spiritual groups, and found myself gravitating to Shiatsu massage therapists, Rolfers, and other types of physical practitioners that were skilled at very deep manipulation. Occasionally there would be breakthroughs or clear insights, such as the *time* a massage triggered twenty minutes of uncontrollable full-body convulsions, but most of the changes were very gradual and incremental. The majority of the shifts were so minor or subtle that they were impossible to describe until, over *time*, they compounded and became more obvious.

Over the years, as I progressed with my exploration of this abuse, I eventually opened to a new level of healing where I could completely forgive and even empathize with my perpetrator. Here I began to see my journey in a much more positive and purely experiential light. It took years of focused attention, but slowly I started to feel stronger and clearer in every way. However, despite all the improvement, I could tell that there was still a piece of the puzzle that I was missing or avoiding. Something fundamental to my wholeness was still not being fully addressed.

Then, just after my sixtieth birthday, I was on vacation in Central America when one of my architectural projects in the U.S. suddenly spun into a critical crisis mode. As I spent the last week of my remote backpacking vacation riding around on buses with broken suspensions, desperately searching for good internet connections, I noticed an unfamiliar tingling sensation that was originating from my sacrum. I made it home, but my second night back I woke in severe pain as my right leg suddenly went into a violent, cramping spasm. This marked the beginning of a very difficult, but spiritually significant, month-long ordeal of deep introspection and *soul work*, fueled by the continuous deep pain diagnosed as severe *sciatica*.

My pain was so extreme and constant that sleep was out of the question; the rare *times* that the pain eased to the point where I could attempt rest, as soon as I fell asleep, the severe cramping returned. For an entire month, I could not sleep or work so every moment, a full twenty-four hours a day, was devoted to searching for solutions. Each day I would visit massage therapists, doctors, chiropractors, nutritionists, other types of alternative practitioners, including psychics, looking for some form of relief; I tried everything. I was told by doctors that my condition was "quite common," would probably come and go, and that eventually, I would learn to "manage" the pain. Some doctors recommended spinal surgery or other invasive treatments, but also indicated that results were very "hit or miss." Even if surgery was "successful," I was told that the problem might reoccur, and my mobility would probably be reduced forever. I tried several prescription painkillers, but at the doses needed to even feel a slight difference, they made me completely nauseous and weak.

Ultimately I ran out of external solutions and having exhausted every possible method or technique that was available, I continued my search, alone and within. I stretched and prayed for guidance, and stretched and prayed, all day and night, fully knowing that my body was desperately communicating that I needed to pay closer attention...but to what?

I found some yogic positions that seemed to lessen the pain for a few minutes. After about a week of this intense and solely internal focus, I developed a mental state where I could fall into a trance and simply "be with the pain" instead of "fighting it." Somehow, just witnessing and allowing the pain seemed to reduce its intensity and sharpness. My single-focused prayer had become "please help me to open to whatever it is I am being asked to receive and understand."

I didn't consciously relate this condition to my childhood sexual abuse because I thought that I had already completed my deep and extensive forgiveness work around this issue. By this point in my larger process, I had become completely accepting of, and comfortable with, my entire abuse experience, my abuser, and the related traumatic events from that period of my life. My perspective had even evolved to the point where I was genuinely thankful for all the growth and insights that my experience had provided. However, I assumed wrongly, for my PTSD was about to push me through another door.

After four weeks of this painful sciatica, one day, in the early morning hours, a brand-new, unexpected, yet very clear verbal thought just popped into my head. ***"You have never forgiven your father for not protecting you when you were that young boy."*** These words were completely surprising; they seemed to come out of nowhere. They were not part of any chain of my thoughts because, except for the intention of my prayer, my mind was empty of thought; the weeks of constant pain and focused meditation had driven all the busy internal chatter away. Immediately, as I focused on and thought about the truth of these words, I noticed a new sensation; a powerful electric, yet pleasant, tingling feeling flowing down my leg and up my back. I quickly discovered that, with this sensation, the pain completely evaporated, leaving even more quickly than it had arrived a month earlier. ***Within those few seconds, by simply hearing, recognizing, and honoring those words that I unexpectedly heard in my head, my month-long painful ordeal completely and miraculously ended.***

Ever since that day I have held a different understanding of my father, his life and the circumstances that shaped his temperament, and our relationship. Although he had transitioned years earlier, with this life-changing realization and the resultant opening, everything changed as I embraced this deeper forgiveness and understanding for my father and our strained relationship. This forgiveness and acceptance have continued to grow naturally to include everyone and everything else in my life. *That particular inner storm originated with a specific event involving my relationship with my father, but its broader lessons extended beyond to include my releasing the general impulse to blame anyone, both others and myself, for any of my experiences within life.*

Over the years, the lesson from that release of blame and resentment has also morphed into a solid appreciation for the unexpected way that life always delivers the perfect experiences at the perfect *time*. Today, even the tiniest taint of judgment or resentment about anything or anyone automatically triggers a deep internal inquiry. For me, this difficult life-lesson marked the beginning of the end of my holding onto any personal resentment about others, or towards any part of life in general. It demonstrated how shadowy resentments hold a hidden power over us, a deep influence that, up until that moment, I could not fully understand. From that moment on, I began to experience a much deeper appreciation for all aspects of life. A new type of lightness began to appear more frequently in my life, which today has evolved into the unwavering joy that fully infuses my everyday *being*.

That particular day also left me with a clear and pressing desire to record my breakthrough experience for the benefit of my daughters. I only intended to summarize my experience into a short letter to be read later when they could better appreciate the profound nature of my experiences. That "short" letter to my daughters kept expanding and ultimately morphed into my first book *The Architecture of Freedom*.

HEALING AND CHRONIC TENSION

Our Western culture and its well-established medical model has yet to fully recognize the influence of these deep, energetic blockages. The current Western model for healing still tends to treat most problems at the superficial and secondary level of symptoms. Rarely do doctors address deeper causes such as diet, stress, and exposure to toxins, even when these are fully recognized or understood. Because of this misplaced focus, the Western population is rapidly becoming dependent upon medications, pain-relievers, and mood-altering drugs, either through prescriptions or as self-medication. We resort to surgery to deaden nerves, relieve pain, loosen muscles, and realign joints – all conditions created because the body has shifted out of its natural balance and harmony. Even when these methods appear to work, addressing health issues at the level of symptoms just temporarily masks the deeper imbalances. Problems compound because many of these medications interfere with other important natural biological processes, often producing serious side effects. Surgery also creates more scar tissue that eventually leads to additional physical blockages and problems. As a result, in our modern Western culture, many people are just "living with" and resigned to chronic medical conditions.

Our bodies, like everything else in existence, are *holographic* and store the complete record of our emotional and psychological journeys. If we learn how to listen within their depths, our bodies will direct us to the hidden keys for identifying and releasing any chronic tension that impedes natural *flow*. While resistances are still minor, they preemptively signal their presence with small aches and pains; at this point, they are only gently signaling or "asking" for further attention. Only much later, if ignored or left untreated, do these early warnings evolve into the diseases and debilitating conditions that require more extensive intervention and treatment.

The only real and permanent solutions to chronic health problems involve mindset, lifestyle, avoidance of environmental toxins, diet, and addressing the causative psychological or emotional issues at their source. *As improbable as it seems from within our Western mindset, the source of most chronic medical conditions or disease, when not directly caused by genetic, dietary, or environmental problems, can usually be reduced to fundamental imbalances in our lives, often involving our traumas and attachments.* At their deepest level, all of our food and diet issues involve *attachments*. An important concept borrowed from several Eastern philosophies, this very important idea of *attachments* is referenced throughout this book and thoroughly discussed in part three.

FEAR OF OPENING OUR HEARTS

"Life begins where fear ends." *Osho*

We all recognize that the expression "open your heart" is describing a positive change for living our lives, but most of us have only a vague idea of what this is asking of us. As we grow and add experience to our lives, we usually discover, early in life, that our universe is filled with many uncertainties and unknowns that we naturally fear. To protect ourselves from these things that we don't understand fully, we build walls. At first, these are just *energetic*, but over *time* they eventually integrate and become fully physical. Einstein taught us that *energy* and form were just two expressions of the same thing. **Through attempting to protect ourselves this way, we begin to learn what it is like when we contract and close our hearts even tighter. We wind up realizing that we are then feeling less of everything – less pain, but also far less joy. We slowly become numb to almost everything.**

As the repetition of these everyday experiments helps our awareness to become more conscious, we develop a sense of what it might be like to do the reverse; **when we open our hearts more, we feel more of everything. Initially, this is a frightening possibility for most of us because why would anyone want to experience more of the world's overabundant suffering and pain? However, what unfolds when we finally open our hearts is exactly the opposite: life becomes less confusing and far more joyful.** The result, which the logical mind would never expect, is the happy result of *energy transmutation* and balancing occurring at unseen levels.

SEPARATION VERSUS ONENESS

Our physicality is a wonderful slice of existence where we have the powerful opportunity to experience both *separation consciousness* and *unity consciousness*. **We live in a special type of space where we have the opportunity to explore the awareness that we are simultaneously one and many. Here we can discover that while we are separated we are also always fully interconnected: we are "oneness," having the very human experience of exploring separation.** The co-existence of our "oneness" and separation has been described as both the paramount *duality* and ultimate *paradox*. **This relationship is the critical glue that holds our entire universe together.** Just as it is with everything else found in *duality*, all extremes must exist together for our physical universe to function. Neither *polarity* is preferred by nature, for the natural world is always seeking both diversity and a balance. Our lives improve tremendously when we seek, find, and dwell within both of these perspectives and eventually allow our entire lives to unfold from this place of *balanced awareness*.

THE DIFFICULT RATIONAL PARADOX

Over *time* we move beyond the logical and difficult paradox of simultaneous *separation* and *oneness* to a more *holographic awareness,* where every cell of our *being* knows that while we appear to be individual and separate, we are also an integrated part of this greater "oneness." As we reach this point of intuitive knowing that we are both one and many, a *holographic awareness* further expands this recognition to include everything in existence. In this *space*, we have become joined with all the birds and bees, plants and trees, rocks and socks, Mars and stars, and all that we label good or bad, and saintly or evil: all of "the seen," and the "to-be-seen," along with the many times greater amount of "the un-seeable." **Everything that could or would exist must also be included – absolutely everything.** Every person brings their own unique and critical viewpoint, a unique perspective for consciousness to explore so that it can come to "know" the greater extent of its "oneness." *Awareness* is never individual; it always understands that it is not possible to ever separate ourselves from any piece of existence, including all that which we once judged as frightening or negative.

I remember the first *time* I was able to embody this collective "oneness." I was enjoying a troupe of Chinese acrobats when I suddenly realized that I was watching with a background awareness similar to "I didn't know that *'I'* could do that." **Of course, this type of acceptance and integration of our collective "oneness" is relatively easy when artists or athletes are displaying their brilliance. Feeling this intimately connected to the homeless, the sick, and the maimed is a much more challenging task. Going even further and being able to recognize this same type of intimate connection to the Hitlers of this world, while impossible-seeming, is a critical and unavoidable step for our collective healing.** As we open to our entire *being*, we must, by definition, open to all of its separated parts, even those very challenging ones that we want to push away the most. The understanding that all *beings* and all their expressions only represent other parts of our divided *self*, many of which we have been reluctant to accept or embrace, does not come easily or quickly. **Despite the seemingly impossible nature of this challenge, until it is directly experienced, the inner transformation that occurs when we finally open to this level of "oneness" is, itself, impossible to imagine or describe.**

For many, this critical step is the most difficult step of the entire process. Allowing everything in existence, including all that we fear and judge to be negative, into our personal space of "oneness," goes entirely against our deeply ingrained human "nature" to want to push the "unpleasant" as far away as possible. At this critical stage, we are being asked to finally move beyond our deeply patterned judgments by embracing and welcoming all that we previously rejected. All that which we have previously judged as "bad" or "wrong" must instead become recognized as the critically important missing pieces of our wholeness. *For many of us, especially early in the process, this is the most unnatural and most challenging step in the entire process of whole-ing.* Just being willing to look at these orphaned parts of our *being* marks a significant breakthrough, so the next step of embracing these once-rejected pieces initiates a major shift and acceleration in the self-healing process. This challenging shift of *awareness* is a critical turning point; it marks the real beginning of profound and meaningful healing within, one that directly leads to a steady stream of beautiful, yet gentle, openings, where our darkness can then become illuminated by the light of *consciousness*. As more of the darkness is *enlightened*, we become more transparent vessels, living to our potential by transmitting the *energy* of love as we now *resonate* fully and freely with all parts of existence.

TRANSMUTATION

We must always remember and continuously remind ourselves that it is our wholeness that we are exploring in this physical world of *duality*, so our journey must include the entire range of different-appearing aspects and all possibilities of expression. This means constantly reminding ourselves that no single aspect of duality is all of one thing and none of the other; every aspect of our lives is built from elements that we might sometimes judge one way or the other, depending on circumstances.

There is a major misunderstanding that each of us must consider before entering the exploration of our shadow world. We only have shadows because some part of our whole *being* or experience has been separated and judged as not right, or not belonging. It was this act of judgment that created the shadow. *What is not fully understood is that the moment we are finally able to re-visit, feel, and re-experience this dark appearing aspect, this completely transforms the way our once-feared shadow appears to us. Once illuminated and experienced in even this small amount of new light, it reveals itself to be quite different from that long-held darkened image of our imagination. Not only will it appear transformed, but when this aspect finally integrates with its other dualistic compliments, this once-frightening form re-emerges to become one of our most powerful allies. The stunning transformational power of transmutation completely defies our logic and expectations.*

When we were children, we could not have known that the fearsome shadow that caused us to tremble was only our loving grandmother bringing tea and cookies into our bedroom. When understood from a greater perspective, our once-feared shadows completely change character. All of our fears and judgments are only reflecting our misunderstanding of *duality*. Once that which we fear is rejoined with its misunderstood and isolated *complementary* parts, our internal darkness becomes illuminated and no longer will *project* those shadowy *images*. We then realize that our fears were entirely imagined by our *egoic* minds as a by-product of internal separation. Instead, this part of our once-darkened interior unexpectedly shifts in quality to present something radically different in appearance: a combined, but now entirely new, aspect of our authentic *being* that fully supports our new life. Once integrated and *transmuted,* our former shadows radically shift character to fully support and sustain our entire lives and well-being.

THE MAGIC OF TRANSMUTATION

"Go into each negative thing and you will find the positive. And knowing the negative and the positive, the third, the ultimate happens – the transcendental." *Osho*

Once our old shadows that were built upon traumatic or difficult experiences are illuminated, re-experienced, and finally integrated, we discover that our awareness automatically expands. Through the reintegration of more of "that which we truly are," we increase our experiential "bandwidth" and simultaneously notice that many new types of experiences have suddenly become possible, normal, and even common. Initially, we should expect periods of adjustment, instability, and mood fluctuation while these shadows are being recognized and integrated but eventually, these shifts and swings settle to be replaced by a steady and solid presence as a wonderful new feeling of wholeness washes through our *being*. The entire universe responds and rejoices as parts of *being*, which were once isolated, feared, and rejected are now illuminated, integrated, and fully expressed through a new and more expansive holistic balance. This surprising shift happens because of the miraculous power of *transmutation*: unexpected *magic* that occurs naturally and organically when we fully embrace all aspects of our *being*.

As we excavate and reveal unexamined aspects of ourselves, parts that have been resisted, rejected, and isolated for a long time, their shadows will often appear to us in a particularly frightening or threatening form. As our blocks reemerge, they usually present themselves as things that clearly should be avoided, and often as the very things we fear or despise the most. This is just our ego attempting to protect itself by maintaining the old familiar state of being through trickery.

Many readers will logically wonder why anyone would ever want to uncover, explore, or relive their most difficult and unpleasant experiences. *If we desire peace within, then some version of this type of shadow work, like the whole-ing process, is absolutely necessary.* There is no way to short-cut or work around this step of the process; we all have to eventually do this "inner work" no matter how frightening its unearthing appears. Eventually, every person will engage in this healing process; the deeper nature of life makes this so. All that we ever can control or adjust is the timing of this process, and even this can't be rushed for it will (and only can) unfold once we are ready. The only real power that we have is our attitude which can only resist and slow down the process, but this option only increases our suffering. Being open to the natural unfolding is clearly our best option, but even this won't speed up the process; it will only unfold once we are fully prepared through our experiences.

By embracing all that we encounter, those once-feared aspects of our being shift appearances; their qualities unexpectedly change to support our lives. Reunited with their long-lost dualistic complements, they can finally be seen and understood through a very different light; their full spectrum becomes more visible. Once these orphans have been illuminated, embraced, and integrated with the rest of our being, these same aspects instead inherit critically supportive roles in our lives. As these abandoned parts of ourselves that we previously tried to push away are welcomed within, they completely transmute to function in unexpected and extremely positive ways. This happens naturally and automatically, once we move past our resistance. It is only the incomplete, divided, isolated, and misunderstood that creates the shadows that arise to disturb our peace of mind.

STEAM AND ICE

Transmutation is a familiar phenomenon in our physical world; it is what happens when the chrysalis that was once a caterpillar, morphs into a breathtaking butterfly, or when the hot pile of rotting organic matter in our compost pile turns into fertile and welcoming soil for seed. It is what ancient alchemists sought in their attempts to turn common metals into gold. Examples of *transmutation* can be found throughout nature with the meeting and blending of many *dualistic* pairs; possibly one of the easiest to understand is the *transmutational* change that occurs naturally with steam and ice. Viewed in separation and isolation, both extremes can appear somewhat threatening to our human form. When we encounter hot steam, we quickly learn to be very careful because experience has taught us that hot steam is something to be feared. The burns steam can cause are often disfiguring, painful, and life-threatening, so naturally, we are cautious and avoid direct contact. Ice may appear to be a little less dangerous, but without the extra protection of warm clothing and shelter, few of us would survive long if surrounded only by snow and ice. While ice can be beautiful to look at, our bodies instinctively push it away because water in that form is too cold for normal physical comfort. *However, when we bring the two (steam and ice) together and allow them to co-join and fully mix, a very natural but significant shift happens; suddenly we have neither steam nor ice, but instead, a third and different element, one that is completely safe and something that our physical bodies require in great quantities: warm water.* Not only has the threatening nature of both extremes instantly shifted in this co-joining, but even the physical state of the *elements* has completely changed. Instead of a gas and a solid, we have a new, third form: liquid water, the most critical medium for supporting human life – a thing we once feared instead becomes an important ally.

As we allow for the complete integration of *dualistic* opposites, many similar unexpected shifts begin to happen. The most fearsome aspects of our shadows, which were once so frightening that we could not even look their way, dissolve to be replaced by surprising new qualities that fully support and sustain our lives. As these hidden parts (once judged as "horrible" or "unworthy") unexpectedly shift in character, they often *transmute* to become some of our most powerful allies. *This integration produces more than just balance; the new wholeness becomes entirely transformational.*

FEARS RELAX

Once we fully understand the transformational power of our darkest shadows, our inner-work becomes an even more important part of our daily life. Recognizing and opening to this new awareness marks the beginning of our actual healing and our incredible journey into this deeper "oneness."

As this powerful idea is integrated, we begin to see our shadows from an entirely new perspective. Now, illuminating our darkness is no longer frightening because, with more of the light of consciousness available to guide our newly expanded awareness, it becomes, instead, a vibrant and joyful exploration of *being*. *The most significant personal evolution that is possible begins with the illumination of our shadow world. The deeper and darker our shadows, the greater the potential for change.*

Discovering Where Our Shadows Hide

We are probably already aware of our shadow world, for it has been revealing its presence over and over throughout our lives. We can witness the shadow's influence by reviewing our fears and life choices. PTSD is usually easy to spot because it will often trigger particularly extreme irrational reactions and odd, unexpected behavior. These PTSD triggered outbreaks can sometimes occur without our conscious minds ever noticing or even remembering. Less extreme shadows reveal themselves through many other types of psychological discomfort, appearing in both social and private situations. A new opportunity for *honest inquiry* becomes available whenever we experience such discomfort. Our discomfort may appear to have a clear and obvious external cause, but it always is related to the psychological echoes of past experiences. We need to be very careful about assigning responsibility to "external" causes, for the ego is very gifted and quite skilled in the art of deflection. As we become more whole, *honest inquiry* becomes easier, clearer, and more truthful.

Our oldest shadows had *time* to become fully physical and these embodiments often become our most persistent guides. When we notice repeating aches, twinges, or sensations that can't be directly associated with a new physical strain or injury, we are receiving *information* about where our shadows are hiding within our physical body. Chronic injury or problems with particular body parts or organs are very clear clues. Because of our deep interconnectedness, there are no accidents in our causal universe, so our "accidents" only provide additional clues for experienced explorers.

However, many of our most shadowed areas have been cleverly hidden, buried within our sub-conscious or deeply interned within our bodies, rendering these completely invisible to the everyday light of our conscious awareness. To discover and unearth these hidden shadow treasures, we must be like miners looking for gold, using all of our best tools for this exploration of our bodies and subconscious. While this book describes many known techniques for this important inner search, many others also exist and we are completely free to discover or even invent different methods. New techniques unfold naturally from within this process. Once begun, this exploration, discovery, revelation, and integration process is destined to become a lifelong and joyful way of living and moving through our world.

Using Trauma as Our Initial Guide

"The wound is the place where the Light enters you." *Rumi*

Due to the human condition and just by living life, every one of us becomes exposed to trauma; it is a given from birth. In today's worldwide culture there is no technique or method for completely avoiding trauma. The birth process itself is quite traumatic, and it only marks the beginning of our growing awareness that our bodies are, throughout every second of our lives, somewhere in the never-ending process of being "consumed" by some other competing living organism and always simultaneously subject to the "decrepit decay" of *time*. Within this physical world, founded and defined through the ideas of separation and competition, trauma will always be a part of life. ***This means that in some strange way, trauma is a "normal" and functional part of life; and, as with everything in life, it offers a most wonderful gift. It offers each of us a clear entry point for the expansion of our conscious awareness.***

Earlier I described how my traumatic past was the entry point for my journey towards wholeness. It is often through the need to heal our psychological wounds that we begin to look within and earnestly deepen our spiritual journey. Our awareness of inner wounds is the most universal entry point for our committed journeys towards wholeness; it is often from our desire to understand and heal our trauma that we first engage this deeper exploration of life. ***Our traumas are not the enemy; instead, they become our most useful guides and teachers, particularly through the early stages of this lifelong journey.***

DISEASE, ILLNESS, TENSION, AND STRESS

My own life experience taught me how trauma and unexpressed emotions contribute to physical illness; but after many more years of exploring, I have become convinced that the actual relationship is even more direct and dramatic. As crazy as it sounds, I now can understand how all disease and illness is related to our emotional and *energetic* history. If the physical world is only a *projection* of the *quantum field*, and our deeper *energetic being* is our direct connection to this first source of all form, then our entire physical experience, including all of the physical circumstances of our body, is communicated this way. All of our physical health issues are related to the same deeper *energetics* that shaped all of our personal experiences throughout our lifetimes, and this includes all diseases and accidents. Disease is never just the result of random exposure to bacteria or virus; those "externals" only come much later, as the actual mechanisms for the *projected* physical expression of the deeper underlying *energetics*.

When we look back at any individual's unique personal history, interesting patterns emerge. We all know people who are often ill, accident-prone, or constantly facing certain types of difficulties such as losing their jobs or their relationships. We also witness those who always seem to have everything go their way; we may even label these people "lucky." One can't help but wonder why these consistent patterns persist throughout an individual's life. Our patterns are never accidental because, once again, they only represent the secondary physical expression of the deep underlying *energetics. Secondary waves* of these patterns persist and repeat until the deeper underlying conditions change.

We have this experience of a three-dimensional realm for a purpose, and that is to discover, reveal, know, and then heal parts of our divided *being* through a living process that merges and integrates *complementary* aspects. Each separated individual has a unique "assignment," "job" or "role" in this integration process, but all "roles" are equally valuable and all contribute to our collective wholeness. For each of us, to facilitate this process, life appears to unfold as an ordered series of related dramas presenting multiple opportunities that are perfectly choreographed and individualized for each of our "assignments." Once we are *awake* to this process, it becomes clear that everything that comes our way, even our diseases and chronic illnesses, are only personalized guides for this journey. Our personal history, no matter how difficult and challenging it often appears to be, will eventually be recognized as our greatest and most accurate teacher.

Ultimately, our unexpressed shadows are impersonal, having roots that reach far deeper than our short, individual lives. Most, if not all, originate far back in our primal evolutionary history and these *primal fears* have then been further deepened by generations of human *karmic* activity. Throughout our lives, reinforcing events have been added to these inherited unexpressed emotions, but because only these additions from our personal lives can be recalled, we tend to believe that our "problems" were caused by things that occurred in this lifetime. Instead, because we entered this realm for the express purpose of *whole-ing,* these energetic inheritances, and the related experiences we encounter, are purposefully aligned to better guide us. Healing only happens when these fearful emotions that have long been unexpressed, repressed, and hidden deep within our collective shadows, are finally brought into the light of awareness to be acknowledged and expressed. The more full and authentic the expression, the deeper the potential for healing.

When faced with disease or a medical issue, we are just being strongly encouraged to address these issues at a "high-alert" level; our bodies are communicating that they have tried the more subtle clues like hunches, aches, and pains, but these have repeatedly failed to trigger our focus and attention. Because we have ignored these early clues, our deeper *being* steps up and "raises the stakes," hoping to finally get the full level of attention that is necessary for our healing.

INSPIRATION, ATTENTION, OPENING, REALIZATION

There are four distinct seasons that we always pass through during the birth of *awareness* that are necessary for successful *whole-ing*: inspiration, attention, opening, and realization. When all of these *flow* in an unbroken and continuous cycle, undisturbed by our thinking minds, there rests *awareness*.

Initially they first appear as divided steps that each require substantial *time* before transitioning. Just like our planet's weather, which might bring an unexpected bit of winter to the fall, these *energetic* "seasons" will intermingle. The first season is the initial *inspiration* for entering the process; this can be relaxed or forced. This *inspiration* then creates a new focus for our *attention* which eventually leads to an *opening*. The *opening* creates the internal *space* necessary for *realization*. Deep within this repeating cyclic process, it can be easy to get confused or lost, so it can help to recognize these seasonal changes as just normal parts of an ever-flowing process.

LISTENING TO OUR BODIES

The most important gift we have been given for our personal healing is our bodies. Our bodies and lives are the *holographic* manifestations of the *energies* we engage and interact with; they are the physical expression of our deepest *energetic* truths. While our bodies never lie, learning to listen to them accurately is not natural or easy. We must first learn about our automated responses (*Default Mode Network*), our conceptual prejudices, our bodies' limitations, and how to move beyond all of these. Once we finally understand how to interpret our body's responses, the body itself becomes our most reliable guide for navigating our emotional depths to discover the truth of our *being*. In this unexpected way, our own body, with all of its "issues," is destined to become our greatest teacher. We always need to remember that our brain, located in an extremity, is only one small part of our body.

Once we begin to integrate the radical idea that "we are not these bodies," the entire process suddenly becomes much easier because it has also become far less personal. Once liberated from this misunderstanding, we no longer have the vested interest in protecting or hiding some frightening truth that we once feared might expose or harm us. Not only do we begin to understand that these deeper truths can never "harm" us, but more importantly, we also start to realize that the "I," which we all protect so fiercely, does not even exist as a separate entity; it exists solely as a mental concept created and held in our minds.

GROWING PAINS

There will be times when we will experience periods of heightened intensity: days infused with a much more dramatic range of feelings or emotions; times in our process where we feel more of what is going on inside. This is nature's way of saying "pay extra attention" and "take a deep and honest look at precisely what is appearing right now." As we grow through this process and consciously make changes, the boundaries of our personal experience will gradually expand, or even contract for a while, if we so choose.

Until we learn about and become more comfortable with this new terrain, there is usually some degree of discomfort and unease associated with these changes. As we change and adjust, temporary disturbances must be expected; these are completely normal. This adjustment period can last months and appear in many different ways: twinges, cramps, aches, vertigo, lightheadedness, tiredness, emotional swings, and other unusual sensations are not uncommon. All these eventually pass to be replaced by a much greater sense of well-being. Our difficulty with the process is minimalized when we don't dwell on these external symptoms, for they are nothing more than our temporary growing pains. ***Throughout this deepening and expansion of being, we are continuously invited to fully feel these new sensations, but at the same time, we must be cautious to not dwell in, overanalyze, or assign these growing pains too much importance.***

As extreme emotions rise within the healing process, it is always beneficial to allow them to be expressed and felt fully; this is a key step for the reintegration of any lost parts of our *being*. Our deeper nature is asking us to allow these old repressed emotions to be authentically expressed, even though the meaning of this expression may never be fully understood by our minds. Our deepest resistances are not "understandable" through our limited thinking minds for what rises may not be even human emotions. It is not our job to understand or judge; our responsibility is only to fully feel and experience. Through new awareness, we are given fresh opportunities to not just repeat our old pattern of pushing these uncomfortable feelings away. Instead, we now recognize that they are only appearing anew so that we can finally experience and feel them, knowing that their full expression is the only thing that will send them on their way. If we allow and just provide witness, with this new understanding of how to "let-go," we can finally fulfill the emotional release that wanted to be expressed so long ago. An energetic pathway long closed is now reopened.

We must always remind ourselves to proceed slowly and safely. Doing this inner work requires *time* for allowing natural adjustments and balancing; this is certainly not a process that wants or needs to be rushed–it can't be. We also recognize that there is nowhere else to go because when living in these bodies, we are already home, right in the center of existence, and connected to everything. We also understand that *time* does not exist in the linear marching way we might have once thought; it is, instead, fully flexible and will always adjust to our individual needs. ***Rushing only gains us new opportunities to explore the nature of our impatience.***

RENEWED SENSATION CAN APPEAR AS PAIN

Our bodies will feel strange and experience new sensations as renewed life once again begins to move through abandoned and unfamiliar pathways. Conscious but neutral observation of these unusual sensations is another useful tool for the facilitation of our expanding awareness: Is the sensation pleasant or unpleasant? Is there an uneasiness or fear? Where in the body is this newly discovered sensation located? Have I experienced this before? How old was I when first noticed? We must not overanalyze or think too deeply; instead, we recognize, look for resonant connections, and become more aware. Every awakened or new sensation, even the purely pleasurable, will initially feel unusual and strange, sometimes even scary. Are we now ready to allow it to be re-experienced fully? We must continuously remain aware to not repeat our old pattern of judgment and shut down; our repressed emotions are only asking for new opportunities to be expressed and manifest.

When energetic pathways that have been silent for a long *time* are reawakened and stimulated, initially the fresh sensation can be overwhelming, and our bodies, through their primitive protective *programming*, will often interpret this sudden new sensation as "pain." ***We might have observed that people who have been losing sensation through neuropathy or similar problems first notice tingling, then later pain, as the nerve pathways fail and deteriorate. When these neural or other closed pathways start to*** *come back to life, the process can be reversed, meaning these same symptoms of "pain" and then tingling can also be the signal of renewed life.* Recognizing and allowing this "pain of rebirth" is an important skill acquired within whole-ing.

Because our strong instinct is to avoid any process that causes "pain," we may initially resist the heightened stimulation of a new opening. However, as we sit with this "pain," it soon morphs into the less threatening "tingling" or other milder sensations, and eventually, we can finally recognize it for what it represents: the "growing pains" of renewed life and feeling surging through some forgotten pathway that is vital to our full *being*.

Sometimes within this process, old traumatic memories reappear. If we have fully worked through our PTSD and feel safe enough, we can allow ourselves to drop fully into this experience, knowing we always have the power to pull out in an instant, because we now fully recognize that it is only a memory. If we reach a point where we don't feel safe because revisiting is still too visceral, then we can back off and just sit with a reduced level of sensation. This is a part of the healing process that involves moving forward into more sensation and then retreating, as we "dance" close to our "edge," always staying safely on the comfortable side. We allow as much feeling as possible, while always being extremely careful to observe any judgments that arise, including our many, persistent, subtle, and clever forms of self-judgment. We also must continue to remind ourselves to not let our minds overanalyze the event or any of our reactions.

Neutral observation and awareness are fundamental and critical, for any judgment or reaction will quickly interfere with the natural *flow* of this opening process. *No attempt should ever be made to explore a painful experience more deeply until the initial strong response has subsided.* Instead, we should practice breathing deeply and just sitting quietly with the sensations being offered. *The reopening of closed energetic pathways is a very gradual process that won't and can't be rushed.*

This is a form of psychological yoga, and just as in physical yoga, best results are achieved by riding close to, yet staying on the safe "edge" of sensation; if we push too far, or too fast, we can create new trauma and set back our progress. Holding the sacred *space* of non-judgment, being present, and simply feeling the new or strange sensations as fully as possible is more than enough. It is only through practice and experience that we can discover exactly where our personal "edge" lies, knowing that this point will also shift as our process progresses.

Throughout this entire process, we casually observe any internal mental analysis, but continue our practice of not dwelling there; we simply witness and observe this mental activity. If feelings of guilt, shame, or anger arise, we then practice observing these from the neutral *space* of *non-judgment,* treating these thoughts as if they are arising from an innocent toddler, because, in the greater scheme, we are all just "toddlers" figuring out this thing called life. Remember that these are only "old" memories of "old" confused reactions, that were formed in the "past" from "old" misunderstandings. What arises is built from old perceptions, but it still fills and drives our minds, leaving precious little room for present moment experiences to unfold freely. These difficult memories are not founded in any deeper kind of truth; they are our perceptions, either inherited through our primal genotype or taught through our culture. We reinforce them whenever we "think" because we are always understanding the world through the paradigm-reducing filters that were built by our *egos* to validate their importance.

We eventually find that we can slowly and carefully move back through *time* and review how these *waves* of similar experiences have reappeared, like echoes, to reinforce and shape our beliefs and lives. We might even discover some aspects of our current awareness that have been shaped this way. We can explore related experiences by gradually moving back through *time* to our youngest memories, where we might even witness the actual triggering event. Eventually, we arrive at the point in this process where we can allow ourselves to safely and fully re-experience these once-repressed *emotions.*

Even if we can't remember our experiences consciously, these memories are still stored within our physical bodies in the form of tension, stress, disease, or injury. Even if memories of our deepest traumas don't surface and become conscious, there will still be some active spillover into our lives: some visible indicator actively signaling the presence of this hidden shadow. More than twenty years ago, when my intuitive friend asked me to get down on my hands and knees in the corner of the room, this instantly unleashed a repeating series of *waves* of disturbing and non-visual energetic feelings that surged through my body. At that moment, I had no conscious memory or picture linked to any of those feelings and, yet, even without actual memories, my healing process began. Twelve years later, while exploring the intense pain of sciatica, I received a clear message in actual words, and this *time* much clearer memories followed. Eight years later another clear, but completely surprising, visual *image* from this same period of my life also announced itself. We need to remind ourselves that even if the process seems to move slowly at times, the amount of *time* involved is always perfect for the process and the gradual unfolding of our *awareness*; the *whole-ing* process is not for the impatient. With every part of our journey into the depths of our *souls*, the nature and pace of the specific unfolding will be different.

Our default mode's habitual reflex almost always includes judgment, so when we probe our buried embodiments we must always be on the alert for this – we must practice never judging ourselves or our experiences. We need to remind ourselves that all experiences, no matter how we or our culture might judge them, work together and contribute to making us more complete and whole. Non-judgment is the critical mindset that is required throughout the entire process of allowing these old, repressed, and painful experiences to resurface. As our healing unfolds and gradually reveals the deeper truth of our *being*, this type of releasing and integrating work becomes easier, eventually becoming almost automatic and even playful.

The *whole-ing* process requires that we develop and maintain an internal environment of absolute non-judgment through all of our witnessing, opening, feeling, accepting, honoring, embracing, and integrating. With each incremental release, our newly expanded awareness will also contribute to a dramatic increase in our sensitivity. Many things will seem different, including the range and swing of our emotions, which might appear to be more extreme and unsettled for a while. Swings in emotion, mood, and sensitivity are normal and expected until the changes become more deeply integrated.

There are two related, but very common, conditions that result in an uncomfortable increase in the range of our emotions. Both are asking for the same response from us: to allow the full experience and expression of the once-repressed emotions. The first of these often occurs right after our awareness has gone through a period of increased expansion and we are still trying to adjust to the new landscape. Initially, these swings in sensation and emotion can make us feel unbalanced or even dizzy, but if we don't push away the *flow* of new sensations, these swings will usually subside quickly and naturally as we settle around a new center. The second condition that can lead to this same type of dramatic mood and emotional swing is when "life" has been asking us to look at some particular issue for a while, but we have been cleverly avoiding doing so. Here, additional pressure is applied by the universe (all things interconnected) to help us move beyond this sticking point.

This can happen when we have been avoiding a critical issue for so long that "life" itself must turn to "nagging."

THE DEEPENING

The healing process can shift gears with the personal recognition that "we are not these bodies." **However, not much will change as long as this important awareness remains only a thought and is not integrated to fully permeate our being and lives.** Once this idea is fully embraced and our lives are lived and experienced through this new expanded vision, we begin to observe that different types of patterns unfold in our lives. As we move through these different experiences, we begin to build reinforcement and support for this positive shift in our awareness. Eventually, sometimes only after years of living life within this expanded vision, a new type of "knowing awareness" gradually emerges. As our lives become guided by this new level of "knowing," deeper *magic* and synchronicities are no longer unusual; they become the norm. Cell by cell, we gradually become "one" with this more universal "knowing." In our world of *time,* this is a process that must unfold at its own pace; it can't be sped up because it requires certain types of new experiences and a series of smaller awarenesses.

While we can't speed up this process, we can, and do, often resist and slow it down. We do this by immersing our minds or bodies in one, or many, of the millions of activities or distractions available for that very purpose. Recognizing and inhabiting our place in "oneness" is inevitable, but we are also free to play and linger in our old concepts of self as long as we like, want, or need. Moving forward in this process is similar to skydiving; our end destination is inevitable, but we also are fully able to control the speed of our approach by choosing freefall, parachute, or something in between. Whatever our chosen pace, our deepening continues with every experience in life.

ENJOYING THIS HIDE AND SEEK

Through living, our bodies gradually become the physical, visual, and energetic record of all of our personal beliefs, hidden or visible. Each of us embodies the entire record: a *holographic* record of all our beliefs and individual responses. **Our physical form represents the holographic image of our core energy that reflects our beliefs as they evolved, changed, and were expressed throughout our entire lives.**

Things appear to happen to and around us, and our bodies respond. We experience physical and emotional traumas, elations, accidents, breakthroughs, illnesses, and health; the record of all of these experiences is incorporated into our physical bodies. As awareness grows, we begin to better notice when we encounter a place or moment where our body is not responding in an ideal way. We learn to go within and ask questions like, "How is the *ego* invested here and creating this tension? What is our body resisting or hiding?"

Clear answers rarely come easily or quickly at first, but as the conditioned responses begin to be recognized for what they are, this method becomes much more efficient and effective. Comprehensive answers occasionally arrive, sometimes as direct verbal responses, but the vast majority of new insights are reached gradually and subtly through incremental and nonverbal emotional sensations, small adjustments, and releases. We might not even notice each small incremental shift; but after years on this path, suddenly, one day we recognize that a great change has occurred through many of these tiny imperceptible steps.

Occasionally there will be dramatic personal breakthroughs, insights, or shifts of awareness, but these larger adjustments may also take years to fully integrate. *Life grants us exactly the right amount of time required for each level of expansion – no more no less. There is never any need or reason to rush, for the unfolding of truth is always perfectly timed. In fact, this is one of the main reasons why time even exists.*

FROZEN MATTER AND AGING

The sense of personal identity of a seventy-year-old person is very different from that of a newborn. The newborn is looking out and experiencing life without a reflective type of self-awareness; it has needs but no directing self-identity. As we age, we gradually become more invested in our bodies and their seemingly independent lives. Over *time* an identity grows and solidifies as families, careers, and personal reputations build. As *time* progresses, a *self* begins to form internally around the powerful cultural idea of being a separate and competing individual. This *self-image* then spends a lifetime shaping itself, sometimes even imagining itself as a commodity that has economic value. Through many years of this *selfing,* life *energy* can become blocked, rigid, and solidified, eventually manifesting as physical problems, including disease and illness. As *energy* freezes within and solidifies, joints stiffen, mobility diminishes, activities become habituated, and the other typical effects of *time* creep in.

While this may be the typical aging process, it is not inevitable. If instead, physical issues are recognized, understood, and then utilized as our most helpful and very personal guides for healing, then the outer physical expression of aging can look very different. When I was in my twenties and living in San Francisco, I was a fit runner who enjoyed racing. A fellow runner was over 100 years old and beat me every race; he also worked a double shift as a waiter. Aging in our physical realm is a normal process, but how it becomes embodied and shapes our lives also depends on our personal approach.

DEFAULT MODE NETWORK – COGNITIVE DISSONANCE

"It's not a question of learning much. On the contrary.
It's a question of unlearning much." *Osho*

Our sensory system, which makes comparisons based upon *contrast,* is a great tool for three-dimensional living. Its full analysis of all the input, decisions, and processing involved would be far too *time*-consuming, especially for the quick responses that are often necessary to preserve the safety and well-being of our physical bodies. To move through *space* efficiently and quickly, and to protect our bodies during emergencies, our nervous systems automatically become trained to shortcut the most repeated steps. The nervous system senses familiar patterns and speeds up the process by automatically responding with the most common and likely conclusions, based on our history of past experiences. This process, which has been named *Default Mode Network (DMN)*, operates much like subroutines in computer programs. It is doing its job well when it is making our activity more efficient and speeding our self-protective reflexes. The *DMN* is a very handy tool if we are trying to avoid a leopard or an oncoming car, competing in a sport, or playing a musical instrument, but this auto-reflexive response can also be problematic because it can become too automatic and miss subtle or different, but sometimes critically important, signals in its rush to react quickly. While our *DMN* reactions are helpful for emergencies and repetitious work, they can also result in inaccurate or inadequate responses and mistaken perceptions. In extreme cases, the DMN becomes our only "go-to" response. One of the characteristic identifiers when diagnosing PTSD is this over-activation of the *DMN*.

Beyond this shortcutting of assumed or "routine" steps, our minds and nervous systems can fool us in other ways. Another common trick is to completely filter and control the input and retention of certain *information. Cognitive dissonance,* mentioned earlier, is the name that psychologists give to our system's natural ability to completely shut out, modify, or rationalize *information* that does not mesh with past experiences or interpretations. Without even being aware, we automatically filter most *information* that has the potential to cause inner conflict or confusion. Even before we begin to process discordant or strange *information*, it can be automatically discounted or discarded so that it never reaches our conscious awareness. When processing bits of *information* that are in direct conflict with our internal vision, to "quell the dissonance within," this new conflicting *information* is automatically shut out, modified, or buried within our shadows. When new *information* does not fit within the familiar *parameters* of what can be easily and comfortably processed, we are prone to adjust or even not register the most disruptive or disturbing parts. We only "see," register, and process new *information* that fits what our system can comfortably deal with.

It is common for *cognitive dissonance* to appear in many small ways throughout our lives, often expressed externally as actions not aligning with beliefs. Small everyday examples include personal justifications similar to, "I only smoke a few cigarettes each day so I am not still a smoker." *Cognitive dissonance* happens automatically at times, such as wanting to do something "right" while being strongly tempted by something we know to be "wrong," or believing an unlikely story because it is from a person we trust, like, or admire.

When I was a child, I could not eat the meat from the animals on our farm that we had named, but I had no conflict when eating meat from the store because I could avoid the direct association of that meat with a living animal. Some examples are more complex, such as knowing you shouldn't get drunk while out with your boss but then ordering more wine to quell the feeling of discomfort you have around that person. The well-known "Fox and the Grapes" fable describes a classic example, pointing to the way we might criticize the very thing we once desired once we discover it unattainable. To quiet internal dissonance, our thinking and expectations are automatically adjusted to match actual or predicted outcomes. At a more significant level, *denial,* in its many forms, is a direct offspring of *cognitive dissonance.* If your income and security are dependent on a product that hastens climate change, you might not believe climate change even exists, despite visible evidence; you can look at the same data as others and still not recognize its clear message. At its most severe, *cognitive dissonance* can appear as a component of PTSD, expressed as dangerous or harmful behavior, such as not "seeing" a red car in your lane because you habitually avoid or block out red, still associating that color with a major disturbance from your past.

The important takeaway from all these interesting ways that our minds function is that we are constantly filtering, disrupting, and altering information for many unexamined reasons. Our more traumatic experiences will modify how the central nervous system works in ways that are tricky, complex, and difficult to recognize. In its most extreme form, PTSD, these neurological changes can include large memory gaps, seemingly illogical but fully automatic reflexes, and a complete rewiring of the way our brains normally process *information*.

HUMAN SEXUALITY AND WHOLE-ING

As one of the biggest drivers of human activity and interconnection, our sexuality is closely tied to, and a part of, every aspect of the entire *whole-ing* process. Our sexuality is fully tied to all of the most critical areas of human activity: biological, emotional, and spiritual. Its biological importance for reproduction is well understood and unquestioned, for without this reproductive function we would not even exist. Beyond reproduction, but still operating at the biological level, sex can be healthy for the body by providing exercise, stress and tension release, fun, entertainment, and general relaxation.

The emotional and spiritual aspects of our sexuality are far less understood or discussed. Sexual attraction has a mysterious way of shaping relationships that are ideal for shadow work, balancing *energy*, and healing. A strong attraction might indicate a good gene match and reproductive potential, but it also can indicate the beginning of a deeper journey into self-awareness – often it is both. Sexual attraction helps to create powerful and long-lasting bonds that facilitate working through difficult issues, those which we might otherwise run from. From within this type of polar relationship, the interaction will often appear to be the least healing precisely when the deepest shadow work is most fully engaged.

The spiritual-sexual journey does not require that we approach it consciously; the deeper spiritual aspects unfold automatically. Deep within the chemically infused bliss of sexual union, we tend to more easily drop our separation and boundaries to experience the deeper type of interconnection that is always possible. With the help of this biochemical mechanism, two separate bodies and *souls* are able to experience a deeper type of merging and sharing in an energetic *space* that can touch "oneness," for a moment. Through these connection experiences, we often have our first taste of "oneness" as we also learn about the possibilities and challenges of deeper interconnection.

MALE-FEMALE SEXUAL POLARITY

An important part of our purpose in this life is to heal ourselves but to even begin this healing process, we must first become aware of our shadows. One of the most powerful mechanisms for ensuring that this happens is a powerful *polarity* that has been conveniently provided by nature: our sexuality. Nature purposefully distributes *energetic* content to individuals in a *dualistic* way that encourages and drives a great amount of human interaction and interconnection. Fully tied to our genetic mechanism for reproduction, our sexuality is one of the great organizers and energizers of human life. It fuels many of our most challenging and interesting experiences, as enormous amounts of human *energy* and drive are unleashed by this powerful, but often confusing, *polarity*. It directly contributes to the wide range of possibilities found within duality, through the engagement and interactions of a particularly dynamic spectrum; one that runs from the extremely receptive pure feminine *polarity* to the much more aggressive focused drive of the pure masculine. Every individual embodies some mutable blend of these two *polarities* and this "spice" flavors all human thought and action, providing a wide and varied palette for human *energetic* exploration.

Just where each of us falls along this continuous *spectrum* will shift and change as we live our lives and continue to evolve. When we open to authentic interaction with those of a *complementary sexual polarity*, pairs of interacting *polar energies* interact, *resonate*, and exchange *information* to help activate, stimulate, balance, and ultimately heal. Healing occurs naturally and automatically whenever we openly interact with others, but *sexual polarity* intensifies and accelerates this process many times over. While any interaction between individuals will trigger an *energy* and *information* exchange that eventually leads to growth and healing, authentic sexual expression engages this process much more deeply and the effects become greatly *amplified*.

The attraction of opposites is always a powerful organizer in the natural world. As we mature, we each discover that we have a particular sexual makeup or a unique mix of sexual *polarities* that is drawn to, and attracted by, certain other mixes. While it is quite rare for anyone to have the experience of being just entirely male or female, our culture attempts to polarize these complex *energetic* "traits" by assigning each individual only one of the two extreme endpoints of the scale: male or female. Despite any intellectual or cultural resistance, we are repeatedly drawn to certain complementary "opposites" and through this process we usually discover ourselves, and our mates, to be much more complex than a simply divided male-female *polarity* might imply.

Through this complex and natural mechanism of sexual attraction, we instinctively find those who *complement* or, more often, stimulate our darkness, and through this complex but powerful process, we are presented with fantastic opportunities to discover and explore our shadow world. Like everything in life, our personal sexual attractions are never random or accidental; they directly guide us to those who will be most helpful or stimulating for illuminating our shadows.

For much of our history, the culturally accepted view of this wide spectrum of sexual expression was focused only on the endpoints of the scale; there are males and there are females based entirely on physical expression (*phenotype)* and all other permutations must be "wrong," "perverse" or even "evil." Over *time* our culture has gradually grown more awareness and acceptance for the great complexity of our sexual makeup. As male *phenotypes* have integrated feminine aspects and female *phenotypes* have integrated male aspects, the culturally recognized range of sexual orientations has widened tremendously. Today, our contemporary culture recognizes that the simple bi-polar male-female definition no longer accurately defines human sexuality, and in some cultures, very complex preferences have even become quite socially acceptable. Many dating services now ask users to sort through over a dozen sexual classifications, and Facebook recently defined at least fifty-two different gender profiles. Because nature always seeks diversity, it will naturally distribute the male-female polar attributes in any, and every, possible combination. *We are all small parts of this greater oneness, each seeking our lost or separated parts for reconnection. Our sexuality is one of the most powerful tools for guiding and activating this process.*

FOUR LEVELS OF HUMAN SEXUAL INTERACTION

Sexuality has always been one of the most powerful biological driving forces on our planet, even supplanting hunger at times. Sex is so important that individuals of many species have been known to deplete the last of their critical resources for a chance to procreate – sometimes forsaking food, water, lodging, friendship, and protection for that opportunity. While the clear biological reason for this extreme behavior is species survival, there are others.

Sex also has other important functions, at least for human *beings*. Sexual interaction offers an extremely multifaceted and complex menu for human communication and expression. As just mentioned, at the most basic biological level, our sex drive is an evolutionary mechanism deeply integrated into our biology, programmed to ensure that our species continues to procreate and propagate. We know that "opposites attract," and there are many powerful biological reasons for attracting potential procreative mates that are *genetically* different. At this level of pairing *genotypes*, *polarity* and diversity are highly valued in the natural world. Differences help create the broad and varied *gene* pool that produces strong viable offspring and more opportunities for novel genetic expressions that allow for better adaption and new evolutionary directions. Maintaining a diverse gene pool is usually an effective long-term genetic strategy for species survival. **Nature loves diversity because it creates new possibilities for expression, and these allow the species to adapt and fill more niches of opportunity.** Chemical *pheromones* and other subtle biological signals are constantly communicating below the level of the conscious mind as they "conspire" to create beneficial biological pairings. At this level, the instant sexual attraction which we sometimes label "love at first sight" might only indicate a chemical communication carrying *information* such as "I recognize, through our *pheromones*, that my *genes* and your *genes* will make biologically strong offspring." At this basic biological level, nature cares very little about our poetic and romantic notions.

At another level (the emotional-physical healing level) for many, sex provides a healthy pathway for the release of built-up tension; for some, this may be the primary avenue for the release of their habitual tension. While this process and the release can be fully healing if the underlying issues that shaped these tensions are simultaneously recognized and addressed, for most of us, most of the *time*, the sexual release only results in a temporary relaxation of tension. As with any "treatment" that works only on superficial symptoms, a dependency on the treatment itself becomes a possible risk. A reliance on constant sexual activity for the release of tension can easily become a problematic addiction or obsession. As this book emphasizes, the only real long-term "cure" for chronic tension is to directly address the emotional and physical "darkness" that creates, hides, and holds this tension.

At a third level, our sex drive also serves our subconscious desire to connect with others to explore that sometimes *timeless* place that lies outside of our normal, day-to-day routine. Sex serves *being's* natural desire to express itself through deep interpersonal connectedness; sexual excitement will relax our ego's usual self-protective constraints, allowing deeper levels of communication. Again, because it is common to experience our very first taste of "oneness" through sex, in our attempts to re-engage this wonderful state of *being,* there will be the natural, but misdirected, tendency to rely on sex. **Of course, what is being missed is that it wasn't the sex that caused this**

memorable experience; it was, rather, the surrender of the divided and separated "self" into a shared space of deep connection.

Beyond these more common human sexual interactions, a more formally organized practice such as *tantric yoga* can help us learn how to deepen sexual connection in a more conscious and, therefore, powerful way. This fourth level of sexual relationship, the spiritual or *tantric* relationship, represents the most conscious form of sexual union. In this very intentional practice, the *energy* created by sexual polarity is co-generated, merged, harnessed, focused, redirected, and then transformed. This powerful activation and redirection of *energy* can be used for many purposes such as healing, communion, expanded ecstasy, spiritual growth, exploration, or just plain fun. At the *tantric* level, this joint intention is utilized very consciously to explore the further possibilities of deep interconnection. Instead of two individuals having separate sexual experiences, the intention is a more conscious union or merging of *being* into the heightened "oneness" experience. Within such a deep and conscious surrender, we find ourselves more present and centered, and sometimes even touching the *infinite field* where "everything is also no-thing": a level of experiential meaning that lies beyond any possible verbal description. *tantric practice, energetically* fueled by a vibrant sexual *polarity*, can greatly expand our awareness of the many powerful possibilities of union with "others."

However, while the unconscious desire for this "taste of unity" is a powerful component of our basic sex drive, tantric practices are not restricted to just our sexual partners; the potential for generating energy through harnessing sexual polarity always exists and can be useful for other types of non-sexual relationships. These powerful polarities always exist even without actual sex, so when we bring this natural force into our conscious awareness, we add many new possibilities for the expansion of our being.

All forms of sexual expression have important functions, and most sexual relationships blend some combination of them. In the woven fabric that is existence, all parts of the cloth are equally important for the fabric's integrity. While the reproductive, biological, and pleasure functions are obvious, enormous amounts of inter-personal *energy* and *information* are also shared, exchanged, and expressed even when a relationship is based only on the most basic level of physical sexual interaction. Since deeper connection can be spontaneously triggered by relaxation, even the most basic of sexual relationships can lead to unexpected and powerful new insights; aspects of

Tantra are a part of all sexual connections, even if it is not consciously practiced as such. The gift of *sexual polarity* is a powerful tool that can help guide us to a fuller understanding of our *beingness*.

SEXUAL UNION

Our sexuality intersects every aspect of our *being* in significant ways and for most of us, there are very few parts of life that are more fundamental, powerful, motivational, disruptive, or confusing. Our range of human response varies from "sex is the devil's temptation and the road straight to hell" to "sex is the most direct path to *enlightenment*" along with every possible variant between. Advertisers have long realized that "sex sells" and many, if not most, movies, websites, and magazines use it as a tool for grabbing our attention.

As discussed, for many of us, the first profound experience of a deep interpersonal connection that includes the dissolution of personal boundaries occurs spontaneously during sex. Within the bio-chemically charged and heightened state that can be reached while engaged in sex, a deep ecstatic moment of awareness can unfold: a moment of *un-selfing*. **There is, then, the very natural tendency to associate that wonderful ecstatic feeling with the sex act itself. Almost universally, we fail to recognize that it is not the sex itself, but rather the deep and open energetic connection with another person that triggered the profound nature of that moment.** These first experiences of deep interconnection sometimes initiate years of attempts to recreate those ecstatic heights again and again through sex, only to eventually discover that this highly desired authentic connection is not automatically triggered by, directly linked to, or dependent on sex.

At any moment, some of us are more focused on exploring our separation and some are equally energized about exploring interconnection; but the complex nature of life means that we are always dancing amongst both. In our separation, when we interact, we often discover the common ground of our intersection. As we "go deeper," we also uncover aspects of the "other" that are unfamiliar, odd, and sometimes even frightening – things that don't always make "sense" when viewed from our perspective. Opening to these unfamiliar aspects begins the deeper healing process of exploring union. *Union* exists as a *space* of *being* that overlays all of our separate *spaces*, a *space infinitely* larger than the limited *space* of our more easily recognized intersection. Only when we are finally able to open to another and "dance as one" in this new and much larger expanded *space*, do we begin to see the real potential of *union*. Another name for this non-judgmental *space* of union is Love.

Imagine how life could be different if this wonderful awareness of our deep inter-connection became our everyday mode of living: an ordinary way to walk through life. This is not just possible; it has become the waking experience of many on this planet. These deeper moments of connection are possible in every type of human interaction when all parties involved approach interactions with an open heart that is free of agenda, judgment, and expectation. We constantly and unknowingly create barriers to connection with others and much of our personal sabotage hinges around our old concepts about sex and our sexuality. Because of this, we continue to create new wounds around our sexuality that only further reinforce our separation.

MERGING AND SHARING OF ENERGY THROUGH SEX

We are constantly exchanging, sharing, and balancing energy with others, but when engaged in sex or practicing *Tantra*, the amount, rate, and depth of energy exchange can increase exponentially. When any sexually active couples are together for an extended *time*, a tremendous amount of *information*, *energy*, and related *soul matter* is shared and exchanged between them. During the heightened and opened state of sexual union, it becomes exchanged automatically and in much greater quantity. Once this *soul matter* has been transferred, each partner begins to "feel" the other internally in a much more intimate way as these energies redistribute to become embodied components of their own *being*. When approached consciously this sharing of *energy* is welcomed and even encouraged, but in most cases, because this intimate sharing activates, exposes, and reinforces both partners' existing shadows, it often stimulates difficult and challenging interactions. Because aspects of our partners have become more present within us, we become much more aware of, and sensitized to, their qualities that we find disturbing or annoying; these qualities are now literally a part of us. Difficult to accept and integrate, these strange new internal feelings often become expressed as flashpoints for anger, frustration, and conflict. While this is an automatic human response, unless we understand what is occurring and respond in a measured way, we can quickly become deeply embroiled in confusion and conflict. Just being aware that this type of automatic *energy* exchange occurs, will help us process our feelings.

The process of opening to our "oneness" involves gradually letting go of all our conscious, hidden, and reflexive concepts about our separate self. While this is never a fast or easy process, intimate relationships that naturally stimulate our shadow world are amongst our most important teachers. If we remain aware and open to the process, our most intimate relationships will act as accelerators and guides for this greatest of all journeys into the depths of our *being*.

SACRED-PARTNER WORK

Despite obvious and hidden risks, there are enormous understandings to be gained by consciously exploring our sexuality with committed partners. Throughout our lives, we frequently find ourselves attracted to the very partners that most stimulate our darkness. This is neither accidental or a "problem" to be avoided because the revelations that surface because of these relationships *resonate* powerfully and help us better understand our shadowed depths. Later in our opening process, we might find ourselves attracted to a different type of partner: a *sacred partner*, one that can meet us in a more quiescent and stable place of "oneness." In many cases, these two types of relationships can be found in the same partner.

When working with a *sacred partner*, the first requirement is to build a deep level of trust and a clear understanding. The perfect partner is one who has extensively explored their own shadow world and is now prepared to meet midway in the powerful and expansive *space* that opens when two or more resonate together without any *interference* from *personal agenda*. The neutral shared *space* that can be created this way allows for an entirely new level of discovery about our *collective being*.

Because "others" are involved, in some ways this type of coordinated inner work is much different from most of our other *self*-explorations, but in most ways, it is just a continuation of the same work we do alone. Here we practice the same skills just in the presence of another, a partner who is prepared to meet us halfway, free of all agenda, including sexual agenda. When open love sits in the presence of open love without an agenda, a *timeless space* of sharing is spontaneously formed. Because, initially, this new *space* feels so similar to our individual private *space*, it might not even be recognized as a new and expanded shared *space*. The experience still feels similar because it is still the same "you' at the center. This is because as we grow and learn how to occupy larger *space*, nothing is ever left behind but our limiting ideas: the awareness does not change so it still feels just like "you." However, as this new *space* is carefully explored it is discovered that it actually is much more expansive because now awareness extends much further to include the physical *space* occupied by our partner. More importantly, even though our vision still sees the other, there is now no sense of two separate *beings* – all awareness of separation has disappeared.

The primary requirement for this type of connection experience is the availability of both partners to meet in a neutral *space* of non-judgment that is free of agenda. Years of prior shadow work are usually necessary for this level of required openness and availability. We are correct to recognize that we are stepping into new and uncharted territory so, during our early shared explorations as we first explore "opening" this way in the presence of "others," we will often feel extremely vulnerable. With experience, we see that the openness, vulnerabilities, and the nature of the *space* that we co-create will vary from partner to partner, but we also discover that we can safely and easily regulate the degree of our openness to protect ourselves as needed.

Within this sacred *space*, one must not expect a sameness from our partners. Our sacred partner's internal experiences will often be very different from our own, and this is what allows them to bring fresh perspectives and insights into this shared psychic *space*. Because they are able to "feel" and experience more aspects of you from their intimate perspective, a sacred partner can function as a unique type of therapist, offering focused suggestions, such as concentrating on specific physical spots, colors, energetic sensations, or experiences. A safe sacred partner will never judge, define right or wrong, blame, or shame; in this work, all participants must always be prepared and willing to adjust, or completely stop, if these, or any other controlling types of agenda enter the *space*. We must also have developed enough personal awareness to not confuse our partner's neutral and helpful suggestions and insights with our old habits of self-judgment or self-shaming. An astute and very valuable partner can observe, comment on, and often activate energies and thoughts that may trigger our inner pain or reflexive self-judgment. If we are relatively new to the *whole-ing* process and don't yet fully understand our own contribution of self-judgment, we might incorrectly feel that our partner is the one "doing" this to us and creating our pain. Later, with more experience, these same types of explorations can be easily understood as helpful guidance for liberation. Patience is critically important. The process *flows* organically and perfectly, once we get out of our own way.

Sacred partner work is a tricky and complex exploration of *being*, so, at least initially, a slow and deliberate pace is always recommended for safe adventures and successful growth. Everyone is different, and in these open explorations, it is natural and normal for missteps and adjustments to be made throughout the process. It is absolutely critical that both partners hold a clear and loving intention while simultaneously remaining flexible and open. Safety is paramount. A "safe" partner is one who has the same goal and respects all personal boundaries to meet us, midway, without any intention to influence or a personal agenda beyond the singular desire of letting go of all self-made boundaries to discover what can unfold in the presence of another. For our journeys into the greater extents of self, a sacred partner can help by providing inspiration, encouragement, support, and protection, but any attempt to influence or manipulate the organic and natural process with either partner's agenda is completely counter to the principles of safe practice. Since the desire to control or manipulate is a ubiquitous and almost reflexive human trait, for this level of exploration, partners need to be chosen very carefully. The right partners are those who will meet us without personal agenda, hold and protect the sacred *space* of opening, and continue to respect and honor the entire process. For these to be healing relationships, they must be open, equal, balanced, and treated as sacred by all partners.

This co-created and sacred energetic *space* is uniquely powerful because of all of the previously unseen experiences it can introduce to our awareness. Learning to navigate and open our sexuality in this new way is an enormous step in the journey towards "oneness." Of course, there are as many ways to explore our sexuality as there are people, and while many of these possibilities would not fit my definition of a "safe sacred sexual container," they all contribute to our broadening of experience. Because all forms of sexual exploration are a part of human expression, they are all valuable; all of our sexual explorations ultimately contribute to our collective experience. *If we find ourselves energized by a particular type of sexual expression, and instead of engaging to explore its deeper message, we repress it, then we have, instead, created more shadow in our energy system: a restriction that will shape and limit our growth.* As Osho (Rajneesh) showed us so demonstrably, it is our direct experience that frees us from our enculturated sexual blocks or kinks.

OSHO – WHY SEX IS SO IMPORTANT

"In the total orgasmic joy of sex, *time* disappears, *ego* disappears – these two things disappear. That is the greatest longing in you – you have known that in deep sex two things disappear: ego and time. You are not aware of time, you move into eternity, and you are not aware of separation, the ego is not functioning at all – that is the joy. Once understood that this is the root cause of joy, you are free of sex because now the whole thing is that you can drop ego and time without going into sex." *Osho*

Osho is known for many things but his wide popularity can be largely attributed to his reputation as the "sex" guru. However, in the deeper truth of his actual teachings, he is more correctly understood as the "beyond sex" guru. He brilliantly taught that sex by itself was not what was so enticing to so many, because the deeper thing that we are desiring is the "liberation from the self" that sometimes occurs spontaneously within the sex act, most often around the *time* of orgasm. *It is often through sex that we first experience this surprising new sense of liberation and freedom. Since our culture has not taught us other ways to achieve this fully satisfying and spiritual connection to oneness, we naturally associate this life-changing experience primarily with sex. It is our overwhelming desire for freedom and its direct but incorrect association with the sex act that fuels our culture's unbalanced obsession and preoccupation with sex. It is the experience of liberation, not the actual sex, which is so powerful and compelling.*

His teachings offered many effective tools, but our cultural media largely focused their attention upon his encouragement of followers to freely explore their sexuality. He instructed his students to "get over it" by fully expressing their sexual desires so that they, then, can "go beyond" them. He also applied this same general philosophy to all of our kinks and neuroses; we free ourselves by "acting upon" instead of "repressing" our desires. Because we are human and have our powerful subconscious reproductive drive directly tied to our sexual polarity, almost all of us have developed energetic resistances that are associated with our sexuality. If these become blockages that are not understood and released, they remain dark and become limiting parts of our *being*. *Osho taught that it is only through first exploring sexual attraction that we burn through our obsessions to discover the freedom that is available with, or without, sex.* He also taught that the ultimate non-reproductive use of our sexual *energy* is to transform it through *tantric* meditation. His students were taught how to effectively express these same desires and needs through *tantric* meditation instead of sex. Many of his most dedicated students eventually mastered these methods and continue to pass these techniques on as they live dynamic lives filled with joy, ecstasy, fulfillment, and freedom.

Sexual Abuse and Healing

Because sexuality is such a powerful driver of human activity when it is not channeled in a healthy way, it will still exude, but often in some damaging form. Unfortunately, this mischanneled expression of *energy* often becomes abusive and far too often young children are the objects of this abuse. Because of this, it is also common to find some form of childhood sexual abuse in the roots of many of our *shadows*. Early sexual abuse creates deep scarring because young children completely lack the power, psychological tools, and perspective to even begin to understand or process these powerful, overwhelming, and frightening experiences. Further complicating any meaningful processing is the fact that, in child abuse, the perpetrator is often a dear and loved family member or friend. As a result, the psychic damage often leads to powerful and very confusing divisions in the wholeness of their *being*. The parts of the whole *being* become separated and divided, stored in darkened and difficult-to-reach corners of their psyche and bodies, creating classic PTSD symptoms. In their isolated and hidden recesses, these *shadows* then automatically build additional layers of darkness that completely obscures the real source of the fear, anger, and confusion that drives much of the rest of their lives.

While the wounds of sexual trauma run deep, they also can lead us directly to life-changing breakthroughs and insights. Our long and confusing history involving our sexuality means that many of our embodied traumas have a sexual component, providing a powerful entry point for our healing. Again, as Leonard Cohen reminds us, "The crack is where the light gets in." Because healing must address the whole *being*, all successful healing must include the sexual component. Our learned cultural practices either completely avoid addressing our sexuality or they address it in a very unbalanced way. As we explore our inner shadows that were shaped by an imbalance of sexual polarity, successful *whole-ing* requires that we fully recognize, embrace, and honor all aspects of both the masculine and feminine.

Since many of our most resistant blockages have a direct relationship to our sexuality, when we finally address this shadowed part of our *being*, we often experience dramatic breakthroughs. Any practices that work to free up blockages built around sexuality will naturally facilitate overall *energy flow* and contribute to our healing and general well-being. **Because the natural expression of our sexuality has been so systematically repressed by our culture, it is often through the exploration of our sexual shadows that we release the final major dam blocking the free flow of our full energetic river.**

SEX AND ENLIGHTENMENT

Within certain philosophies and religions, *enlightenment* is seen as a final objective of the spiritual search, but my personal view is somewhat different. *Enlightenment* is not a particular state of awareness as many have imagined, but rather, it describes an ongoing process. Not only is this an ongoing process but it is also one in which everyone is always engaged – we are all gradually becoming more *enlightened* as we gain life experience. Every person and other living *being* is fully engaged in this process whether or not they are aware of this. No person is exempted and it always results in the expansion of awareness – at least eventually. The process of *enlightenment* is automatically driven by the accumulation of experiences from our day-to-day lives. Life is *enlightenment!* **Life's "theme" dramas repeat many times until each experiencer understands that this physical manifestation is, in fact, only their "story," which has been projected on the "big screen" called life simply to reveal the hidden truths that lie deeply buried within their psyche. Life happens as it does so that each may better "know" our deeper selves. This gradual illumination is enlightenment, a process that, from our perspective, appears to take time. Many lifetimes of repeated experiences may be required for a simple small change, but when the experiencer is finally ready, the shift will then appear to unfold in an instant.**

Enlightenment is the ongoing and continuous process of illuminating our darkest places so that the light of consciousness can shine through us more easily. Because almost all of us have blocks or "darkness" that are specifically associated with our sexuality, our sexuality can act as a reliable guide to help us discover the most direct connections to some of our deepest wounds – those that most interfere with the free *flow* of *energy* and *information* through our *being*. We cannot clear these by hiding from the shame, fear, and guilt that are tied to aspects of our sexuality. We can *enlighten* these parts of *being* only by consciously exploring them with fearless integrity.

This is one reason why, for most of us, the path to freedom must include the exploration of our sexuality. This is also why some groups, therapies, and technologies that address our sexuality will advertise *enlightenment* as a potential benefit of the practice. Since our potential is always limited by kinks or blocks which are often tied to our sexuality, sexual *yoga,* and *tantric* technologies can fairly advertise their potential for *enlightenment.* By exploring our sexuality without shame, these techniques help us to open specific areas of resistance that often can slow or stop our much broader journey towards personal freedom. **While tantric techniques cannot guarantee the liberation that some promise, if we have been on a healing journey and our remaining internal resistances are primarily those tied to our sexuality, then techniques that directly address this part of our being may help release the final blockages to allow a more illuminated state of existence. The converse, however, is fully true; without first and fully addressing human sexuality, enlightenment is impossible.**

When resistances that were interwoven with our sexuality are illuminated and healed, we become more energetically transparent and, therefore, more *enlightened*. While these therapies and practices can help us with the process of *enlightenment*, it must be clearly understood that it is not the sex itself that helps to liberate us; instead, it is our release of resistance that allows the light of awareness in to illuminate our once hidden darkness. Spiritual growth involves opening and integrating all aspects of our *being*, so *tantric* practices, alone, can't promise complete healing. However, by addressing life's universal but confusing sexual component, sexual healing will always contribute to *energy flow* and overall well-being.

As with everything else in life, sex involves expression at levels we cannot directly experience or easily understand. Our sexuality is tied to the biological reproductive imperative, the powerful human desire for release and pleasure, and a deep-level drive to connect with others in the *space* of "oneness." With its powerful energetics, sexual expression reverberates through every part of physical existence, which is why so much of our healing work in three-dimensional *spacetime* is connected to our sexuality.

PERSONAL EXPRESSION

We each bring our unique perspective for shaping the whole of *beingness*. Without the contribution and direct expression of all our unique qualities, the world itself remains incomplete and divided – it too becomes shadowed. Because our uniqueness and potential unique contribution to diversity are what makes each and every one of us so valuable, expressing our gifts is important for everyone and the entire world. Our most important personal mission is to fully express our unique offerings by being true to ourselves, no matter how we might personally judge these actions or our individual gifts.

To be able to fully express our special gift, we must be whole. We can only do this by first discovering who we are, and once recognized, claim and express our unique role within existence. From our limited human *viewport,* we usually lack the larger perspective to recognize the value of our unique "gifts" or those of all others. Nature never wastes *energy* on pointless projects, so everything that unfolds in our lives has some deeper purpose, even though this greater purpose is almost always invisible from our *time*-focused realm. Learning to trust our process of opening is also a process, and for this, we have been given *time*.

SAFETY AND TIME

Whole-ing is a very deep level of *soul work* that leads to internal change, so it is, by its very nature, destabilizing. When first initiated, it almost always disrupts the rhythms and patterns of our everyday lives, so it is also very important to proceed slowly and safely. It is our shadow world that is being explored in this work and there is no clear roadmap for this terrain; it is all of the unexpressed "darkness" that makes it so important to always proceed slowly and carefully. There is nothing to be gained by rushing this process for many reasons; the most immediate is not being properly prepared for the challenges that will arise, while the most abstract is that all the *time* required is created as needed. Because *time* always unfolds with as much "room" as required for any natural life process, we gain nothing by trying to "get it done" faster. Ultimately the healing process involves our learning how to let go of all "goals," and this includes those involving *time*. A temporary guiding "goal," might be to "feel and experience more of everything in existence," but at some point, in every opening, even this type of agenda must be released. The critical skill is allowing it all to unfold easily and naturally so that we never rush our body or *soul*. Each one of us must discover our personal rhythm for this exploration, the natural pace that works best for each of us as we learn how to better occupy our bodies and enjoy this realm of *time* and *space*.

As we proceed to "open" by releasing the tension held in an area of our body, such as a muscle group that has been long bound and rigid because of past emotional trauma or injury, we must treat this new opening as tenderly as we would treat a newborn baby. This renewed mobility and flexibility of our muscles, vertebrae, ribcage, or hips will unleash unfamiliar movements and new physical relationships, and this introduces many opportunities for injury. A joint suddenly mobilized has little or no muscular memory for the structure needed to support this new range of motion; all the muscles, tendons, and nerves that relate to the once bound-up region need to relearn their jobs. For this, the newly mobilized body needs *time* to build and coordinate its structural system; we must take the *time* necessary to relearn how to move in our newly freed body. No one else can or should define our program or pace, and it is critically important not to attempt to compete with ourselves or others; healing is never a competition or race. When unsure, we must always adjust our pace by going slower and being more cautious. The only thing that can be gained by rushing this type of work is injury; but if that happens, even this injury can be used as a positive learning experience to help refine our pace and practice.

As we begin to experience the enhanced living possible through this type of deep healing, we might get very excited about the process – often over-enthusiastic. Because of this, we must constantly remind ourselves to proceed at a slow but steady pace, one that maintains physical and emotional balance at all times. As we uncover restrictive kinks within, we might think it best to release these as quickly as possible but, instead, we must always remind ourselves to slow down and proceed carefully. Patience and caution are necessary because rushing this process runs a significant risk of further injury and damage. If we attempt to release our habituated kinks too rapidly, we only create more problems. Instead, with the gained wisdom of experience, the wisest explorers have learned how to open in smaller incremental steps so that the body and nervous system have more *time* to adjust. While the "go for it" attitude is helpful at times during our journey, this phase of clearing of our bodies of its old habitual restrictions is not one of them.

WHAT SHOULD I BE DOING IN THIS MOMENT

The most powerful thing that we can ever do is to give the very things that appear before us in each and every moment of our life our full and focused attention. We are all points for the propagation of conscious light and we illuminate whatever it is we focus upon. Whatever we focus upon becomes activated and contributes its *energy* to help shape the forms that appear in our daily lives. Our focused attention is life's "steering" mechanism, and by correctly using this "tool" we reshape our lives to look entirely different from what came before. *Through our attention, we create and regulate the flow of energy and form through our individual and collective lives.*

At first, this new level of awareness might evoke internal worry or pause because it also implies an enormous level of personal responsibility for the appearance of our lives. Here we must come face-to-face with our awesome power and recognize that we are the ones creating the visible appearance of our world. With this sudden recognition of our true power, shame and guilt also re-appear. We must remember that self-criticism is always our *ego* speaking, and that shame and guilt are the *ego's* most powerful tools.

So then, knowing our "terrible" power, how can we ever possibly know what is the best thing to do at any given *time*; where is the best and right place to focus our attention at any given moment? *Fortunately, the answer to this universal and ubiquitous query is astoundingly simple and always the same: it is to focus our attention on whatever is showing up in our*

personal life in each and every moment. With its perfect intelligence, life itself always presents for us exactly what we need through the coordinated unfolding of our fully interconnected lives.

When we begin to live our lives this way, engaging with whatever shows up in our lives, giving it our full and open attention, there is a different and expanded type of wisdom that starts to take over. We begin to recognize two extremely important but normally hidden truths about our existence. ***The first realization is that we all, together, create our entire experience of life.*** We gradually discover that our creation is not enacted by conscious or willful acts, but instead, it manifests through our *energetic* interactions operating outside the sphere of our conscious minds. By allowing each moment to unfold, naturally and organically, we begin to sense the presence of these *deeper* energies and how they are unconsciously shaping our entire existence. We have not recognized this before because so much of the creation occurs through the momentum of our collective *Default Mode Network* – most of our collective co-creation has been purely automatic and hidden.

The second realization is that we eventually discover that our most beautiful creations are those which occur when we completely let go of our desire to control the outcome or situation. When we drop our egoic desires and instead allow ourselves to *flow* with life's river, the greater wisdom of co-creation is fully revealed. There is a completely unexpected but beautiful and transformational shift that occurs upon the integration of these twin hidden truths. ***Without the ego's agenda controlling our actions, unexpected events and opportunities just seem to pop into existence and completely change the potential of every outcome. This is what happens continuously once we open to the infinite and interconnected space where this type of impossible-seeming magic unfolds every day.***

The healing process expands enormously when we openly and patiently address the issues that arise each day in our lives. We learn to trust that the issues appearing in each of our present moments are only being presented because they address our most critical work. ***Being fully interactive, the universe always presents the perfect experience for each moment, so there is never a problem with our knowing what we need to be doing at any given moment; it is always whatever shows up in our life.*** This is a beautiful transitional moment when this special type of wisdom takes hold. As we begin to realize that we have the ability to create anything, we simultaneously discover that the most beautiful creations are those which occur when we let go of our desire to control the situation. *To understand where we should be focusing our attention, all that we ever need to examine is what life is presenting to us, in each and every moment.*

PART THREE – DUALITY AND HEALING

INTRODUCTION

"We live under the impression that in order for something to be divine it has to be perfect. In fact, the exact opposite is true. To be divine is to be whole and to be whole is to be everything"
Debbie Ford

This part of the book contains further discussions of various topics that relate to this vision. This part expands upon Duality and the dual nature of our world and lives, the fundamental *energetic* nature of our existence, the necessity of maintaining balance, and the hidden interconnectivity or "oneness" that is built into the universe. Here, I re-examine basic principles and include some historical perspective to help the reader better understand the life-changing implications of this different way of *being*.

MEANING AND PURPOSE OF DUALITY – CONTRAST

"A certain darkness is needed to see the stars."
Osho

We live in a world that I have described as *dualistic.* Understanding the nature of *duality* is fundamental to a deeper understanding of many ideas that are presented in this book. *Duality* has three definitions, and all relate to important ideas that are discussed throughout this book.

Duality can be defined as the dual nature of human existence: our simultaneous immersion in both the physical world and the "spiritual" (or non-physical) world. This meaning of *duality* is often expressed through quotes similar to the following: "We are spiritual *beings* living physical lives." or "We are not these bodies, but they allow us to experience physical life." While these all describe an important and very relevant use of the term *duality,* and they also accurately describe how our lives appear within this physical realm, this particular definition of *duality,* while very significant, is not the most important definition for understanding the ideas of this book.

Duality is also a term used by physicists when describing the connection between two things where the properties of one define or describe the properties of the other. One of the simplest examples is that when standing in front of a mirror there are two objects – the body and the reflection – and both have a *dual* connection and relationship to each other. The *holographic principle* describes a much more complex *duality* and possibly the most powerful one recognized by modern physics. This principle describes the strong mathematical correspondence, or *duality*, between a volume of *space* and the outer surface of that volume: the inside and the outside are fully interconnected. The *holographic principle* states, "The *information* contained within a region of a *space* can be determined by the *information* on the surface of that *space*." With *black holes,* this implies that we may be able to understand what is going on inside by studying the more "visible" surface of the outside. Physicists are also referencing this same meaning for *duality* when referring to the *dual wave-particle nature* of all physical things in the universe. Depending on the perspective or intent of the *observer, radiation* from light or objects that we *observe* can appear to act as *particles* or *waves,* but never both at the same *time.* This *wave-particle dual nature* lies at the very root of *quantum physics* and it leads directly to many observations and questions about the very nature of our existence and just who or what we are – the primary subject of all of my books. Since this *wave-particle dual nature* is foundational to modern physics, when I am referencing this specific technical use of the term, I will refer to it as *"the wave-particle duality"* or *"the wave-particle dual nature."* **It is my conclusion and the conclusions of more and more physicists that there is an even more profound example for this second definition of the term duality. Our entire physical world is only a direct and dual reflection of the current state of the quantum field.** Even though this interconnected *dual* relationship is critically important to many of this book's ideas, this is still not my most significant use of the term *duality.*

When I use the term duality unless it is clearly stated to be one of the above definitions, I am referring to a third use: the understanding that everything sensed or experienced within our three-dimensional world is only manifest and recognizable because of having been shaped and built from contrasting pairs of polar opposites. Our entire sensory system is built upon contrast. For our neurological systems to function, we must have some sense of what "hot" is to be able to understand "cold," while "short" gives meaning to "tall" and "bad" is necessary to recognize "good." For our sensory and neurological systems to be able to gather and process *information* about our environment, we must be able to compare. This means that all the "things" in our physical universe, by necessity, must be built upon *contrast.* "Up" must always be understood in comparison to "down," "in" requires contrast to "out," "on" has "off," "soft" needs "hard," and of course "light" only has meaning against "dark." Without this type of direct comparison or *contrast,* we would not be able to see, hear, or feel. *Contrast* allows our brain to compare, register differences, and "understand."

Our most direct interface with the physical universe is our *neurological sensory system*. This system which is responsible for our tasting, hearing, seeing, smelling, and feeling, is completely dependent upon the existence and measurement of *contrast*. This *contrast*-dependent system not only defines the range and extremes of our senses, but it also controls or limits the input of *information,* which can then be processed by our central nervous system and ultimately interpreted by our minds. Our entire experience of this world is understood through this process that begins with our sensing of *contrast* and finishes with our unique and personal filtering and interpretation. Without *contrast,* we would lack the discernable *information* necessary to understand or interact with our physical world.

This *space* that we understand as our physical universe, the realm that most of us think of as our "reality," is completely dependent upon the existence of opposing energies and attributes. Our physical universe can never be just one-sided; it requires the "other side" for us to even perceive its very existence. **Duality is the contrasted way that the universe must present itself so that we can have physical and meaningful human experiences. We live in a universe defined by a duality where extremes of expression are necessary to create, define, and maintain the wholeness. The push-pull of duality also determines the extent and the center balance point for our ever-widening range of experience and expression. At the level of the individual, if we deny or reject any aspect of our existence, the result is greater imbalance, more separation, increased isolation, and wounded access to our wholeness and healing.**

This third definition of duality describes one of the most critical and intrinsic qualities of our reality. This polar division of attributes makes our world, and experience of it, possible. The structural foundation of our familiar three-dimensional world is entirely based upon the separation and comparative analysis of time, space, and things. Our reality is completely dependent upon *dualistic contrast*, critical for the creation of our unique and separated experience within *time* and *space*: an experience that is expressed and can be interpreted as a *time*-ordered series of individual things and discrete events. Without *dualistic contrast,* our type of experience could not even exist. *Our universe requires dualistic contrast so that we can sense our environment, measure or quantify its divided appearance, and then operate within it. The necessary contrast is created by pairs of polar extremes that must always exist together for our world to have any meaning: pairs like hot and cold, light and dark, up and down, in and out, good and bad, and male and female are what make the world real and meaningful for us.*

BOTH EXTREMES MUST ALWAYS EXIST TOGETHER

"Sadness gives depth. Happiness gives height. Sadness gives roots.
Happiness gives branches. Happiness is like a tree going into the sky,
and sadness is like the roots going down into the womb of the Earth.
Both are needed, and the higher a tree goes, the deeper it goes,
simultaneously. The bigger the tree, the bigger its roots.
In fact, it is always in proportion. That's its balance."

Osho

When we look at something in the sunlight, we see illuminated areas and shadows; bright and dark together form the *images* that inform and please our visual senses. Both the darkness and the brighter highlights are necessary to shape the forms that make our visual world. The light and darkness work together to define the shapes that inform us, and together they create beauty that can take our breath away. Without both extremes working together through *contrast,* we would not be able to register visual *information* and detail.

Our physical universe is a special space within existence that can only be experienced by us through the separation and division of "oneness" into many different discrete things and divided incremental time. It is only through this separation that we are able to have such a wide variety of individual experiences. To function in this environment and sort through all these divided aspects, our brains and nervous systems require *contrast* so that we can measure, compare, and process this wide variety of *information,* things, and experiences.

Our particular type of *space* (one that was historically, but incorrectly, assumed to be the only kind of *space*) is structured to be perceived by us as a collection of "things" and moments in *time* that appear to us divided and separated. All of these individual "things" (people, planets, *particles*, objects, ideas, experiences, passing *time*, phenomena, etc.) exist and seem real to us only through the ability of our nervous systems to measure and compare – this is how our brains and nervous systems must function. We only learn and process by comparing new experiences to those that we already "understand." "Hot" has no meaning to us without comparing it to something that we have already experienced and understand as "colder." What we are sensing and then analyzing is the degree of difference between our current experience and that which we have already experienced and think we "understand." Observing, comparing, and processing our internal measurements of *contrast* is how we categorize everything in our world. Our entire "understanding" of a physical existence completely depends on our ability to measure and compare *contrast*.

Contrast shapes our entire world for us. A world without polarities, or opposites, would not be experienced as "physical" in any way that our bodies could sense or understand; without these extremes, our experience of the world would quickly disappear in a blended and undifferentiated "soup of sameness." Because *duality*, the deeply divided polar nature of our three-dimensional universe, is responsible for shaping our *space*, these mutually dependent opposite pairings must always be present; our perception and functioning depend on this. **There can be no understanding of "good" without the simultaneous existence of the "bad." Both extremes that create the necessary contrast must always be present for our perception and comprehension.** *Contrast* is the keystone of *duality:* the critical property which allows our physical universe to be perceived, experienced, and enjoyed by all of us. **For this life-defining contrast to shape our world, the entire range of duality must always be present and available.**

A WORLD OF ONLY GOODNESS IS IMPOSSIBLE

"God is not only good, or else he could do better."
Meister Eckhart

The noble desire to live in a world defined entirely by goodness is completely natural and understandable; but, unfortunately, such a one-sided world is not realistic, useful, or even possible at the level of our physical experience. Many well-meaning people anxiously await the day when "consciousness has finally expanded to reach a critical *mass*, where the entire physical world transforms together so that peace and human goodness replace war and selfish behavior." **This hopeful expectation of a future physical world built only upon peace and love, a common and deep human longing, is unfortunately misguided and ultimately impossible for at least three important reasons. First, such a place would lack the necessary contrast to even exist at the physical level; second, this change involves waiting for an outside solution (waiting for others) when all real human change can only happen within; and lastly, it**

misinterprets the greatest gift that our physical lives have to offer: that this physical world we experience as external is always the perfect reflection of the deeper energetics stirring within; they are only being expressed outwardly, for our benefit, so that we may better see and know our deeper nature and ultimately heal ourselves.

Fully embracing the awareness that all possible expressions, all the positive and all the negative, must be equally available and manifest for the perception and functioning of our *dualistic* world is a difficult, but critical, step in our growth and expansion towards personal freedom. However, once this idea is fully understood and embraced, the groundwork for *magical transmutation* is firmly in place.

UNITY AND SEPARATION

Many of the long-term "problems" of our planet would quickly disappear if we only understood and embraced two foundational truths about our existence. The first of these is that while all living *beings* seem to exist and act separately, we are, in fact, deeply interconnected – so intertwined that several leading physicists have summarized their conclusions with phrases such as, "It is as if everything in existence is all one thing." The second truth is that our primary purpose is to "know' our greater self, and this deeper type of "knowing" can only be achieved through the direct personal experience of all the possible permutations of life. Our appearance as many separate things is only a trick of nature that allows us to fully explore and experience every possible nook and cranny of our ongoing co-creation. This apparent physical separation within our special "spread-out" *time* allows us the needed *time* and *space* to learn about, know, embrace, and then incorporate divided aspects of our greater *self*. We have been given the precious gift of *time* so that the true *self* can be gradually revealed through this staggering variety of *dualistic* experiences that we think of as life. Through this collective and deep exploration, we have multiple opportunities to experience all forms of fear and love (including self-love). We also discover that all of those that we consider to be "others," including those with whom we experience the most conflict, are only unrecognized or rejected parts of our greater *self*.

It is only through the lens of our divided three-dimensional expression that these smaller truths appear as many, for they are just different ways of describing that single unified *Truth* that can be most easily understood as "oneness." The separated nature of our language and conceptual thought is what has defined, shaped, and solidified our *human viewport*: a belief system built around the persistent appearance of separation between individuals and things. Within our physical *viewport,* we always experience the fundamentally simple idea of "oneness" as a multitude of *infinitely* complex parts. This division and separation is just the way our *dualistic* physical realm appears to us as a result of the functional limitations of our minds and sensory systems. How we then understand, internalize, and act upon this appearance of separation is entirely up to each of us.

When our worldview is only the powerful concept of separate, competing selves within a framework of limited *time*, we easily lose sight of our deep interconnection. From this viewpoint of extreme separation, relationships can seem complex, tentative, difficult, and ephemeral. **However, humans are innately blessed with a wonderful gift: the ability to simultaneously be aware of the complementary perspectives of separation and connection. As we expand our worldview to learn about, recognize, and embrace more of our deep interconnectedness, all this separation becomes, instead, a playground that provides endless energy, excitement, joy, and inspiration.**

TRANSMUTATION THROUGH WHOLE-ING

The deeper nature and purpose of our inner darkness is universally misunderstood. Culturally, we have been trained to "push away" both the unknown and that which we have judged as "bad." Because our inner *shadows* are, by definition, "unknown," it is quite natural to fear and avoid probing these "unexplored and dark" parts of our *being.* Internally, our shadows are usually being judged as "bad." However, as discussed, what is not being recognized in this avoidance is the *transmutational* change that occurs when once divided parts are again reunited in wholeness. **In wholeness, neither extreme will rise alone to present its lopsided and frightening face. Instead, because wholeness requires that all extremes always co-exist together as a unified entity, what emerges is an entirely new quality that looks and functions very differently from the feared and divided elements of dualistic contrast.** As these once-feared elements merge and integrate, their appearance and meaning completely change through the *magic* of *transmutation*; they combine, instead, to create entirely new forms that often become our most powerful gifts or allies. This is what occurs during integration.

NEED FOR EXPERIENCE

While the feelings that we universally classify as negative (such as envy, hate, greed, and lust) all challenge our ability to reside in peace and joy, they still must be experienced firsthand before they can become fully understood, integrated, "known," and thus made whole. This open wholeness is what allows the light of consciousness to *flow* freely and illuminate our entire *being*.

When we are first introduced to this world as newborns, we only experience the "now." We recognize this special perspective every *time* we witness and admire the bright openness of a newborn's eyes as they gaze onto their new world. For them, every moment is fresh and full of wonderment – but this is a form of "openness" that is only available through their innocence of life, duality, our culture, and its ways. Later, as their polar experiences in life lead to inner conflicts and disappointments, and the formation and solidification of traditional societal concepts and patterns reshape their *viewport*, that automatic and natural openness is lost. However, this is normal and healthy because after the right and perfect amount of experience within life, *whole-ing* can be initiated and the expansion of awareness that follows allows us to be reborn on the other side of our culturalization. The open and "agenda-free" living that then becomes possible is built on a different and more mature mindset, one that is still aware, open, and free like the newborn's, but also completely immune to the closing and limiting influences of our dominant culture.

THE HERO'S JOURNEY

The great mythologist Joseph Campbell named this journey of uncovering and knowing our true *being*, the "Hero's Journey" because of the endless challenges and obstacles our lives and culture present. This conscious *whole-ing* journey is not for the meek, for it requires a clarity of vision that often conflicts with our "common sense." The journey demands a willingness to tread through a dark wilderness that resists *illumination* at almost every step. We can only begin to engage the conscious part of this journey after a certain depth of experience that involves an honest and *aware* examination of all aspects of *beingness*. Ultimately this journey organically expands beyond the bounds and limits of our *self* to recognize the full *illusionary* nature of the *self* and all of that which we think of as our reality. **While such a profound journey might initially seem to be unnecessary, radical, or completely abstract, once understood and engaged, it becomes completely obvious that it is the best and clearest path through life; nothing could be more** *natural. Once this journey is undertaken and this new perspective is integrated, it even becomes difficult to remember what all our earlier resistance was about.*

Our human nature arrives with a built-in fear of, and resistance to, the unknown: a deep *limbic* survival mechanism that perfectly served the evolution of our ancestors. However, as with all useful tools, there comes a *time* or situation where that tool is no longer the best tool for the job, a *time* when it must be set aside to allow progress beyond. Our fear of the unknown is a sometimes useful survival tool that no longer serves us once we begin "the hero's journey."

TWO LEVELS OF HEALING TRAUMA

There are two distinct levels of healing within this journey: the personal, and the universal. The first is when we follow each of our aches, pains, unease, kinks, diseases, and sufferings back through the events in life that built, shaped, and hold the physical embodiments of these traumas. One by one, as we allow the emotions to freely express themselves, our traumatic life events become seen, accepted, and then healed to become integrated parts of a new wholeness. This level is the sequential and gradual approach that most successful recognized therapeutic techniques employ. This is the level at which every individual begins their journey.

This incremental approach to unearthing and healing personal and cultural layers of traumata initiates many openings and new insights. However, no matter how much of the trauma acquired within this lifetime is processed, we discover that deeper levels of trauma continue to surface. This is because our embodied traumata transcend even these physical lifetimes, so eventually, we find ourselves unearthing the more impersonal archetypical trauma that we inherited at birth. In our depths, we discover a never-ending supply of cultural traumata to explore and clear: the entirety of human history is available. Because of this, the incremental level of the *whole-ing* process, while very healing, will eventually be recognized as limited. While this "trauma-by-trauma" path to our healing is a wonderful way of walking through the world, there is another deeper level of healing possible, a more encompassing technique that becomes evident and available only after the lifetime specific incremental path has been explored for some *time.*

As I introduce this broader approach, two things must be kept in mind. This could be seen as a "better" pathway because it may seem to be more "efficient," but this is a *time*-based concept that is meaningless within the deeper levels of *being*; as we have learned, there is never any need or reason to "rush" this process. The second point is that once this approach is described, our *egos* will automatically attempt to turn our mastering of this technique into a goal: an agenda item. Our egos will quickly attempt to hijack this process to create the appearance that we are more "spiritual." This is *spiritual materialism.*

When an individual explorer is fully ready and feels attracted, there exists another level of clearing, a more thorough or bulk approach. While not a well-worn path, in "one fell swoop" this ancient technique with deep roots addresses the common source of all of our shadows, the ego-created *image* of *self* that we all hold so tightly. Throughout the *whole-ing* process, we gradually become more aware of the critical role that our concept of *self* plays in all of our dramas and traumas. Eventually, our long-held idea of *self* is recognized as the single common element in every trauma that surfaces. This is the same "*self*-maintained" identity that we believe has been "wronged" by "others" and "outside" events. It is the same *self* that appears again and again in all our most challenging experiences. ***Because our self-image is the leading character of all our drama and trauma, when we learn how to fully release our attachment to it, our path begins to completely clear.***

THE COMMON ROOT OF ALL SHADOWS

Through our explorations of the step-by-step approach to *whole-ing*, we gradually learn how to discover our shadows, address them, and then open further. Through this incremental process, we also discover that whenever we are successful in the illumination of one dark corner, another issue will always pop up; there appears to be no easy way to put an end to the continuous stream of newly discovered shadows and *attachments.* While the step-by-step approach is radically life-changing, at some point it can also feel like a game of "Whack-a-Mole" where every *time* we discover and address one aspect of our shadow, another just pops up.

At this place in our journey, we begin to realize that all of these shadowed and obscured parts of our personalities are fully related and have a single common origin. ***After more work involving deep exploration of our being, we eventually uncover the single connecting root issue: our reflexive and persistent misidentification with our separated body and ego.*** Deep within our journey of exploring our shadows, we discover that we each still harbor a small belief in the importance of the *self-image* that we have been creating and nurturing. After years of *whole-ing,* our self has certainly changed, but now it has just become a more "spiritual" *self-image*; we are still being influenced by our *ego's* agenda and storyline. We likely have also become convinced that these more "spiritual" roles that we are now claiming are real and important to the rest of humanity. We may even imagine ourselves as "healers" or "lightworkers," but such judgments of *self-*importance are just another of the many ways that the ego's trap of *spiritual materialism* can, and will, appear.

This is a critical epiphany, and with this shift in awareness, we can begin to address our darkness at its common root source: the *image of self* that is constantly and relentlessly created, maintained, and protected by *ego*. As we will discuss in more detail, the deeper issue is not the existence of our *egos*, for they have been, and will continue to be, very useful and functional for the maintenance and protection of our body. *Ego,* by itself, does not create our shadows; they are created by our investment in, or *attachment* to, the *self-image* that *ego* automatically creates and protects.

Albert Einstein and many of the gurus of Eastern spirituality describe our perception of our reality as "a stubbornly persistent illusion." Contemporary *quantum gravity physicists* describe a similar perspective using words such as, "A primary disturbance in the *quantum field* causes secondary *particles* to form and arrange themselves to appear as the physical expression of that which we call our reality." What all of us consider to be real, solid, and immutable is now being discovered to be more like an interactive "three-dimensional movie": a secondary expression of physical form to express the deeper interplay and resolution of fundamental *energies*. As we dive deeper, we realize that because what we think of as our "reality" is actually more like a "movie," nothing ever "really happens." Inside this illusionary *image* of reality nothing physical ever really "happens" to any of us, because all "actions" are just *images* or secondary *illusions* (*reflections* or *projections*) of the interplay of deeper level energies. ***However, it seems so real to us because, just like when we are engrossed in a 3D movie and duck as the sword swings our way, we can become completely "lost" within our "story."*** It seems Shakespeare was far ahead of even today's best science when he wrote, "All the world's a stage, and all the men and women merely players."

The deepest, but still understandable, *truth* is that we do not exist, as we universally believe, as separate and defined individuals confined to our bodies. The more scientific, but less understandable, wording of this *truth* is that the *quantum field* is first excited by a hidden form of *energy* and this "disturbance" produces the *particles* that take the shape of our world and its inhabitants. Only then are there actors to perform their roles in this "movie" that perfectly expresses these underlying *energies*. **Because of this, there is nothing that is ever actually personal about any aspect of our temporary lives; that which we think of as self does not even exist as a real or separate entity.** The separate *self* only persists as part of the *ego*-created individual identity; its appearance as something real and enduring is entirely illusionary. All parts of this separation, our entire physical universe, exists only as a temporary, but very useful, *image.*

Once we recognize and integrate this new level of understanding, then all of human history can be recognized as just a "story," that has been created, learned, and remembered by our *minds.* Because it is nothing more than a story held by our minds, it is also one that can be freely rewritten, or disappear in its entirety once our old concept of *self* is released. Upon this *un-selfing*, a bright new awareness of our *infinite* creative freedom is unshackled and it suddenly becomes very easy and natural to illuminate any remaining embodied darkness. We recognize that the entire "story" of trauma has been built and fortified with our always-active *self* continuously creating and adding more layers to help the illusion appear more "solid" and "real." **We suddenly realize that because our remembered history is only a story held tightly in our minds by the invested ego, we have always had the power to change the entire plot! This is an amazing new gateway to freedom.**

When we free ourselves from our persistent identification with the illusion of a self, we finally undam the river of life to find ourselves flowing in true freedom. We can look at each trauma sequentially on a "piece-meal" basis, or open to and embrace a more global method of healing: the realization that the self, who might suffer and be harmed, is only a made-up, fictional character that has no independent existence. The most complete level of healing possible involves this dramatic and bold step into our oneness. Without the old distraction of self, we naturally and easily become the "oneness."

When this is first recognized, it automatically initiates a renewed commitment to exploring the deeper nature of our *being.* Through this energizing expansion of our process, we recognize that we have always been presented with opportunities for witnessing the illusion of *self*, but now we can finally step back and observe our lifelong relationship with *self* with much more clarity and personal detachment. This is a transformational part of the *soul's* journey that has been described within the most esoteric writings of many of the world's philosophies and religions. The ancient spinoff of *Hinduism: Advaita Vedanta,* is an excellent resource for this level of *soul* work because it offers specific tools for helping us recognize and move through this important transformational step. Its method of direct *self-inquiry* teaches us to recognize and understand our misidentification, as it also helps us become more familiar with that part of our *being* that is eternal and transcends form. Today, in the West, our current awareness of this spiritual philosophy is largely due to the Indian mystic Ramana Maharshi.

RAMANA MAHARSHI AND ADVAITA VEDANTA

Ramana Maharshi was an Indian spiritual philosopher and teacher who was a precise contemporary of Einstein, though it is likely that they never met. Self-taught, his developed sense of *time*, *space*, and an individual's relationship to the world often made him sound as if he had been fully trained in the philosophical aspects of *quantum physics*. He developed his own style of intuitive reasoning or *self-inquiry*, but his overall philosophy stood squarely on the shoulders of the three-thousand-year-old *Advaita Vedanta* tradition. While he fathered the modern branch of this Hindu lineage, it was his direct protégé, Papaji, who had the special talent for attracting serious Western students to India. There, Papaji trained a generation of Western students and many returned to Europe and the United States to become influential spiritual leaders themselves. Once this Eastern export reached the fecund medium of the Western world, it evolved many modified, cross-bred, and interesting variants.

From his mountaintop in India, where he spent his entire life, Ramana shared these deep contemplative understandings that contain more than a hint of the contemporary *quantum* mystery:

"Investigate the nature of the mind and it will disappear."

"The universe exists within the Self."

"Apart from thought, there is no independent entity called world."

"When your real, effortless, joyful nature is realized, it will not be inconsistent with the ordinary activities of life."

"The universe is only an object created by the mind and has its being in the mind. It cannot be measured as an external entity."

"The world is an idea and nothing else."

"There is neither past nor future: there is only the present. Yesterday was the present when you experienced it and tomorrow will also be the present when you experience it. Therefore, experience takes place only in the present and beyond and apart from experience nothing exists."

"Even the present is mere imagination, for the sense of time is purely mental."

"The feeling of limitation is the work of the mind."

"The ultimate truth is so simple; it is nothing more than being in one's natural original state."

Covering one eye with his finger, "Look, this little finger prevents the world from being seen. In the same way this small mind covers the whole universe and prevents Reality from being seen."

"Reality lies beyond the mind."

"Environment, time and objects all exist in oneself."

"Good or bad qualities pertain only to the mind."

"The numeral one gives rise to other numbers. The truth is neither one nor two."

I find it extremely interesting that Ramana's lifetime unfolded at exactly the same time, and yet worlds apart from the breakthroughs in relativity and *quantum physics*. **Advaita and quantum physics are as different in background and methods as any two schools of thought can be, and yet the deepest conclusions from both methods of inquiry are remarkably similar.** This is only natural because both are methods that sincerely strive to understand what ultimately must be the same universal *truth*.

WHO IS HAVING THIS EXPERIENCE

Eckhart Tolle is an extremely popular contemporary philosopher, author, and speaker. In his first widely distributed book, *The Power of Now*, he states, "When our consciousness frees itself from its identification with physical and mental forms, it becomes what some call pure or *enlightened consciousness*, or simply presence."

However, it is clear from living our lives that we cannot just decide to be unidentified or non-attached; this way of being can only evolve over time and with a great amount of direct experience. It requires deep, personal,

open-minded, and broad experience with the constant overlay of honest self-inquiry that is focused upon discovering just who and what we are. As we will learn, releasing our attachments is the end result of a long process, not a decision that we can simply make; but because this process can be very difficult to maintain, we must constantly remake the decision to stay with this process.

Self-inquiry is a process that deeply examines the question, "Just who is having this experience?" Is it the "body–brain–ego" that is having this experience or is it possibly "something else" entirely different?" Through this type of inquiry, we begin to observe the unwavering existence of this "something else." We start to recognize an *observer* that has always been present, but up until this point, has existed mostly unnoticed: possibly this newly recognized presence is consciousness itself. Through this process, over *time* and practice, we eventually learn how to experience this presence without our *egoic* desires directing the show. These ancient questions about "who we really are" sit at the very heart of our lives, *Advaita Vedanta, quantum physics,* and contemporary research into consciousness.

OUR "STORIES"

Ultimately we can recognize this "presence" as a true, or at least truer, *self.* With this recognition, we also understand that everything else that we believed to be true about our *self* and our life is just a remembered "story" that our *ego* then keeps alive within our minds. **We ultimately discover that we created this story, repeated it in our minds, and have now become attached to its image.** This new awareness also serves to inform us about the creative potential of our "story" for, through this adventure, we also discover that we can also "create" a different, and better, story for ourselves; we are learning about our awesome power to manifest. Then we can practice this creative power by shaping and testing new or better "stories" before eventually, one day, we suddenly become thunderstruck by discovering the enormous value of dropping all of our "stories." Once we understand that this "*image*" or "story" only exists because we actively keep it alive in our minds, then we also realize that we, and only we, have the power to release our *attachment* to all of our self-created "stories" and free ourselves from the artificial bounds that any type of "story" automatically defines.

We are able to live in the present moment and fully experience the natural flow of life only when all of our self-manufactured and self-maintained "stories," even the "good" ones, are dropped. Until this life-changing moment, we are too busy living in our *"image" of* the past and creating new expectations for the future to be able to experience and enjoy what is always unfolding in the "Now." We eventually also reach an awareness that this "Now" is the only moment that is real; the past and future are only interpretations of experiences that form and exist in our minds. ***And while we are free to change our story at any time, we are also free to drop all stories.*** Living any "story" is entirely a personal choice. ***To liberate our minds and live in true freedom, we first must no longer be bound by our past experiences or future hopes: our stories.***

INQUIRY – EXPLORING SELF

Whenever we encounter a difficult decision, conflict, or unexpectedly disruptive event, it is always valuable to "inquire within" through questions such as, "What do I want? What is my *ego* desiring?" If we then just watch and observe, we will find that these simple questions will open doors to new insights. The answer might be as direct and clear as, "I want more money," "…a new house," or "…a new lover"; or it might be as diffuse as, "I want to feel different about myself or life." Either way, the next question in our process of *self-inquiry* could be something like, "Just who is having this desire?" At first, this deeper level of inquiry is almost always deflected with the obvious and reflexive, "I am having this experience!" By this point, our direct inquiry is probing the roots of our *being*, so the next logical questions should be, "Who is this I," or "Where is this I?" These are simple-sounding questions where the answers require a deep level of understanding. For any issue or tough decision, this type of direct and simple sounding exploration can lead to unanticipated breakthroughs in our understanding because all of our personal issues hinge upon the same fulcrum: our unexamined *attachment* to a non-existent *image* of *self*.

AWARENESS OF THE UNCHANGING "PRESENCE"

As mentioned above, the first reflexive answer to the internal exploration "Who is having this experience?" is usually, "*I am having this experience*"; but then we quickly re-inquire, "Who is this I?" As we gain experience and are able to more carefully observe our inner landscape, one day we begin to recognize something very subtle that has not been previously noticed: a "presence" that is always watching and different from the egoic "I" (the one that believes that it is real and having this experience). This presence has always been there, but it has just been completely obscured by the constant chatter and distraction of our egoic minds.

If we have been granted the additional blessing of many years lived, we also gradually become aware that this part of us has always been with us and still feels exactly as it did when we were teenagers; the presence that is being revealed seems to be ageless or "outside" of *time*. We begin to recognize that this "presence" has been a constant observer from our first conscious memories, but one that has always been hiding in the background. We have not noticed it through all the chatter and clamor of our daily lives because it sits so quietly, never imposing or complaining. At first, this new perspective appears to be extremely subtle, and possibly even just a trick of semantics or rhetoric; but over *time*, a steady awareness grows and eventually this "presence" is recognized and acknowledged as a "real" part of us that has always been quietly observing but never reacting, interfering, or aging. While appearing to be ageless, it is also infused with a powerful but quiescent type of *timeless* wisdom.

This is the nature of the self-inquiry that *Advaita Vedanta* teaches and it is the foundation of a process that gradually and slowly guides us into an entirely new level of understanding about our inner *being*. The individual "I," which once existed in separation, becomes recognized to be only a creation of our separated egoic minds and is, therefore, impermanent and non-existent in any real sense. Initially, the deep always interconnected presence (the "I and I", representing our access point to the greater *collective consciousness* that we are all a part of) may still seem strange, unapproachable, and unfamiliar. This newly opened *space* can even seem frightening at first, but it is an invitation not to be feared because this *space* is sacred ground and precisely where the great human adventure, devoted to exploring the expansion of our consciousness, must unfold.

PREFERENCES, JUDGMENTS, AND ATTACHMENTS

In a world that is entirely shaped and defined by *duality*, it is, of course, very easy, natural, and completely normal to have preferences. We may prefer a warm day over a cold one, corn over potatoes, and sunshine over rain; but these are all just preferences, which by themselves do not create disturbances in our inner lives. We all have our individual preferences, they are just a part of nature's gift of diversity. It is completely natural and healthy to have many evolving and changing preferences throughout our lives, as long as we consciously realize that a personal *attachment* to one of these preferences is quite a different thing. It is our *attachments,* and not our preferences, that interfere with the open *flow* of *information* and our journey towards freedom. When preferences are explored they guide us to new experiences. ***Issues*** *only arise when preferences manifest as judgments or we become attached to one outcome or another.*

When enjoying nature, many of us will prefer a bright sunny day over a rainy day. This is completely natural and has no impact on the *flow* of our life. However, if it does rain, and, instead of adjusting to the experience which is presented, we resist by believing that the weather has "ruined our day," then we have clouded our experience of that day through our *attachment* to a particular outcome. If we also go further and internally experience this rain (which is a necessary and integrated part of vibrant life) as negative or "bad," then we have also moved into *judgment*. **Judgments and attachments are two of the most common self-created obstacles interfering with our flow, sense of well-being, and our search for freedom.**

If we recognize that *judgment* and *attachment* are only the universal and automatic response of *ego*, we then receive a spectacularly powerful gift. Once this can be recognized, we develop the new ability to more easily sense or spot when *judgment* and *attachment* arise. This new skill quickly becomes one of our most faithful guides for the rest of our inner journey. **After this important realization, when we recognize a new disturbance, our checklist usually needs only two items: judgment and attachment.** Personal freedom completely hinges on recognizing and fully understanding the deeper nature of these two common, and very human, responses.

JUDGMENT

> "Out beyond the ideas of wrong-doing and right-doing
> there is a field. I'll meet you there." *Rumi*

As we start to recognize the full range of expression that must exist for our world of *duality* to manifest, it is critical that we look more deeply at all of the various forms of *judgment* that have separated and divided our complete *being* into its many separate and incomplete parts. While judgment is a natural and common response to our living within *duality* and separation, it is also the major contributor to human suffering.

Almost every aspect of our culture trains us to judge. We were all taught about "good" and "bad," right" and "wrong," "saints" and "sinners," "winners" and "losers," along with "enemies' and "friends." We learn to embrace and welcome some things, while other things are to be pushed away or feared; our culture encourages us to create and maintain these types of separation. Throughout our lifetimes, a tremendous amount of internal tension is created through this unnatural separation. It is all created by our *ego's* reflex to judge and reject one side or the other of opposing *dualistic* aspects: two opposites that actually must always exist together to maintain wholeness, both within and without.

To be able to fully understand and experience "good," we must first recognize and understand that everything that we separate and label as "bad" is an equal and balanced component of "all that we are." Both extremes are needed to make the whole, and both extremes must always be present for our world of *duality* to even exist. To only see and accept one side of a person, object, issue, or experience is to be in denial of the fundamental, complete, and holistic nature of our physical existence. **To not recognize and embrace all extremes equally as equal and integrated parts of our wholeness is to "judge" one extreme as better or right; and through this, our common, reflexive, and often unconscious judgment, we also create an unbalanced and unnatural separation within.**

The division of the whole, within and without, is driven by our very common, logical, and natural-seeming personal judgments about one particular part of the pairs of opposites that shape our world of duality. The eventual result of any judgment will always be more separation, less understanding, more confusion and conflict, and a significant restriction of our clear insight and potential as humans. Most of the chaos we witness in today's world is a direct result of this type of one-sided judgment.

WHEN BEHAVIOR HARMS OTHERS

Non-judgment does not mean that harmful anti-social behavior, such as theft, abuse, murder, or warfare should be ignored or encouraged. Instead, the perpetrators must be seen, understood, and rehabilitated as broken, isolated, and confused parts of our wholeness that are signaling their need for love and support. Instead of more isolation, rejection, torture, or execution which, today, is our society's standard responses, these confused *souls* require, instead, the experience of love and support that has been sorely missing from their lives. Only by understanding, experiencing, and sharing unconditional love, support, and care, can these confused *souls* begin their healing and be brought back.

However, for this radically different treatment to be effective, or even be recognized as a helpful solution for today's fractured world, our entire system of values needs to be re-examined. For this level of healing to unfold and spread, our current culture, which largely fosters, encourages, and rewards competition and greed, must be replaced by an entirely new culture built upon the recognition of a loving and shared interdependence. Because such a radical change would take generations and require a profound and resilient level of guidance, trust, and understanding by everyone, it is not reasonable to expect or hope that the entire world will shift and open to such radical change. Instead, it is the smaller local demonstrations that are more likely to succeed since they can better identify, engage, and inspire those individuals, families, and smaller communities that are ready for change.

Humankind has proved, *time* and *time* again, real and meaningful societal change can't be forced or engineered from the outside. Dramatic shifts in our way of *being* can only begin within each individual; it can't be created through forced external means, such as laws or institutions. At both the individual and cultural levels, what we experience as the outer manifestation is only the physical *projection* of our inner psychic shadows. If we are not peaceful within, our world will not look or feel peaceful without. **Peace must originate within; there exists no other way.** We can't fix our culture if we each are still hiding from our shadows and holding fear inside; this inner darkness will only project and re-manifest, again, and again, and again. **"As within, so without"; it is as clear and simple as that!**

SELF-JUDGMENT

Once it is fully recognized that the outer experience is only a reflection of the inner *being*, then it also becomes clear that all *judgment* of "externals" is only a *projected* form of *self-judgment*. Analyzed from a completely different perspective, once we fully understand that "we are all one," then it becomes self-evident that any *judgment* of others is tantamount to a judgment of some part of our greater *Self*. *Self-judgment* is found at the root of all *judgment*, and all *judgment* completely inhibits our ability to *flow* and recognize the deeper truth of our "oneness."

As we carefully watch our internal dialogue, many of us might observe that we often are critical of ourselves, *judging* our actions and responses with internal thoughts similar to "I never do it right," or, "I am so stupid." We blame ourselves for past "mistakes" and carry this shame about past behavior forward into our present moments. Because we have been so thoroughly conditioned by a culture that teaches that this judgment is healthy, most of us grow up believing that our internal self-criticism is accurate and helpful for keeping us on the "right" track. **What self-judgment and inner criticism do is divide,**

separate, and then isolate important parts of our complete, unique, and more powerful whole being – the very treasure that is our most personalized gift we each bring to this world.

Self-blame, and its inbred cousin shame, may first appear as an occasional nagging voice in the background chatter of our minds, but over *time* its message will solidify and become firmly recorded into our physicality. Early on, this embodiment can be witnessed as small quirks, such as an inability to easily share direct eye contact, or it might express itself internally as an ache, stiffness, pain, or a deep feeling of discontent within our *soul*. Later on, in its more embodied expression, its influence can be observed in the individual's posture; there might be increased slumping, dropped shoulders, or a broken gait. If not addressed, self-judgments compound, deepen, and often manifest as serious chronic conditions or disease. Whatever its form, self-judgment always sends powerful energetic ripples into our present moment experience.

Once we fully recognize the inhibitory nature of self-criticism, we might attempt to adjust and correct this destructive trait only to find that the habit persists, despite our most noble attempts. Einstein's famous quote, "We can't solve problems by using the same kind of thinking used when we created them," describes the dilemma we are facing. A more tailored warning might read, "We can't solve problems created by the thinking and doing egoic mind through a process that involves more egoic thinking and doing."

Einstein even offered the solution to our dilemma with another of his famous quotes: "The only source of knowledge is experience." *Self-judgment* divides our wholeness and this limits our depth of experience, an action that interferes with acquiring self-knowledge. It is only through experience and authentic living that we can illuminate and reveal those parts of our wholeness that have been hidden from our mind and external senses. Once we open to a fuller experience of life, its revealing light automatically begins to illuminate our inner darkness, highlighting those parts that were once banished into the deepest of shadows. Only when revealed, recognized, and embraced, can these lost parts of ourselves be reclaimed and reintegrated; through this process, we restore our wholeness as we also come to "know" ourselves. As we integrate these lost pieces of *soul*, the inner criticism will quiet naturally and without any additional effort. **When there is nothing left to judge or divide, the always present "oneness" once again becomes the center of our awareness; but this time its influence is fully sustainable.**

OUR ATTACHMENTS

Several ancient Eastern spiritual traditions teach that it is our *attachment* to aspects of this illusionary and *dualistic* world that directly leads to most or all of mankind's suffering. We have become so *attached* to the illusion of our bodies, our ideas, our possessions, our riches, our loved ones, and our creations because we have been convinced by our senses and cultural paradigm that all these things are real, solid, and critically important to our lives. We have been taught that these are the important things that make and define meaning in our lives, but as we grasp and cling to these ephemeral illusions, we suffer.

In the West we have become very *attached* to our possessions; at times we even start to believe that our lives "depend" on our "things." In my architectural practice, I have heard clients utter phrases similar to, "I couldn't possibly live without a five-bedroom house, Italian travertine floors, or a four-car garage." Our possessions can become closely associated with our sense of well-being and used by the *ego* as "assets" in its attempts to establish a *self-image*. For many, our possessions have become the most important measurement of our sense of self-worth.

While we might own the fastest car, the biggest boat, and the largest house in the most expensive neighborhood, and even feel "successful" for a while, we probably also will experience an increase of stress and problems as we also must worry about maintaining this unsustainable level of "self-worth." Later, when we experience financial reversals and lose some of these prized possessions, we might start to imagine ourselves as "failures" and become unhappy or depressed. We might also worry that others will judge us as "unsuccessful," even if we have secretly discovered that we are actually enjoying our newly pared-down lives. If we are weighing and assessing our self-value through these temporary physical things, our inner peace will always be subject to dramatic fluctuations and disturbances.

The issue is never about our possessions themselves, for they are inherently neutral in their meaning. The problem is our perception that we need them; that they are important or add meaning to our lives. What we are observing is our *ego's identification* with these things. Once analyzed and understood, we can quickly see that this "need" for a bigger house or faster car is certainly not a survival issue, for humans survived in caves without any mechanical means of transportation for eons. It is also not a requirement for happiness because anyone who travels widely quickly recognizes that happiness does not require an elevated standard of living; a higher level of happiness than that of developed nations can be witnessed almost every place in the world where basic minimal needs have been met. The problem is that our possessions have become an important part of our *self-image*: an artificial and temporary idea about ourselves that is then carried forward as a mental construct in our minds. We then devote a great amount of our physical and emotional *energy* to protect, shore, and maintain this idea. At some point, only after the right amount of experience, it becomes clear that our "things" weigh us down and are not critical for our leading healthy, productive lives; they have no relationship to "who we are." Our self-created concepts about our possessions block and *dampen* our natural *flow* of life.

ATTACHMENT TO LOVED ONES

Our attachment to others (friends, family, and loved ones) is a much more difficult issue to examine honestly and reconcile. **Through our latest physics, it is becoming more and more apparent that we probably having this experience of a physical life without a solid or objective reality existing outside of our internal mental constructs. Our physical body along with all our friends' bodies are just another piece of this great illusionary experience. We mistakenly believe that our subjective internal interpretation is the real and objective reality, but what we are calling reality might be more like a well-coordinated dream. Our misunderstanding inevitably leads to suffering because, when we believe only in the illusion, it appears that our loved ones are constantly dying and leaving us. We are completely immersed in the mistaken belief that we are these bodies, just as our loved ones are also their bodies. The illusion seems so real because we only experience and understand the outside world only through our senses and limited neurological processes; it is almost impossible for us to see that our awareness of temporary separate bodies aging through time is just a subjective view that we formed from a lifetime of internal processing.**

Of course, within our minds, we are fully convinced that this physical world and our bodies are "real," because this is our direct experience. We feel the weight and hardness of the solid stone wall that we just ran into, and our broken leg is extremely painful, so how could all this not be "real?" **While this is our most direct interpretation of our experience, it misses the important lessons of modern quantum physics, our most revered spiritual traditions, and many of our unusual, but extensive, personal experiences. All of these indicate that the external physical expression of life is not so solid or fixed.** Instead, our world and life may be better understood as a secondary manifestation, or *image* that originates in, and is *projected* from, a much deeper level of reality.

Our direct sensory experience automatically and naturally includes our bodies as part of our interpretation, and this association leads to our constant worry about, and a desperate clinging to, these temporary forms. It is nearly impossible for most people to recognize, or even seriously consider, that our unwavering belief in the absolute and fixed nature of the three-dimensional physicality – the deep-seated belief that "we are our bodies" – is the pivotal misunderstanding that prevents us from experiencing the full potential of the infinite possibilities that life is continuously offering.

As impossible sounding as this new vision initially appears, it is starting to gain traction as it increasingly is recognized to be supported by our latest science. For me, it is also the only description that helps to explain the human condition. *Eventually, all of mankind will once again shift paradigms and realize that it is the egoic human mind, so strong in its conviction of a real and independent self, which is the actual pretender, desperately struggling to protect its long-standing, but internally created, concept of self-importance.*

For the vast majority of those raised within our Western culture, the *attachment* to our bodies is nearly absolute. While this belief persists in most human sub-cultures, as just mentioned, some ancient traditions, particularly Eastern ones, teach a much different mindset. *Buddhism* reminds us that anything that can die, decay, break, or get lost is impermanent and, as such, is "not real." *Hinduism* speaks of the Maya, the illusionary nature of physical life. *Advaita Vedanta* teaches that our everyday concept of *self* is a complete misidentification. Despite our resistant and persistent internal perception, these ancient spiritual traditions continue to spread the strange idea that our physical world, bound by *time*, is only a temporary dream-like *image,* and our real *Self*, which lies hidden beyond our senses, is eternal.

ATTACHMENT TO OUTCOME

Thoroughly integrated into our Western culture is one of the most pervasive types of *attachment:* our *attachment* to an outcome. As previously mentioned, this is also one of the greatest difficulties for the Western mindset to overcome because we are trained to take charge, analyze situations, find solutions, and take action. If our first solutions do not work, we are encouraged to try again, and again, until we eventually meet our goal. My father's message to me took this one step further by adding a layer of blame and shame: "If you are not successful, it is only because you have not tried hard enough." This is the "American way" and it certainly can produce short-term "results." However, because all of these "results" reflect the diverse and often conflicting wills of many competing individuals, they often create even more chaos and division. Throughout our culture, we witness a steady stream of societal problems created by our short-sighted "doing" because this "doing" reflects only our division and separation.

Our "do-ing," by itself, is also not the problem. The Western mindset produces a great amount of "do-ing," and through it, we have achieved a bewildering variety of stunning technical accomplishments. Through all our "doing," we have also demonstrated the incredible power of the focused human will. Western civilization has, without any doubt, proven that humans are powerful creators; when we focus our intent, we can manifest many things that were once thought to be impossible. *Space* travel, computers, powerful weapons, transformational surgeries, skyscrapers, precision manufacturing, and our highway and rail systems are all evidence of our powerful ability to manifest our will.

However, as we have also learned, all our sensational achievements come with their own set of new problems, and more and more these introduced problems are presenting enormous challenges for our very existence. Nothing that we build, develop, or construct will be all "good" or all "bad" because everything that manifests in our world must reflect the full spectrum of *duality*. However, when we are driven by a mindset that divides the world through economic competition and differing nationalistic or religious principles of "right" and "wrong," then imbalance and problems are automatically built into our manifestations. When we create from a place that is still filled with unexamined shadows, the results will always reflect, and even magnify, this darkness and confusion.

We are exploring and witnessing our enormous powers of creation; this alone is a valuable and expansive experience, but at the same *time* we are also discovering that we usually lack the greater wisdom to understand or predict the full impact of our creations. ***New and sometimes greater expressions of darkness will always be part of the result when our "do-ing" originates from a place that continues to be obscured by unexamined shadows.***

As an integrated part of our conditioning and training as "do-ers," we learn to expect, anticipate, or at least hope for certain future results. Not only are we conditioned to controlling outcomes through our "do-ing," but we also become invested in the rewards that we expect later: a *time* that only exists in our minds, far outside of our present moment. Many readers will ask, "What is wrong with expecting results?" and the deepest answer is, "Absolutely nothing." This is normal Western thinking and behavior and there is nothing "wrong" with this way of thinking. However, this type of thinking will consistently produce certain kinds of results: outcomes that express the separation, competition, and struggle of our collective psyche. We already know what this looks like, for these are the types of results that shape the world that most of us experience every day. We have a long history of these types of results, and we can only expect to witness more of the same if we restrict ourselves to this familiar path that largely reflects our *attachment* to outcomes.

If instead, we first illuminate and heal our shadows through individual and internal whole-ing, and then create and act from this more illuminated new awareness, a completely different type of manifestation will then unfold: one that better expresses our interconnected wholeness.

ATTACHMENTS DAMPEN OUR VIBRANCY

While our *attachments* to something, someone, some *time*, someplace, some way, or some outcome are a natural part of human life, they are also the most glaring source for our common difficulty with experiencing the full *flow* of life-force. When we are too *attached* to form, things, or concepts, our *core vibrational resonance* (CVR) responds to the weight of this resistance by vibrating with less amplitude (power or volume) and fewer harmonics or overtones (expression and expansiveness).

In the *classical physics* of *vibration, damping* is the technical term used to describe the restraining effect of a weight or *force* that opposes and slows or stops the natural *vibration* that is normally shaped by the object's *natural resonate frequencies*. This added weight or *resistance* then effectively drains *energy* from the system to reduce the *amplitude*, or *power*, of the *vibration*. As an example, imagine hooking a bungee cord between two trees or columns on a porch and then snapping it like a big guitar string. While it is still *vibrating*, throw a wet towel over it and watch what happens; the vibration will quickly slow, then stop. This is *damping*. *Damping* is a good thing to design into bridges, tall buildings, and car suspensions because it reduces movement by absorbing *energy* to keep them from swaying or bouncing too much. In certain circumstances, *damping* can very helpful.

We can understand attachment as "holding fast to the temporary forms of this life." This includes our memories of the past and our hopes for the future. By doing so, we disrupt the creative present moment and our ability to *flow* through life. In my architectural practice, if I present conceptual ideas to a client too early in the process, there is a very real risk that the client will get *attached* to some aspect of one of the designs. If that occurs, the project will then cease to evolve freely as it becomes bound by this *energetic,* and very influential, expectation that then constrains the free *flow* of ideas: the open creative process has been *damped.* Having learned that this happens easily and quite often, I rarely present my ideas to clients until they are almost fully developed.

With our energetic bodies, there are certainly times when a controlled amount of *vibrational damping* may help provide the necessary calm in our lives. When not *damped,* however, the natural *resonant vibrations* of life can, and will, swell into something that may be completely unexpected and *infinitely* much greater: a free and loving expression shaped by vibration from the deep river of our collective *being.* This unrestricted *flow* of natural *energy* will then be felt and recognized by the individual as the powerful feeling of well-being, strength, and joy, a feeling that can also be described as unbridled ecstasy.

Attachments of all kinds will *damp* or restrict our experience and prevent us from discovering life's freedom and the ecstasy that is always possible and always available. **Personal freedom does not require that we avoid the objects of our attachment because the actual person, idea, or thing is never the problem.** Instead, we must reach a better understanding about the nature of our *attachments* to things, people, and ideas, how *attachment* works, and how this weighs us down. **What is blocking our full vibratory potential is not the person, thing, or idea; it is only our mental concepts about them.** These people, objects, and ideas to which we become *attached* are often the most beautiful gemstones within our lives. These may include our children, our homes and prized possessions, our accomplishments, and our lifestyles. Freedom does not require that we avoid or distance ourselves from these. Neither does it require a simple or ascetic lifestyle, although there are many unexpected insights to be gained by living simply. **Instead, freedom requires that we reach the point at which we no longer identify ourselves through these external things and people.** When we don't *identify* with them, we can still enjoy, care for, and love them, but at the same *time,* our deep internal feeling of joy or love will not be dependent upon them. Witnessing, experiencing, and understanding our *attachments* is a critical step in our common journey towards the realization of permanent freedom.

Freedom can only be built upon a psychological state in which our peace, happiness, joy, and love are not a direct function of, or dependent upon, the very things and people that we enjoy and love so much. *In freedom, as outside circumstances and people in our lives shift and change (as they always will), our inner peace, love, and joy remain steadfast and continue to deepen.* We enjoy and are fully grateful for these loved ones and objects, but at the same *time*, we understand that these do not define our happiness because, like everything in the physical world, their existence is only temporary: they all will come and go. The people and things of our lives never need to change, but our ideas about them must if we desire inner peace. *This may seem subtle, paradoxical, or even illogical, but a critical step along our path to true individual freedom is when we fully understand that our release of personal attachment to the very things that we enjoy most in life, is what allows us to experience even greater love and joy.*

ATTACHMENTS, DISEASE AND WAR

Our *attachments* lead directly to associated worry, anxiety, and depression. We worry that we might not earn enough income to make the big house, or car, payments. We are anxious about what the neighbors are saying about our Christmas decorations or our new landscaping. We have concerns about the popularity of our children and their future financial prospects. We are constantly concerned about our bodies, our children's bodies, and our friends' bodies: "Are they growing up fast enough?" "Are we growing old too fast?" "What if we become ill?"

These habitual anxieties create emotional and physical tensions that eventually manifest as both energetic and physical blockages in our bodies. *These energetic and physical expressions are fully equivalent because, as we already understand from Einstein, energy, and matter are just two forms of the same thing.* As these blockages begin to manifest in our bodies, they often first announce their presence as knots, kinks, muscle spasms, twinges, weakness, anxiety, or tiredness that can interfere with our motion, comfort, physical or creative expression, and emotional well-being. At this stage, these small discomforts are just the early and relatively gentle signals that let us know that something is out of balance. The issue could be psychological or physical in origin, but ultimately both are completely related;

all dietary, addiction, or over-consumption issues have energetic and psychological roots. If we are tuned to recognize their message and respond by earnestly seeking to address the root psychological or lifestyle issues, the body can and will guide our search, and as soon as an issue is recognized and addressed, it quickly corrects itself. If these early symptoms are ignored or only treated at a superficial level, the disturbance will progress to interfere with our natural ability to restore balance until it eventually appears as crippling physical and emotional pain, often leading to immobility, sickness, and disease. Once they become fully expressed at the physical level, these manifestations of buried tension are usually assigned official-sounding clinical names, such as carpal tunnel syndrome, sciatica, arthritis, auto-immune disease, diabetes, heart arrhythmia, anxiety disorders, depression, or ulcers.

Our deepest level of *being* does not want to be prescribed pills or other forms of temporary or symptomatic relief; it desires real growth and healing. The psychological discomfort, aches, pains, and finally disease are just the body's way of shouting for attention, asking that the issue be realized and addressed at its source. Pain and disease are nature's way of signaling that our body is out of balance and is asking for our focused attention. The deepest message is rarely about the body and how it is to be treated; it is, instead, usually about the underlying *energies*.

At the individual level, our unexamined attachments express themselves as physical disease, injury, or deep emotional "suffering," but on the broader societal scale, these same types of attachments lead to struggle and conflict between different cultures that eventually manifest as war, mass starvation, and the ongoing destruction of our environment.

Naturally balancing and self-corrective in its nature, life always creates experiences for us that will signal, reinforce, and deepen our awareness of any unhealthy imbalances or blockages. Whether or not we fully recognize or appreciate this quality in the short term, life's processes are always very helpful and directive. *The physical issue highlighted is never accidental or about punishment; it is always the result of a fully interconnected and interactive system attempting to restore balance. Through this natural balancing process, the unfolding of life will always illuminate our most profound areas of darkness as we are simultaneously presented with new opportunities to expand and grow our enlightenment.* Healing or *whole-ing* is a natural and organic process and if we are fully open to whatever presents itself to us in every moment, our healing will always unfold in the most unexpected, yet wonderful and miraculous ways.

ATTACHMENTS AND HEALTHY NON-ATTACHMENT

Learning to live without creating attachments helps clear a path to health and well-being. Because our *attachments* are so naturally formed, living a life that is free of them is an almost impossible challenge for the culturally trained Western mind-set; this entirely new way of living life is extremely difficult to understand, and even more difficult to incorporate. For the Western mindset, this shift of perspective requires a long process of re-exploring and re-learning, and because of our deep cultural conditioning, Westerners, especially Americans, are likely to confuse this process with their ingrained ideas about "detachment." Because of our preconceptions, we are prone to reflexively assume that this release of our *attachments,* or *non-attachment,* is similar to the often witnessed destructive numbing of personal "detachment." Despite sounding similar, these two words describe radically different, even opposite, ways of experiencing life.

We are correct when we understand this Western form of "detachment" as "dropping out" of life itself; typically this contractive state is entered through desensitizing oneself with drug use, engaging in some form of busy and distracting activity, or just wasting away. This type of detachment represents a "shutting down" that inevitably moves us even further away from the interconnection and vibrancy that we likely desire. Many of us have personally experienced this Western version of "detachment," or, if not, we have certainly been close witnesses to this type of destructive behavior in others we love. This type of "detachment" never promotes health or healing, even though it often relies upon culturally approved methods such as prescribed anxiety medications, alcohol use, excessive exercise, or work addiction. Here, it is life itself that one is detaching from, and because this always leads to isolation and desensitization, it is almost the polar opposite of what is called *spiritual non-attachment.*

In the healthy spiritual form, the goal is non-attachment, not from life, sensitivity, and experience but, instead, from the relentless grip of the ego and the drama that it continuously creates. We accomplish this by learning how not to identify with the things that appear in our illusionary or temporary expression of life, including our bodies and our egos. To understand and benefit from this positive form of non-attachment, we must first reach and embrace the deep "knowing" that we are not these bodies: we are not what we do, own, or create in this physical manifestation of life.

DESIRE ITSELF IS NEVER THE PROBLEM

While a deeper understanding of human desire is important for any spiritual path, the need to free oneself from all types of desire rests at the very core of *Buddhism*. This is clearly expressed in Buddha's primary teaching: *the Four Noble Truths*. According to *Buddhist* tradition, all suffering stems directly from our desires and, therefore, letting go of desire is critical for ending our suffering.

My own experience has been that desire, by itself, does not directly lead to suffering, so desire is not the actual problem. Instead, my experience has taught me that desire can be a very positive motivator, one that naturally adds an exciting dynamic to energize, add vitality, and provide focus. If understood and used correctly, it becomes a wonderful tool for the enjoyment of life. When channeled correctly, desire is a natural energetic source for facilitating human expression. At the most primitive biological level, desire is what drives the reproduction and continuation of our species. If expressed through the mind, desire helps produce interesting intellectual and scientific works that shape and energize our culture. When expressed through the heart, desire fuels a wide range of spiritual explorations and expressions. Working through our deep creative well, it can help produce great artistic expressions of beauty or guide the development of technological marvels that change the world. Desire is responsible for initiating, driving, and sustaining many important forms of human expression.

However, if through our desire we become attached to a specific outcome, object, or individual, then we will inevitably suffer from this attachment. At first glance, the desire appears to be creating the suffering when actually it is only our habitual attachments to persons, things, or results that lead to our suffering. To avoid desire simply because it can lead to attachment is equivalent to avoiding automobiles because they can lead to accidents. It is bad driving, not the car itself, that causes most automobile accidents. Similarly, it is not the desire, but our attachment to a specific outcome, that leads to our suffering.

Desire will stimulate, influence, and engage the *ego* which then tries to "control" outcomes, and as already discussed, the *ego* is also not the problem because our *ego* is necessary for protecting our physical form. However, when we become too *attached* to some idea, *image*, or outcome, a critical shift occurs. Instead of simply allowing natural *flow*, our *attachment* to some specific outcome signals the *ego* to take over and control the outcome, ensuring our suffering. *Regardless of how wonderful and lofty any*

personal goal might be, if we are attached to a specific outcome or result, we will inevitably suffer.

Throughout our lives, real learning only happens through direct experience, and nature always balances everything. This means that even our suffering will eventually guide us, through experiences, towards deeper insights into the true nature of desire, control, and *ego*. We will eventually discover that it is only our *attachments*, and not *desire* or *ego*, which form the chains that keep us bound to our unhappy or unhealthy situations in life. Once we learn to dance with the *flow*, by understanding the real nature of our *attachments*, our suffering ends.

We are here to gain experience in all its different forms, and the natural world uses desire as a tool for making sure we do just that. Desire is one of nature's most wonderful and powerful facilitators for three-dimensional living; it sharpens our focus and motivates a fuller expression of life. Awareness is always the key; when we are just reacting to unexamined shadows, we are creating more obstacles and closing down; but when we are consciously and openly exploring our desires, we are continuing to *flow* and grow.

AGENDA OF ANY KIND – AGENDA-FREE LIVING

Attachment can be a difficult concept for many to grasp, so possibly a better way of addressing the same issue is, instead, to just monitor when we witness any personal agenda. Living without *attachments* may also be understood as "agenda-free living." We can learn to ask ourselves the clear question, "What is my agenda or motive at this moment?" If we explore this authentically, we begin to see that we usually operate with some type of agenda, either obvious or hidden. Each *time* we face a decision, this question will help to make us more aware of our actual motives. Once we recognize the habits and signs of *egoic* agenda, when life disturbances rise, we can quickly see where we have allowed *ego's* influence. Keeping this watchful eye on agenda provides illuminating revelations about our always-evolving *self*, and when we practice "agenda-free" living, we serve our greater *being,* instead of just fluffing our *self-image*.

Our agenda is always built on our *attachments*. Once our agenda can be witnessed and illuminated, it no longer can disrupt our *core vibration* through its hidden darkness. With this shift, deep *soul*-level healing or *whole-ing* accelerates dramatically, and the entire way we see and move through life changes. While every step in life eventually leads to the expansion of our awareness and contributes to our *whole-ing*, the process will remain incomplete and chaotic as long as our lives are still "agenda-driven" by all of our individually conceived and competing *egoic* desires. Once we learn how to meet each moment from an "agenda-free" perspective, our entire experience shifts. ***"Agenda-free" living is just another way to describe and differently understand the experience of living in the "now."***

THE "NOW"

Eckert Tolle describes the significance of living in the "now" throughout his bestselling book, *The Power of Now.* This is not a new idea; it is an ancient spiritual concept that has always been an integrated part of several Eastern philosophies and religions. Today, our Western culture is beginning to recognize the life-changing value of this idea, even though its meaning can't be fully understood through words alone, and the nature of our modern lives makes its practice particularly challenging. This method has been taught and practiced around the world as a valuable technique for reducing personal suffering and living a more peaceful life, yet there is no place on this planet where there is more lack of understanding and resistance to this idea than from our *time*-ordered, Western culture, particularly the goal-oriented United States.

In the West, this orientation was clearly described by the mystic Meister Eckhart in the Thirteenth century but it still took many centuries for this idea to spread widely through the "New Thought" movement of the late nineteenth century. The 1960s and 1970s saw this ancient idea reformulated by several contemporary Western spiritual pioneers, after their personal experiences in India. One of the most influential from this wave of Westerners was Ram Das, a Harvard psychology professor, originally named Richard Alpert. With his revolutionary 1971 book, *Be Here Now,* he, along with others such as Alan Watts, introduced many young Western minds to the strange-sounding idea of "being in now." Today the phrases, "living in the now" and "living in the present moment" are almost fully integrated into our Western lexicon but, at that *time*, this was a brand new concept to most readers. However, despite numerous books, videos, and workshops on this topic, including Eckert Tolle's blockbuster best-selling book from 1997, *The Power of Now*, along with his equally popular second book, *A New Earth,* popularized by Oprah Winfrey, this critically important idea is still far from understood or integrated.

What does this unfamiliar way of experiencing life really involve? What are these words actually asking us to do differently? How can we utilize and integrate this principle into our extremely busy and competitive Western lives?

Because this idea is so foreign to our culture, we completely lack tools and models to help us with the process of integration.

The first major challenge is to understand exactly what the phrase "living in the now" is asking us to do differently? Our modern Western lives are perpetually busy, filled with what we perceive to be an endless series of necessary and required tasks. Most of us wake up every day with a full and often overwhelming "to-do" list that we believe must be completed. For relaxation, we often choose activities like reading, TV, exercise, the internet, or socializing with friends: activities that allow us to forget the extent of our endless "to-do" lists. In our busy, but supposedly "productive," Western lifestyle, we usually are operating in a "full-on" or 'full-off" mode, allowing ourselves little or no *time* for balanced, quiet, or deeper reflection.

Living in the "present moment" requires a monumental shift in our usual day-to-day focus. Throughout our lives, we have been programmed by our family, culture, and experiences, and much of this programming is important and useful; it controls our breathing, our heart's beating, and our body's functions, and all of this without conscious thinking being involved. This natural programming is constantly helping us to function efficiently in our culture and our physical environment; it allows for the quicker automatic responses that are necessary for day-to-day living and survival. However, because our brain's Default Mode Network automatically learns and recycles our most repeated patterns, these will quickly become routine, habitual, awareness-reducing patterns that we rarely reconsider or even notice. These automatic "shortcuts" are helpful tools for efficient day-to-day functioning, but because these unexamined reflexes also unintentionally close the doors of conscious awareness, we lose our connection to all the miraculous possibilities that are available through our universe's *architecture of freedom*.

Living in the "now" means that every experience and interaction is approached as a fresh, innocent, and brand new adventure. We are being asked to step into unknown territory, to participate in life without first engaging our automatic reflexes or logic-limited minds. If we are thinking about the outcome, attempting to control, or responding in automatic or habitual ways, we are no longer living in the "now." This way of *being* is similar to how we interacted with the world when we were newborns. Then, we entered every experience free of the programmed preconceptions that gradually formed throughout our lives. With this shift, we are again being guided to experience life anew, raw, and free of our learned concepts and thinking. We are being asked to drop all of our preconceived and acquired ideas along with our entire personal agenda. In fact, to fully embrace this way of *being*, we must ultimately learn how to drop our belief that we even exist as a separate individual having a body. "Living in the now" means being free of any concepts or agenda so that we can be open, available, and engaged in whatever life presents. Living in the now must even include dropping all concepts that anything real is actually "happening" or there is a separate "us" involved. Newborns are not yet aware of being the experiencer: without this concept, whatever is presented simply unfolds for them. For the busy contemporary adult human *being*, this release of our patterned and "efficient" ways of moving through our daily tasks and interactions does not come easily or quickly.

Can such a shift actually be achieved and then fully integrated? Are there real positive benefits? Can we fully participate in our lives and "live in the now" at the same *time*? Will we then become happier or healthier? The answer to all these questions is a resounding yes, yes, yes, and yes! Not only can we fully understand and integrate this revolutionary idea into our lives, but doing so will forever change our lives for the better. This shift benefits our entire health and well-being and there is no other single change I know of that will produce such a dramatic and permanent reset in our personal lives.

FLOWING WITH THE RIVER

Some of these discussions about *attachment* may sound strange and unfamiliar to some readers, but they all describe an understanding or perspective that radically changes lives, and leads to the sanest and most joyful life that is possible. Individuals who have mastered *non-attachment* do not just live full lives within this world, they actually live much fuller lives: they experience life more completely, engage with more of life, and *flow* through changes more easily. By dropping all personal agenda they have much more internal "room" for the incredible experiences that suddenly have become visible and available through their newly expanded *viewport*.

For this same reason, the non-attached individual also does not become giddy when things appear to be going particularly well; they have an ever-present inner understanding of balance in all things. Those who are *non-attached*, living their "agenda-free" lives in the "now," will still make plans, yet always remain flexible, knowing that within the powerful *flow* of this river, all plans must always be free to evolve and change. They no longer become disturbed or upset because things do not go as planned: while they still make plans, their happiness is no longer tied to them. Instead of always being directed by ego's desires, they have learned how to navigate the river of life by allowing and following its natural *flow*, as it winds along its undammed and free path, filled with unexpected, but joyful, adventures. They also fully understand what the American mythologist Joseph Campbell was expressing when he said, "If the path before you is clear, you're probably on someone else's."

This way of moving through life leads to an even-tempered *presence*, a constant awareness of being a "passer-by," even while fully participating in a joyful life. Those who discover how to live in this freedom experience an expansion of joy and a reduction of stress, as they also engage more of life, and enjoy more of its flavors. Once a person can release their *attachments*, they discover that their involvement in life is more like watching a good movie or playing an exciting video game; they can be fully engaged and enjoying the experience but not be really stressed or deeply worried about the outcome. They have come to fully realize that they are not the character or *avatar* that they play in this game. They have released all attachment to their *image* of *self*.

With a clearer understanding of the issues surrounding *attachment* and the eventual mastery of *non-attachment*, we completely free ourselves from the endless repetition of being stuck in our own, self-manufactured quicksand. **With this one change, we learn how to get out of our own way.** Now we are fully engaged through our direct and intimate connection to the ever-evolving *present moment*. Our lives become smoother, more *flowing*, interconnected, and fully tempered by an always-present, underlying sense of peace. *It seems completely ironic, but as we learn how to let go of our need to control and life assumes the less threatening qualities of a more impersonal adventure, it then becomes much more fun, enjoyable, interesting, and exciting.*

WHAT IS FLOW?

When we learn to live our lives without an agenda, we begin to notice something unusual and a little *magical*. More and more, interesting interactions unfold in surprising ways; all without our conscious involvement. We notice unexpected coincidences and well-timed intersections unfolding naturally and organically. We also may notice something else – something that has always been a part of our lives is suddenly missing from our interactions: our over-thinking. We discover that all these encounters are now happening without any conscious control or input from our thinking brains; we are just participating in all this without any thought of "making it happen" racing through our minds. It is as if these perfect decisions are being made elsewhere. We may also notice something else – the moment we have a flicker of a thought about "not thinking about it," the *flow* and *magic* instantly stops.

Flow is what unfolds when the separated mind is not engaged and the experience becomes direct; this is living in the "now." Every musician, athlete, and performance artist knows both the feeling of being "in the zone" and the very different feeling when the mind re-enters the process. Professionals are those who have learned how to extend this *flow* for as long as possible and have also practiced their skills so thoroughly that when their thoughts re-enter, causing them to drop out of their *flow*, they are proficient enough to still produce satisfactory results.

SUBROUTINES

Computer programmers rely upon *subroutines* for efficiency. Complicated tasks are broken down into a series of sequential smaller tasks and each of these smaller tasks is accomplished by a specific series of commands called a *subroutine*. Typing a few letters in a word-processing program can input entire phrases or completely change the look of the entire document and a single click in a drawing program can add a very complex *image*. These often repeated actions are accomplished by simply recycling smaller sets of instructions called *subroutines* making them an extremely helpful tool for maximizing the efficiency and speed of our machines. If the task is to be repeated later, the *subroutine* is simply reused making it unnecessary to relist all the individual instructions.

In our day-to-day lives, we use *subroutines* all the *time* because *subroutines* do the same for our brains, making our often repeated tasks faster and more efficient. For example the *subroutine* "wash the dishes" has many steps but after a little practice, only those three words are needed to communicate the entire task. In the morning we can easily make our coffee before we are even aware of being awake. We can drive our cars in busy urban traffic only because the standard tasks of steering, braking, shifting gears, and accelerating have been incorporated into standard routines that are repeated over and over again; this leaves our brain's computational power available for reacting to new, unexpected and dangerous events. Without any new or additional code (new "thinking") we can go through most of our day by just repeating previously learned steps.

We all welcome the *energy* savings that subroutines provide; the tasks of daily life would be exhausting without them. However, because many of the situations we encounter in our day-to-day living are so routine, when we encounter a situation that seems only slightly different, we will often miss or ignore these small differences. Reliance on this reflex makes it very likely that we will miss a new experience or special opportunity that may be presented through this slight change. For example, since an object moving rapidly towards us often represents a critical danger, as an automatic response to this type of stimulus, our bodies develop *subroutines* for ducking or quickly moving away. If we are playing baseball and we respond by ducking when a ball comes our way, we will either strike-out or make an error, so ballplayers must relearn and re-pattern their reflexive responses to specifically allow for a new type of response. They must learn and practice how to be open to many different, and possibly unnatural, responses.

SUBROUTINES AND OUR DEFAULT MODE NETWORK

As we mature, the complexity of our lives increases. Below the surface of our conscious awareness, our body and its nervous system automatically create numerous unseen shortcuts that make our repeated tasks more energetically efficient. Certain tasks, like learning how to play a musical instrument or driving a car, always begin with a significant conscious component, but through repetition, a deep level of programming inserts itself and takes over.

Our bodies are incredibly efficient and automatic at this kind of learning. We move through our complex environment by making all kinds of internal assumptions and approximations about the nature of that which we are experiencing, allowing us to respond much more quickly. As discussed, the name that neuroscientists use to describe the brain's mechanism for reusing these previously recognized patterns, which then become the automatic reflexive responses that are fully ingrained and integrated into our lives is *Default Mode Network* (DMN). For example, when we sense danger while driving, it is our DMN that quickly applies the brakes of our automobile without our having to first think about and analyze all of the different possibilities. This automatic patterning is a positive contribution to our lives because these programmed routines allow us to quickly respond to challenging, and sometimes dangerous, situations.

However, as we mature, the repetition of many daily activities also means that a large part of our lives becomes reflexive and automatic; long before we become adults, much of our daily activity has been pre-programmed into our

DMN. While this is fully normal, it is also clearly not the most expansive or expressive way to experience life.

Living Without Subroutines – The Now

Like the baseball batter facing a pitcher, we all can learn how to better recognize our conditioned responses and begin to train our minds and bodies to move beyond these ingrained subroutines; but is it possible to live without any *subroutines* at all? Probably not, and it is also not a wise idea, at least while we are living in these bodies because basic biological functions such as breathing and heartbeat also depend on subroutines. However, living without *subroutines* always controlling our thoughts and actions is certainly possible. Again, this is just another way of describing what "living in the now" means. Living in the "now" means living a life that is always looking out for and then examining our habitual *subroutines*. This vigilance also must include the most powerful *subroutine* of all: the egoic *subroutine* that is constantly at work re-generating our *self-image*.

To enjoy personal freedom, it is not necessary or desirable that we eliminate all of our automatic responses (ducking flying objects will always be useful). What we can achieve is a state of awareness where we become skilled at instantly recognizing our conditioned responses; we learn to quickly assess when a reaction is the result of old conditioning and adjust accordingly. We also develop the ability and tools to more effectively examine the roots of this automatic patterning, so if our old conditioning does not serve our current situation, we can then turn it off. The result of this new type of awareness is a more authentic life, filled with richer experiences and more personal growth. The ego is still present, but now it has been reassigned to focus on its real job, that of protecting our body.

Ego Is Still Important

The stated goal of several popular religions and spiritual philosophies is to eliminate or "destroy" the *ego*. Because the *ego* tends to create more separation through its clinging to familiar forms and behaviors, it is sometimes described as a major obstacle to our potential for living in freedom. The "elimination" or "destruction" of *ego* is taught within several spiritual liberation traditions as a necessary step in any individual's journey towards more awareness.

However, working towards the elimination of ego is not a good way to expand awareness because the ego is critically important to our

physical existence, and without its helpful functions, we also will shorten the wonderful opportunity that these bodies present for acquiring experience. Operating on many levels, the *ego* is an integrated and very functional part of this physical manifestation, and like every other contributing part of our physical body, it has an important role. At its most fundamental level, operating through the most primitive part of our brain, *ego* has served us well as the primary protector of our physical bodies. Our evolutionary history would be short-lived or non-existent without the previous protection and drive of our *egos*. This is *ego's* critical biological function; its job has always been to watch for, and alert us to, possible physical danger to ensure the continuation of our species. Analyzed at this primal level, the destruction of *ego* would expose our bodies to constant grave physical dangers. If we desire to continue our exploration of the physical realm, this particular *egoic* function is still a necessity.

The real problems arise at the other end of the *ego's* range of operation, where *ego* often rises to defend our *self-image*. *Ego*, while functioning as our great protector, becomes over-enthusiastic and tries just as hard to protect our made-up, and experience-limiting, *self-image*. At this level, the *ego* can be recognized to be overreaching and problematic because this expression directly leads to stubbornness, resistance, inflexibility, and a myopic vision. Here, our *ego's* activities interfere with the *flow* and expansion of *awareness,* to create major obstacles for our journey towards true freedom.

As with everything in dualistic life, our egos are a "mixed bag," useful sometimes and problematic at other times. When the problematic aspects of *ego—self-image, and attachments—*are successfully redirected through *whole-ing* and more *awareness*, we reap positive benefits because its important activities at the primitive, body-protective levels can continue to serve our physical form quite well. If we succeeded at completely "killing" or eliminating the *ego*, our entire physical experience would be threatened and our wonderful potential for maximizing our experience within this physical existence would also disappear. *Ego remains critical to our experience in this physical plane, so instead of eliminating our egos, we only need to tame them.*

LETTING GO OF "SELF"

As discussed earlier there is a more efficient way to release all of our resistances because once we can identify and release our self-created *self-image*, all our other hidden resistances evaporate at the same moment.

Buddhism, Sufism, Advaita Vedanta, and other related Eastern philosophies present useful tools for this level of deeper work.

When we explore the *self* through any of these methods, we eventually discover that the common link to all our traumas, fears, worries, neuroses, and habits is our *self*-created and *self*-defined *image* of who or what we imagine ourselves to be. When we explore this idea more deeply, we discover that this *ego*-created *self-image* is the single container holding all of our "issues," personal and cultural. Instead of addressing every issue in order as it arises, if we can recognize and release our *attachment* to our *image* of *self*, then all of our separate inner resistances instantly melt away. In this one single release – the letting go of our long-held concept of a separate *self* – our entire shadow world vanishes. Here we discover that, where there is no longer a *self* that exists in separation, there is also no one who has ever been wronged by another, or who has done something wrong to another. All guilt, blame, shame, and judgment completely disappear along with the *self*.

This *self* that we are now discovering and exploring is the container for all of our shadows, yet it is not who we are, for it is only something created by the *ego*. Because the ego is involved with both this illusionary *self* and the important function of our basic survival, this level of *whole-ing* needs to be approached carefully—it can't be safely rushed. To not proceed carefully would be similar to jumping out of an airplane without a parachute; some might survive, but clearly, this approach is not recommended. If we are fully engaged in the therapeutic healing process that I have named *whole-ing*, the right *time* to shift gears and move into this deeper mode is revealed naturally. Attempting a full release of *ego* is never healthy, helpful, or recommended; a more gradual "taming of the ego" is the more balanced and preferred approach.

TAMING THE EGO

> "Here is my secret, a very simple secret;
> it is only with the heart that one can see rightly,
> what is essential is invisible to the eye."

Antoine de Saint-Exupery, The Little Prince

"*Ego* death" is never a requirement for shifts in consciousness; while *ego* is usually far too bossy, some part of it must remain to serve as the protector of our physical body. With *ego* in its correct place, the heart can assume its proper role to become the center of our communication and our most direct connection to everything else in existence. ***Instead of "killing" or***

"destroying" the ego, a very different approach is preferred: taming the ego. What is "taming" the ego? It means teaching our ego how to serve consciousness instead of the other way around. A tamed ego responds first to our heart, instead of our mind.

This shift can't be forced; the only way to accomplish this (whips, chains, and self-blame don't work) is by quieting the mind to the point where the much more quiescent heart becomes easier to hear and therefore more clear. Unlike the dominant *ego*, the heart's inner voice always remains non-verbal and soft. When the mind and body have been quieted enough to hear the heart's softer signals, a surprising new type of wisdom is unleashed, one that can move mountains and change entire worlds.

Each of our *egoic* minds has built a resilient and powerful *image* of *self* that is held tightly by our minds. This *self-image* is so loud and dominating that it prevents our hearts from ever being heard. There are meditative techniques to help quiet the mind temporarily, but we can only permanently tame the *ego* through illumination: shining the light of consciousness on our *ego*-created and well-hidden shadows, a type of darkness that, because of just being born and living our lives itself, is found within all of us. We must first shine the light of our conscious awareness upon our hidden shadows before the *ego* can be effectively tamed and permanently quieted. Only after extensive illumination and clearing of our unconscious shadows can this soft voice be heard, listened to, and begin to serve all of *being*.

In a series of very thought-provoking studies, published in part by the Heart-Math Institute, it has been determined that the critical parasympathetic Vagus nerve of the human body has more ascending nerve pathways engaged in sending *information* from the heart to the brain than it has for sending *information* the other direction, from the brain to the heart. This nerve pathway imbalance is one powerful clue indicating a natural organ hierarchy in which the brain has been designed to be regulated by the heart. In addition to this pathway of *neurological* communication, the heart also communicates with the brain *biochemically* using *hormones, biophysically* through pulse *waves*, and *energetically* with *electromagnetic fields*. It is also recognized that there are likely other forms of intercommunication between these organs that we have yet to discover.

Another indicator of their relative importance is just where the brain and the heart are located in our bodies. The body protects the heart much more carefully than it protects the brain. Unlike the brain, which is located in a vulnerable extremity, the heart is physically more central where it is very well protected in a soft and resilient cage; while head injuries are common because of falls, direct physical heart injuries are rare from everyday living. One more sign that our heart is more central to our existence is its location means it can more easily communicate with all parts of the body. ***However, the biggest single indicator that the heart is intended to be the master of mind, and not the other way around, is the way our lives radically change for the positive when we learn how to carefully listen to our quiet but always fully inter-connected and therefore well-informed hearts.***

While the heart is designed to normally control and regulate the brain, the *ego* is designed to quickly override this natural ordering in the event of an emergency. For whatever reason, the *ego* really enjoys this temporary promotion and from this level of elevated self-importance, when left unchecked and unconscious, the *ego* will automatically try to assert control over everything—its natural self-centered nature makes it a constant bully. Since our competitive human culture is consistently stressful, our *egos* are always being stimulated to rise and assume this ultimate level of control and through this constant repetition, they become far too accustomed to this role.

Largely because of the constant competitive stress in our lives, our heart/brain relationship has long been turned upside-down to favor ego/brain dominance. Easily activated through constant stress, our *egos* have gained an enormous amount of control over the visceral nervous system, influencing almost everything about how our lives appear. ***Instead of being willing servants to our hearts, our minds have been hijacked by our egos as they take control of almost every aspect of our lives. This fundamental misunderstanding of our heart's central importance is directly responsible for most of the confusion and problems in our contemporary lives. Because of the way our modern society greatly undervalues heart and interconnection, many of us have forgotten how to listen to our hearts.***

QUIETING THE MIND – MEDITATION

The heart, central in our physical body, is built for communication and interconnection, while the brain, located in an extremity, only understands its world through separation, division, and contrast. These are also the same qualities that define the primary *dualistic polarities* that shape our physical lives. ***Our brain is a tool for three-dimensional living while our heart quietly guards the portal to the hidden realms beyond.***

When the heart is heard and can function as our center, harmony and balance are naturally restored. Maintaining this internal relationship in our loud, crowded, fast-paced, and action-oriented contemporary society is extremely difficult because of all the competing distractions and noise. Listening to our hearts is only possible once we have first learned how to quiet our loud and busy minds, and this always involves the shifting of deeply ingrained habits, so reaching this point requires patience, vigilance, and trust. However, with practice, the "quieted mind" eventually becomes a life-long way of being that benefits everyone and everything.

Learning how to become "heart-centered" involves two fundamental areas of work: first learning how to meditate to fully witness the noise and chatter present in our minds, and second, *whole-ing*, or the illumination of the darkness within to shine more light on our inner shadows harboring unseen resistances. ***Ultimately we discover that it was largely from these shadows that ego has been driving all the chatter. To better hear the message of our hearts, we must first illuminate our shadow world to quiet our busy minds. As our minds naturally quiet through whole-ing, meditation will function differently as it begins to reveal the heart's message instead of the mind's chatter. Both meditation and whole-ing are necessary and must work together if we wish to hear and understand our hearts' deepest messages.***

Once the heart reclaims its correct position as our center, disturbances will continue to appear, but they no longer swell to dominate as they once did; they can, instead, be understood in the context of unexpected, but illuminating and interesting revelations about our always-evolving self. Eventually, through this more conscious process, our ego can finally relax to become the joyful servant of our hearts and, therefore, Being itself.

UNDOING OUR HABITUAL "SELFING"

Paul Hedderman is a contemporary speaker and workshop provider with a unique and entertaining take on the ancient *Advaita Vedanta* philosophy. Part of his method involves discussion of the self-explanatory term, *Selfing,* a word that describes our constant habitual reforming of the psycho-physical *image* of ourselves that we all create, build, maintain, believe in, and hold tight. Throughout our lives, we are continuously adding new clay to this self-imagined "sculpture" of ourselves, and as these new layers of clay age and harden, this made-up idea of our "self" becomes more rigid, solid, and fixed.

Freedom is only possible when this reflex is recognized, checked, and the old *image* is disassembled. Initially, *selfing* can be a tricky process to recognize because it is completely automatic and it often operates in very subtle ways. Once we truly understand the deeper implications of this powerful yet fully automatic *selfing* process, we can begin to recognize all the subtle ways that we continuously reinforce our old *self-image*. In doing this work, we have to remember that positive *self-images* are just as problematic as negative ones. **Positive self-images may actually be more inhibiting than negative self-images because it is usually our "problems," rather than our successes, that drive us towards the necessary deep inner work.** While those with less than glowing *self-images* can more easily become motivated to explore and release theirs, those with very positive *self-images* are, instead, more likely to want to maintain, grow, and protect a *self-image* that they judge to be advantageous. **It matters not what the quality of the specific self-image is, because any image of self, even a very positive one, will have a quality of rigidity that automatically impedes our growth and natural flow.**

GRACE, TIME AND HEALING

"Truth waits for eyes unclouded by longing."
Tao Te Ching

Our egoic desires, wants, and needs are the primary drivers of most human activity, and the real reason our culture appears the way it does. Our problems and difficulties always repeat themselves until we each have enough personal experience to be done with these endless appearing cycles. Only when we are each fully ready can we consciously release all of our *attachments* to outcome, agenda, and form.

At some point in the cyclic repetition of our dramas, we start to realize that most of our wants and desires are just the *projections* of our ego's ideas about how to elevate and enhance the separate and isolated *individual self*. Through focus and practice, we eventually come to the critical point where we know that our *attachments* are only the ego's way of "keeping us in its game." These patterns appear over and over in our lives until we finally "get it" at the deepest level of our awareness; we are neither our *self-image* or body – their "realness" is entirely *illusionary*. This realization requires *time* for the right amount of direct experience, but once it becomes internalized, change can unfold very quickly.

While it is true that this realization about the "true nature of self" or "self-realization" often happens in an instant – some say "in a flash of grace" – that instant can only unfold after the perfect amount of repeated experience and preparation. Most ancient traditions believe that this preparation takes many physical lifetimes or incarnations. While we can, and do, wakeup "in a flash of insight," this only happens after a tremendous amount of repeated experience and then after this insight, it must still be integrated before one can embody this new way of seeing the world; this step takes even more *time*. *Time* is always our most constant and faithful ally.

BAKTI YOGA

Bakti Yoga is the yogic path where students practice moving beyond their separate *self* by focusing their love, attention, and devotion on another or something outside their *self*. The new object for their attention might be a historical figure or icon, but often it is their *guru* or teacher. This is the same practice at the heart of many religions, including Christianity and *Buddhism*. Instead of only focusing within, these religions guide practitioners to direct their attention, love, and worship towards another, such as Buddha, Christ, or Mohammed, or possibly another loved one or even an inanimate object of worship such as an altar, icon, or statue. By directing attention and love to someone, or something, outside of ourselves, our focus automatically shifts away from our relentless *egoic* desires. Also, because this type of practice indirectly relaxes the inner chatter about ego and all its wants, it can initiate a transformational level of peace, clarity, and relief.

Also, when a larger group gathers, all with the same intention and focus, a much higher level of interconnection is automatically created. Because of this built-in connection, these group experiences initiate moments when participants can almost "taste" "oneness" or interpersonal *transcendence*. However, fully committed explorers also realize that, at some point, all temporary external methods, devices, and tools must be set aside to travel further, and this must include even their guru or this revered object of worship. The real journey is always within, not without.

KILL THE BUDDHA

Buddhists often relate the story of the raft no longer being carried on a monk's back after all the rivers have been crossed; a raft that has finished its usefulness to become, instead, a burden for the narrow foot-path ahead. At a certain point, we realize that all of our external objects of devotion must be released because they only inhibit the deeper journey within; the *time* for relying upon external objects to help focus our mind has passed. We must now "kill the Buddha," our most revered external object of worship before our *soul* can continue on its journey to experience true freedom.

Through our journey, it becomes increasingly understood that all external objects of worship, even those that have been useful for diverting our attention from the self, are subtle forms of *attachments,* which are, ultimately, not necessary or even helpful. The expression "kill the Buddha" refers to this near-final step of liberation within the *Buddhist* path. This is also the point in the *whole-ing* process where all the differences of various spiritual paths begin to merge and become one. **All individual attachments, even a wonderfully spiritual one, must be released as an integral part of the process of becoming truly free. All seekers eventually encounter a time when they are being asked to "kill their Buddha," in whatever form it appears to them.**

OUR ATTACHMENT TO FREEDOM

Within *Advaita Vedanta,* new and revealing insights can be gained when we inquire, "Who is desiring this freedom?" All our desires, even those of spiritual intent are created by the separated egoic *self,* and the *ego* will continue trying to claim its place of special importance – this is the always present temptation of *spiritual materialism.* At this point in our process, we discover that we are still caught in a trap of sorts because we still have a personal "agenda" item. We suddenly realize that our long-guiding desire for obtaining freedom from mind and suffering was just another disguised form of *attachment.* Because it drove our thinking and actions, this particular desire has been a gift: a very useful tool that brought us to this special point in our journey. However, like all tools, its utility is limited and only useful up to a certain point. While our *desire* for liberation helped bring us to this critical point in our process, once here, this same *desire* then becomes a restriction and an unwelcome burden – just another heavy raft to carry. At this point, even this once very handy tool must be released and shed.

As taught within a few Eastern traditions, including *Advaita Vedanta,* the focused and specific "desire for freedom" is a great teacher because it provides motivation and focus for much of our spiritual path; and for many seekers, it has been the very specific "desire to be free of suffering" that has guided and directed their journeys. However, these same traditions also remind us that one of the last steps in the process of any individual's *enlightening* is, by logical necessity, letting go of all *attachments*, even our once very useful *attachment* to the idea of becoming free.

At some stage, all seekers recognize that their desire for freedom, which has long propelled their spiritual search, is itself just another form of attachment and therefore a very real impediment to the actual freedom sought. Every individual on this journey is eventually invited to explore this critical and extremely interesting threshold.

THE NATURE OF "DISEASE"

Our muscles, tendons, and joints function best when there are flexibility and symmetrical muscular support for both the joints and their movement. After we experience emotional or physical trauma, the body adjusts and compensates. It usually does this by swelling joints and tightening certain muscles to protect us by reducing the movements that might cause pain and further injury. If left untreated, day to day living adds even more strain to these compromised body parts so the original trauma becomes reinforced. Over *time*, these effects will compound to become deeply embodied, integrated, and very visible, but even more important are the less apparent, at least initially, parallel restrictions to the movement and *flow* of our *energetic* system.

Older kinks continue to build new layers of resistance as they adjust for local weakness and imbalance by strengthening certain muscle groups in order to temporarily compensate. This over-compensation creates an unnatural form of temporary balance that, over *time*, causes the body to become even less supple and more vulnerable to additional injuries. New emotional wounds are easily shadowed below and within these restricted areas that begin to interfere with our lives in every way. After further reinforcement by the inevitable additional trauma of all types, this complex of layered wounds eventually surfaces to fully manifest as some serious problem that soon becomes chronic. By this point, these conditions involve a powerful multilayered emotional and energetic component so the physical injuries that may have started this build-up are long forgotten. Lacking any clear understanding of their origination, these chronic lifestyle conditions are then identified as a "disease" by the medical establishment and assigned official-sounding names like "Arthritis," "Fibromyalgia," or another from the long and rapidly expanding list of chronic "diseases" classified by Western medicine. All of these names are simply descriptions of the physical symptoms that are expressed at the surface when our energetic resistances and blocks are not addressed at their emotional roots. Our unattended shadows, if unaddressed, eventually rise to become fully expressed externally through what we now call "disease." In the bigger picture all "dis-ease," is just nature's way of saying that we have ignored all of its gentle hints and we must now pay full attention.

HEALING AND "THESE" BODIES

Most people associate the healing process with their bodies and assume that healing means restoring them to a healthy or more functional state. True healing operates far beyond our temporary bodies engaging an extraordinarily deep process that works to balance fundamental energies. This common association of healing with our physical body represents a significant misunderstanding of this deeper process and its meaning and is the direct result of our habitual misidentification with our bodies. *It is entirely possible, even common, for complete and deep healing to occur even as the physical body transitions from its current form. Because we always are so much more than "these bodies," the deepest healings have little or even nothing to do with the physical state of the body.* "Nature", which always keeps itself lean and efficient, will discard a body when its ingrained habits and physical blockages have become so embodied that it no longer serves the expansion of consciousness. At these times, "nature" has determined that a "full reboot" is the most efficient option. While deep healing will often have a very positive effect on the body, this is never guaranteed; these bodies are fully subject to the invisible but powerful influences that always exist beyond the narrow framework of *time* and *space*.

An important person in my life who was a generation older was struggling with his body after a debilitating stroke. Every one of the few remaining things that he still enjoyed in life – walking in his garden, driving his car, reading, and dining in nice restaurants – were instantly taken away from him by this stroke, and suddenly he was reconsidering his once-powerful desire to survive in his form, which had served him well for 92 years. On the evening of my plane flight to visit him, I experienced a particularly real-seeming lucid dream, involving an extended "conversation" between the two of us that objectively looked at all his options. In the dream, he chose to move on, so I was not surprised but found it interesting that he passed early that very morning when just the day before he was adamantly informing friends that the stroke was not going to slow him down. At the end of my dream, I observed his obvious relaxation as he reviewed how fully he had lived and enjoyed this particular physical life while reaching a clear understanding that this infirmed body was not who he was. It was completely clear to me that his unexpected and very easy passing indicated a profound shift in his attitude. I often wonder if his sudden change could have been related to the particularly deep healing that I witnessed in my dream. I would have enjoyed more *time* with him, but his passing was as perfect as that healing was deep.

THE NATURE OF ACCIDENTS

As more of our life begins to unfold in the "now," we better see how our movement through our life resembles the flow of a powerful river. We begin to "feel" the current and discover that swimming against this deep "current," while possible and very common, is extremely difficult and exhausting. Suffering is always the result of resisting the natural *flow* in the "river of our life."

The pain that emerges from human suffering can appear in either of two fully intertwined egoic forms: the physical and the psychological. As expressed in different ways throughout this book, *I firmly believe that all suffering and illness have egoic and, therefore, psychological roots.* For this discussion, the psychological form is defined as all the mental, emotional, and spiritual forms of suffering before they become expressed externally in the physical. The distinction between physical suffering (disease, pain, hunger, and injury) and psychological suffering is, therefore, largely about "timing," which we already understand to be "illusionary."

When our physical body is injured, it signals with pain. If the injury is the result of an accident, then, to almost all of us, the cause of the problem appears to be simple and direct; it seems clear and obvious that the physical injury was the result of this "accident," which is often assumed to be random or chance. Common examples might be injuries sustained in auto accidents or falls. Our culture teaches us that most of these accidents and injuries are just the results of "bad luck," but when we reflexively make this assumption, we are completely overlooking the deeper and usually hidden messages of our "accidents." Our reflexive assumption that "accidental" also means "random" completely ignores or misses the deeper cause-and-effect relationships that always spring from the vast inter-connectedness of all of life's processes.

Hidden connections and interrelationships are constantly unfolding in the invisible realms of the *universe*, and these shape our lives. An "accident" may appear to be completely random, but it is never "accidental" because it is the result of the natural cause-and-effect process that is always operating at levels that cannot be accessed by our rational minds. External events may seem unexpected or unplanned from our rational perspective, but they are always the direct result of this invisible process. Events are never actually accidental – unexpected, yes, accidental, no. *There is no such thing as a random accident in the multiverse; events only appear random or accidental because the actual causative interactions are unfolding beyond our vision and understanding.*

Once we have learned how to open and live lives that are not driven by darkness and fear, we begin to find subtle ways of accessing this usually hidden *information*; through our inner sensations, we learn to recognize and start to understand the deeper meanings of these clues. Only later do we fully recognize how our "accidents" are clear communications of a deeper need; they are the dramatic "wake-up" calls that point us in new directions. **When recognized and seen from this perspective, accidents and injuries become guiding lights that illuminate the very places where our lives need attention.**

"Injuries" might linger to become chronic conditions. Some chronic problems may appear to have started with a single "accident"; others may be the result of long-term imbalances, but most have components of both. Recognizing the limitations of our *viewport* and the deeper interconnectedness of everything, we understand that healing a chronic issue always requires that we open to its deeper origins. Carefully chosen therapeutic approaches can help us to identify the specific imbalances and patterns of physical tension. As we continue to explore our depths, we eventually uncover the root of our imbalance; all our chronic conditions are founded in our repressed subconscious fears and emotions.

Once they can be witnessed from this equanimous place of experiential wisdom, our injuries and accidents become an extremely valuable source of *information* about our emotional holding patterns and the state of our *being*. Here we fully understand that our body's challenging conditions are only late-appearing manifestations of our psychological and emotional state. **Accidents and the resulting injuries appear in our lives only as a "last resort" in life's process of guiding us to self-discovery; they only arrive on the physical plane as a desperate scream from the depths of our being.** Our accidents and chronic injuries are direct communications from *being* itself, telling us "because we have been avoiding the more subtle signals, we now must be persuaded to pay more attention." Through accidents and chronic conditions, we are specifically being directed to explore and address issues at their deeper source.

If we simply medicate the pain away in our attempt to regain function, we are disregarding one of nature's fundamental and most honest teachers. Pain is an extremely powerful attention-focuser; it is nature's way of helping us grow, change, and get the most out of our lives in these bodies.

EMOTIONAL SUFFERING AS OUR GUIDE

Our emotional/psychological body is a powerful tool because we can use it for locating our resistances early in our process, long before physical symptoms emerge. Compared to physical pain and disease, when emotional resistance first appears it can be very subtle; we might not even recognize it, and, if we do, it is often easy to ignore. It typically arrives obscured, hidden by the many layers of old and dense psychological and intellectual defenses. Over *time*, as this restriction builds, its signal gradually grows stronger and more obvious and more of its expression becomes physical. Recognizing these signals early in the process is a skill developed and integrated as part of *whole-ing*. As we gain wisdom and tools, we learn how to identify and address our more subtle resistances long before they rise to appear as pain and disease.

All human suffering has this emotional aspect because all suffering has its roots in our rigid identification with our bodies and ego. When we finally understand, at the deepest levels of our *being*, that we are not these "bodies" and not these "egos, the emotional component of our suffering begins to release and evaporate, the physical pain may continue but it will also become tempered by this dramatic shift of perspective. Through this one encompassing "let-go," emotional and psychological pain quickly loosen their tight grip. When we no longer are creating new psychological darkness, there is less darkness for any new forms of suffering to grab and hold onto. Once free of this self-created internal turbulence and resistance, we become more like empty vessels that are wide-open for life-force to *flow* through. While our old physical issues may still show up because of tissue damage and old remembered habits, because we are no longer adding new layers of tension and imbalance, we find that we now have more *time* and *energy* for repair. Once we are empowered by learning how to recognize and release our *attachments*, we arrive at another critical crossroad; the place where we finally recognize that all emotional suffering is completely optional.

By its very nature, all of the *information* in the *multiverse* is *holographic* and this means that everything is intimately interconnected. It takes practice, training, and patience for the human mind to recognize and trust these typically invisible but very real connections. They may not be readily visible or seem accessible, but when we are living our lives close to the "now," we begin to move to the rhythm of these connections naturally and easily.

THE VALUE OF CONSTRAINTS

We often encounter our most powerful opportunities for growth when new constraints are imposed on our lives. Changes like loss of hearing, sight, or a limb, paralyzation, economic reversals, loss of loved ones, or any major shift in the external *parameters* of our lives, can unexpectedly provide the stimulus for great leaps of growth.

The famous and recently deceased Stephen Hawkins discovered that he had ALS in his early twenties while studying physics at Cambridge. He credits the ALS, which quickly confined him to a wheelchair, for helping him gain the necessary focus required for theoretical physics; before the disease, his *time* and *energy* were spent chasing drinks, parties, and women. His unique and evolving constraint also motivated the world's most skilled computer scientists and accessibility designers to explore and develop new types of special equipment to facilitate his communication with others. Today, many others with similar constraints are directly benefiting from the technologies that were harnessed because of Hawking's condition.

Ram Das (*Still Here*) and Jill Bolte Taylor (*Stroke of Insight*) both credit their strokes for expanded insights into the deeper nature of life. Ray Charles might never have explored music as a potential career without the focused musical education he obtained as a child because he was blinded in an accident. In my architectural practice, it is the constraints (codes, budget, zoning, structural, clients' needs) that shape the solutions for every project. Without constraints, there are often so many possibilities that it can become difficult to even identify the best solution. Constraints, by themselves, are neither good nor bad; how they impact our lives depends on our attitude and approach.

Our physical lives will always be defined by constraints; they shape the very *space* and *time* we live in. We are born into a specific place and socioeconomic situation. Where and when we are born shape constraints that set the tone for the beginning of our lives. The sensitivity of our nervous system, diet while developing, size and convolutions of our brain, height, sex, way we look, and body type all define parameters for our lives. Our confinement to a particular body can easily be viewed as a constraint. We all recognize that we only have a limited, but unknown, amount of *time* in this adventure of life, but even this constraint in *time* can be seen as a fantastic guide and motivator.

Our entire lives are the product of constraints that define the *parameters* of our lives and shape our world. Like everything we encounter, if we view constraints as blessings and guides, their entire nature changes and they then begin to work with us instead of against us.

AGING IN THE WEST

One of the most powerful and defining constraints for the experience of these physical forms is the short and always uncertain timeframe that we are granted. Within the physical realm we all age, and this means we all will experience certain *time*-driven biological changes. The rate and order of these changes will vary from person to person but there will always be inevitable and somewhat predictable processes in every person's physical birth-to-death march.

We recognize the changes of youth and puberty as normal and healthy even when they are uncomfortable, but beginning somewhere in our thirties or forties we start to interpret some of the natural changes as problematic: the bone that breaks from the same activity it had completed successfully a hundred times before, the joint ache that does not disappear within a few days, the new growth on the skin, less available *energy*, and the extra rest day that is required after a strenuous event. Moving into our fifties and sixties these changes continue to pile up: decreasing sexual appetite, more unexplained aches, gray hairs, drying and bruising skin, eyesight changes, loss of hearing, tasks taking longer and, what is most distressing to many, a clear change in the way our brains work. We start to notice little things like forgetting an old acquaintance's last name, not remembering a meeting, and not having the same ability to hold and rapidly complete complex mental manipulations. These very real changes in our thinking and *energy* are one reason why many, if not most, famous scientists and inventors complete their major technical contributions early in their careers, often in their twenties. There are additional factors involving culturalization, the traps of expertise, and the increase in personal responsibilities, but these physical changes to the aging brain and body are real, consistent, and, when witnessed through a narrow perspective, they can be extremely disturbing.

It is our natural reflex to resist these changes. In the West, we express our resistance to these natural changes through our constant attempts at regaining youth. We endure stressful and damaging plastic surgery, Botox (injected neurotoxins), hormones, hair dye, make-up, and enormous emotional stress, all to satisfy our misguided desire to appear or act younger. In our modern culture, we usually treat our naturally changing mental function as an illness or at least a problem to be fixed, even though, with a more open-minded outlook, this change can be viewed as a gift that only arrives late in a long and successful life. We are almost universal in the way we miss the deeper significance and value of aging.

I propose a completely different approach. ***What if, instead of viewing these changes as problems, we view them as "life directives" or "universal guidance" about how and where we might best and more fluidly focus our attentions.*** Instead of fighting or resisting these changes, we can begin to view them as the "perfectly timed gifts from nature" that they actually represent. With this change of perspective, not only does the problematic nature of these changes suddenly shift, but aging and its gifts can become extremely advantageous and liberating. ***A great relaxation occurs when we finally realize that aging, like all of life, is not something that happens to you, it is, instead, something that happens for you.*** If we adjust our lives to gratefully allow for these natural changes, this newly carved path helps guide us to the new insights that will open us to a life that we never could have imagined in our youth.

While our fast-moving Western culture is fundamentally ageist (and regards aging as a problem), many, if not most, traditional human cultures recognize and honor the great gifts of aging. In Fiji, where I lived for many years, as villagers age and lose their physical strength, they gradually evolve into the village decision-makers and wise keepers of the culture. In Fiji, a death initiates a three-day celebration honoring the life and transition with dancing, drinking, and singing right along with the weeping and mourning: a balanced response to aging and death that fully reflects their balanced approach to all of life. In the Hindu culture, it is understood that all the different ages of man have their distinct purposes. After youth, career, family, and business phases, sometime in their early 60's, Hindu men, who up to this point have been steady family providers, are encouraged to enter a more introspective and deeply spiritual period. This may even include leaving their home to spend the rest of their lives as wandering *Sadhus*, fully dependent on the charity of strangers for food and shelter. The new role helps the *soul* to complete itself through an entirely new setting and perspective, a phase for healing or *whole-ing* that we entirely miss in the West.

WHAT HAPPENS WHEN WE DIE

There is a single individual realization that changes everyone's life for the better. This one realization, by itself, can end depression, reduce tension, eliminate fear, promote joy, and mark the end of many diseases; and once it becomes fully integrated, entirely shifts our perception of life. Again, this is not a new concept because it is the same awareness that shapes many ancient spiritual practices. I will not be able to prove its veracity either logically or rhetorically, because it is a deep-level *truth* that can only be grasped through direct personal inner experience. Its meaning resides outside of language and our rational mindset, so it can only be realized and integrated through direct experience. For the Western-trained mind, a shift in consciousness is required before this *truth* can be experienced.

This realization springs from two seemingly separate existential questions. What happens to "me" when my body dies? And what is my deepest relationship to everything else in the world? Only after the deepest of experiential explorations will the answer be revealed; it may gradually unfold, or it can arrive in a single-appearing burst of insight. *The clear realization that arrives and makes such a difference is discovering and knowing that the deepest and most fundamental aspect of our being is awareness itself, something which is not dependent upon these bodies; instead, our awareness transcends all the temporary forms that exist within the realm of time. The realization that changes everything is our clear knowing that we exist and will continue to exist beyond all form, including these temporary bodies.*

It is important to understand that today there is no clear rational argument or proof that will ever convince anyone about the truth and meaning of this deeper knowledge that recognizes that we never are, or were, just these separate, short-lived bodies. Philosophers have been arguing about the "seat of the *soul*" for thousands of years without ever producing a substantial rational proof. Definitive proof may not exist because our true existence lies far beyond the limits of rational thought, logic, language, and proofs. When we first begin to explore this existential truth, a host of new thoughts, questions, and rational arguments immediately appear. Our minds are always attempting to control and limit exploration to the more familiar and comfortable territory of divided, separated, *time*-ordered conceptual forms that *ego* understands: the terrain where it still has influence and control.

This life-changing shift includes two critical realizations about our realm of duality. The first is that "we are not these bodies," and the second is that "all separate-appearing souls and things are completely and fully interconnected." Even though these truths can't be discerned rationally, they can be verified and completely "understood" through open, direct, and broad experience. This type of "experiential proof" requires *time* and life's hidden processes; the only way to reach this deep, integrated level of awareness is through repeated direct experiences involving the full range of possible physical manifestations. *We learn about our eternal nature and oneness only through the repeated experiences of life itself.*

REINCARNATION AND ENLIGHTENMENT

Reincarnation is a fundamental belief and defining part of many Eastern philosophies and religions. They also teach that it is only after "direct" experience with all aspects of our *dual* world, that we can learn how to open, embrace, and integrate life's deepest *truths*. Being born again and again is an effective mechanism for creating a wider variety of life experiences.

Our opening to the deeper truth of our existence is a never-ending process that builds on experience, so reincarnation may actually be a mechanism that nature likes to employ; at the very least a belief in reincarnation can encourage a relaxation around *time*, individual identity and current life circumstances, and this relaxation alone can trigger powerful new levels of awareness. Even in the West, it has become common to hear someone speaking of having been "this or that" person in a past life, more often than not a famous historical figure like Cleopatra. There is a lot of truth and wisdom contained within the idea of reincarnation because regeneration occurs throughout the natural world; nature recycles everything. However, the divided individual *self* that reincarnates again and again to just continue its compartmentalized and separated personal set of experiences is not a necessary or even very efficient process when viewed from the perspective of "oneness." Here, where it is only awareness that is having all these experiences, once we open to this "oneness," we recognize that we all embody the direct experiences of Cleopatra but, in our wholeness, these must also appear with the difficult experiences of the downtrodden slaves she kept. From a place of "oneness," we all embody the collective experiences of mankind.

THE IMMORTAL SOUL—AKASHIC RECORDS

This Christian concept of *soul* also involves a more subtle form of *attachment* that may be particularly tricky or difficult to recognize: our attachment to the *ego*-derived idea of an eternal, but still individual, *soul*. Today, most Westerners are familiar with the Eastern concept of reincarnation, yet we might miss that our Christian culture also promotes a similar belief; that of the immortal but personal *soul* that later must atone for all its accrued misdeeds. Believing in a personal *soul* identity that stays intact through some longer journey is a variant of linear reincarnation. Our having a divided and individual *soul* is a concept that fosters even more separation by encouraging *attachment* to this particular, yet somewhat ethereal form: the "eternal, but personal, *soul*."

When reincarnation is imagined in the West, it usually includes the *egoic* idea that we somehow manage to maintain the same individual *soul*, intact and continuous from lifetime to lifetime. It is only our relentless *ego* that could have dreams of continuing our separated personal history this way. Nature does not recycle like this; it likes mixing, integrating, and composting everything together for it always wants to shake things up and start anew. It is clear to me that we all have some Cleopatra within, just as we all have some of the poor peasant or slave. Ultimately, because all of our collective experience is stored *holographically* within, and throughout all of existence, we are a part of everything, as everything is also a part of us.

This idea of a separated and individuated *immortal soul* is just another product of *ego* that only serves to create more separation and attachment to a physical form. The *ego,* completely dependent on the body, fears nothing more than the inevitable transition of our current physical forms. Ultimately this Christian concept of an *immortal soul* is just another not-so-subtle trick of *ego;* an attempt to extend its *self-image* beyond even this lifetime's experience to "death-proof" itself.

The existence of a separated self of any kind, including an individual reincarnated soul, is entirely illusionary. Soul exists, but not in that divided and separated form; because everything is so intimately interconnected, whatever is being experienced always involves the much greater *universal Soul*. Once we learn how to connect with this universal record, we can start to experience being "all that is" and this taps into all the collective experiences from the lives of all "others," including our Cleopatra moments. Some have used the term *Akashic records* to describe this complete *holographic record*.

GRATITUDE FOLLOWS FORGIVENESS

> "If the only prayer you ever say in your life is 'thank you,'
> it will be enough" *Meister Eckhart*

There are no greater allies in our journey towards freedom than *gratitude* and true *forgiveness*. *Gratitude* becomes an easy and natural state of *being,* once all "others," including oneself, have been *forgiven* for all actions and thoughts: past, present, and future. Once our vision has expanded enough to realize that, because there are no "others," the once-powerful realness of "others" doing things "to us," can instead be seen as nothing more than an illusionary story that is being held and reinforced by our minds. *With this new level of awareness, many unseen things quickly become clear; we now understand that because there are no "others," our holding blame and resentment towards "others" is just a disguised form of self-blame. We also begin to see how this resentment and blame that is carried forward in our memories and bodies alters our personal experiences and interferes with our openness and our next present-moment experience.*

Now, with this new level of understanding, we can inquire within and ask a different type of question; "once we let go of all our memories, including any resentments that we are holding, what remains?" Suddenly all sense of our separation evaporates. *Within each of our journeys into expanded awareness, there will be special times when we reach through the veil, a point like this where we finally connect with our being at a deeper intuitive level as we suddenly see that no one else is ever involved and no one else could have ever done anything to us because no separated ones even exist.* We recognize that all our holding of blame, shame, and resentment towards others is a complete misunderstanding of our "oneness" and therefore our own *being*. We have only been blaming and shaming aspects of our own self. Self-blame and self-resentment have been shaping our personal experience of life and interfering with our joy. Life, and the separated world, has just been appearing to us as an externalized graphic of our inner divisions. Our world of separated things is just a brilliant illusion generated for our benefit, so that we can better see and come to "know" our true deeper nature.

Once our anger and blame begin to be released, there is a new lightness of *being* that replaces the old weight of resentment. A newly emptied *space* appears, waiting to be filled and occupied by refreshed content that reflects our more open state of *being*. *Any newly opened space could automatically be filled with more of the old type of blame and anger, but this time around we possess enough awareness to fill it, instead, with something very different and much more empowering: deep gratitude and love.* In the *infinite* stream of moments, over and over, we are presented with many opportunities to choose gratitude and love over the limiting reflexive choices of times past.

We are only able to make this our clear choice once we have become fully prepared through the profound experiences of life. *It is only through gaining deep personal experience that we can uncover the true healing power of gratitude and forgiveness.* First, we must reach the point where we fully understand that any blame or anger directed towards others is always rooted in some form of self-blame or judgment. If we have not successfully understood and then addressed blame and anger at this root, then we will remain unable to freely choose gratitude and love. Our old familiar and reflexive *egoic* responses will continue to be fully automatic and act as our *default mode*. We may practice gratitude through daily affirmations, but until we have cleared this shadowy ground that sits deep below our conscious awareness, our gratitude will feel forced, unnatural, and any healing will remain a fleeting thing. *The deep level gifts of forgiveness and gratitude are only available through complete whole-ing: the internal clearing, integration, and healing process that is built upon the recognition of our eternally interconnected "oneness."*

EMOTIONS COMPARED TO FEELINGS

For this discussion, I will use the standard definitions of *emotions* and *feelings* as they are defined in psychology and psychotherapy. *Emotions* are defined as the primary raw signals; they appear before any cognitive imaging or processing. They initiate the quick reactions to threat, reward, and everything in between that help our species survive. Representing lower level responses occurring in the subcortical regions of the brain, the *amygdala,* and the *ventromedial prefrontal cortices*, *emotions* create biochemical reactions in our body that alter our physical and mental states. These are physical, electrical, and chemical changes that can be monitored and measured using medical equipment. *Emotional reactions* are coded in our genes and, while they do vary a little from individual to individual depending on circumstances, they are generally universal and similar across all humans, and they even initiate parallel responses in other species. For example, humans smile and dogs wag their tails when pleased. The *amygdala* plays a critical role in emotional arousal for it regulates the release of *neurotransmitters* essential for memory consolidation. This is one reason that our memories become bound with strong emotions and can be so long-lasting. *Emotions precede feelings and are instinctual as well as physical. Because they are physical, they can be objectively measured by blood flow, brain activity, facial micro-expressions, and body language.*

Feelings only come later. They originate in the neocortical regions of the brain and are secondary mental associations and reactions to the more primary *emotions*. Feelings are *subjective* because they are completely influenced by personal experience, beliefs, and memories. *A feeling is a mental portrayal*

of what is going on in the body when we have an emotion. Feelings are the secondary byproduct of our brain perceiving and assigning meaning to our emotions. *Feelings* are formed later after having an emotion and they always involve cognitive input. They usually originate in the subconscious and, unlike *emotions*, cannot be measured precisely. **Many responses that we think of as emotions emerge only after some mental processing, meaning they are actually processed feelings.**

TWO TYPES OF FEELINGS

The common psychological definitions describe *emotions* as primal and originating from deep within our bodies (primitive brain stem), and *feelings* as that level of interpretation we experience after our *emotions* are processed by our rational egoic minds. This definition of *feelings* needs more clarification because there is a polar range to any interpretation of experience and, therefore, there exist two extremes to the types of *feelings*, along with everything in between. There are common everyday *feelings* where the *ego-mind* is always involved but unconsciously, and authentic *true-feelings*, where these same emotions have been processed by an aware and conscious mind and the raw emotion is not as subject to manipulation or alteration by ego.

The *emotion* is the raw material or *information*, but the resultant *feelings* depend on how these *emotions* are processed. As a result, our level of awareness determines the type of *feeling* that we tend to experience: *ego-based* or *authentic*. *Emotions* moving through a clear, open, and free pathway produce the *true-feelings* that eventually will become our most reliable guides throughout our new and transformed lives. **This clear and open type of feeling, true-feeling, produces clear psychic intuition and is a gift that facilitates a new and profound ability to flow freely with the entire multiverse.**

THE SHUTTING DOWN OF OUR EMOTIONS

> "To avoid pain, they avoid pleasure.
> To avoid death, they avoid life." *Osho*

When we are afraid that we have done something "wrong," what often happens inside, after, is a bigger problem for our living full and expansive lives, than any possible "wrongdoing." Reflexively, we attempt to mitigate the intensity of any pain arising from this self-effacing choice by shutting down: subconsciously restricting our *energy flow* or *life-force*. We usually do not realize that this pain is an important but temporary messenger and that its lingering is the product of inner tension and resistance to the natural *flow* of *life-force*. In our attempts to avoid this pain, we subconsciously throttle down our entire energetic system. Unfortunately, closing our energetic system to one emotion also means restricting our ability to sense or feel everything, and this completely reduces our ability to connect with life itself. **When life-force is restricted or shut down this way, the body/mind will, initially, experience less pain, but there is an enormous price to pay for this relief.** The body/mind becomes less sensitive and less receptive in every way, which means it will not feel as much of anything; we become numb to emotion, and therefore all types of authentic feelings. **A system restricted this way becomes less vibrant and therefore less capable of also feeling the "good" stuff, including joy.** In the bigger picture, this is a self-defeating behavior that does not make long-term or logical sense, and yet this "shutting down" remains our most-common response as we continue this unconscious and reflexive reaction to fear and pain.

The only way to expand the flow of joy in our lives is to feel more of everything, instead of less, and the most direct way to feel more is through opening ourselves through the discovery and release of the hidden chronic tension that is interfering with our natural energy flow. The ultimate source of all tension is the *fear* that is generated and propagated through our cultural idea of separation – the primal and very "real-appearing" *fears* about dying alone and isolated after a short existence that is entirely shaped by struggle. Instead, our healing hinges upon understanding that we are always fully connected with everything in existence, even with the death of these bodies. It requires our deep "knowing" that we are not just these bodies and that our existence is part of a profound and much greater *timeless* process, one which does not end with the transition of this particular physical form.

Healing requires that we feel more and not less, so the path towards wholeness, and ultimately freedom, must involve dropping all the common judgments that divide and separate our more whole being. By learning how to embrace and welcome everything that appears, within and without, life can be seen differently. Every part of life, even what was once considered "bad," then becomes recognized to be valuable, important, and equal. The ability to open to the very things that we fear or reject the most requires deep insight that can only be built from broad personal experience. **Our development of this nonjudgmental approach to all aspects of life is often the most difficult step in the entire process of reclaiming our gift of freedom.**

NEUROPLASTICITY AND HEALING

For the facilitation of healing, we need to understand that it is our mind that is responsible for our *feelings*. Our *feelings* then produce additional chemical and electrical changes, which further alter the brain's function and structure. *Neuroplasticity* is the scientifically proven ability of our brain to change form and function, based on repeated *emotion*, thought, and behavior. While *emotions* always just appear, *feelings* shift and change with conscious thinking, and cycles of this repeated action will change the structure and function of our brains. ***How we choose to respond to our emotions will, over time, physically change our brains and, therefore, our lives.*** Brain *neuroplasticity* is one of many biological mechanisms that is becoming better understood through the rapidly expanding science of *Epigenetics*.

Personal freedom requires that we learn to understand, interpret, and authentically respond to our *emotions* instead of repressing them. For better health, our emotions want to be directly expressed without the many psychological layers of modification, interpretation, and judgment. This is *authentic living*, but we can only live this way if our shadow world has first been illuminated.

KNOW THYSELF

Many of the world's troubles, most of which we create ourselves and then exacerbate through our culture of separation, would be eliminated or greatly reduced through one simple understanding. This is a *truth* that we observe daily, yet we live in a culture that teaches us to ignore it because its integration would require a total change of our belief system. ***This understanding is that each of us is a small part of a single, connected, and fully inter-dependent organism called life.*** This *being* is widely recognized, usually externalized, and has been called by many different names: "all that is," awareness, consciousness, or any of all the many traditional religious names like God, Allah, or Jehovah. This world-changing shift would unfold automatically once it was widely recognized that every single person and living thing, without exception, is a vital and meaningful part of this *being's* full magnificence.

When an individual consciously harms "another," this action is always motivated by their fears or desire for personal gain, advantage, or power. This reaction is the inevitable result of a belief system built upon deep fears of separation and death. What we consider to be "evil" is often only the misguided result of this deeply conditioned and continuously reinforced fear of isolation and aloneness – a desperate terror that has been hardened around our cultural beliefs about our existence as competing and separate selves. Because of its deep *limbic* origin, this primal fear is easily stoked by others seeking political or market advantage. This type of massive fearful *energy field* can be amplified by "groupthink" to stimulate the anti-social impulses and actions that eventually result in wars, human atrocities, and the gross mistreatment of other forms of life, including our entire planet's ecosystem. For example, what we now call "terrorism" is a reactionary result, triggered by earlier events, usually involving the oppression and subjugation of one group by another seeking more power, influence, and riches. The chaos of today's world is not "evil"; it is, instead, a dysfunctional wounded reaction to centuries of interference, repression, and warfare. *The solution is clear and it centers on the recognition of our interdependence and interconnection: "Treat others as you would treat yourself for they are you, as you are they."* This is a foundational principle of several major religions, yet because of our cultural conditioning, we continue to create wars in the name of these same religions.

The path beyond this eternal chaos has only one real and lasting entry point: a deeper understanding of the "self." "Know thyself" is the primary mission of our lives, and despite all our complaints, the often difficult but fully revelatory life that we lead on this Earth is the perfect vehicle for fulfilling this purpose. *Because everything that we witness and observe as external is just the visible revelation of our personal and cultural shadows, by exploring our external lives fully and honestly, we are also exploring our inner being at the deepest of levels.*

By remaining open to any and all *truths* revealed throughout this personal journey, we expand awareness to fully love ourselves, and therefore all "others" equally. This paradigm-changing practice of treating others as we would wish to treat ourselves will initially function as a very difficult, but always revealing, mirror into our feelings about self. Later, from the perspective of greater awareness, this same practice becomes a natural yet sacred loving ritual – one built on the recognition of our eternal oneness.

Loving all aspects of the complete "self" is the polar opposite of the narcissistic and separated form of "self-love." It is a love built from a much broader and more encompassing understanding of "self." This is the realm of a deeper type of love, where all the separate "selves" merge into an experience of the single "Self."

EMBODIMENT, PTSD AND SURFACING

Every single thought, action, and experience from our lives is remembered and stored *holographically* throughout the universe, including within our bodies. All of our experiences are stored and imprinted throughout our bodies, but the most disturbing ones are parked secretly in our most shadowed and least accessible places, where they can more easily remain unexamined and add to the layered darkness. This hidden embodiment is why the deeply traumatized have unconscious and irrational responses, and why this condition is so difficult to reverse. The condition only gets worse when, over *time*, the disturbing memories begin to make physical changes and become integrated parts of our bodies. Those with PTSD act out unexpectedly because what gets stimulated has been buried deep within and hidden by darkness, where it then manifests in a way that is fully separated from the rest of our conscious *being*. When the wound is inadvertently stimulated, it acts alone, powerfully, and out of context because there are no conscious and integrative connections or processes available; it has become completely isolated from the aware parts of our *being*, interacting only with other hidden and separated trauma that also lurk within these shadowed depths.

To heal trauma, it must be recognized consciously, and this often involves re-stimulating the very memories that we devoted so much of our *energy* to keep hidden. This step of the healing or *whole-ing* process will always appear to be extremely frightening because of the very nature of the process: our opening to the darkness that is hidden inside. Why would anyone in their "right" mind want to re-stimulate their most painful memories?

Successful treatment requires this conscious integration, which can occur only in the clear light of awareness. We don't need to recall all the details of trauma for effective healing, but we must bring the unconscious into our awareness. At a minimum, this requires recognizing the "presence" of the trauma, its influence on our wholeness, and witnessing our automatic reflex. We must also be fully willing to experience and express whatever arises; our "goal" is to make all aspects that surface fully conscious and visible. Of course, this will be frightening, since the full face of our shadow is unfamiliar and its previous stimulation always caused difficult problems in our lives.

While direct stimulation will trigger our *Default Mode* reaction, in the right therapeutic environment, this stimulation can facilitate a more conscious recognition, a critical step in every healing process. However, because constant re-stimulation deepens the wound, some method of breaking the automatic cyclic pattern is necessary. Deep and energetic body-work can be very effective to break this cyclic pattern when the right therapist is matched to a client fully prepared for this deeper exploration within. Intentional stimulation through bodywork can be used to identify, track, and reveal the physical areas of the body that are most directly associated with the wound. It increases awareness of what is happening within and this allows for clearer observation and, eventually, more understanding and integration.

As our shadows become more consciously illuminated, the automatic *Default Mode* response weakens, and our ability to apply rational oversight gradually returns. Progressing safely requires *time* so this must be a gradual process, and each person's ideal path will be unique. As I became more committed to my *whole-ing* journey, I was particularly drawn to this type of deep energetic bodywork. Through this process, I was able to unearth, deep within my *being*, reactions and sensations that had been long hiding completely invisible to my conscious awareness. I was completely overcome with powerful sensations, uncontrollable shakes, or full body convulsions multiple times during this many-year process involving multiple therapists. Gradually, over *time*, these hidden terrors and little fears seeped into my awareness until they surfaced and were finally "seen." Once visible, they then could be acknowledged and consciously released. Through this type of repeated stimulation and witnessing, our darkness "lightens" and then rises to the surface to become a more integrated part of our physical expression; this *surfacing* of our hidden emotions is a difficult but very important step in the process of our gradual *enlightenment*.

However, for many, the fear of their shadows is so great that they won't be able to proceed with this self-revealing type of bodywork. For these, and most patients with overt PTSD, a more tender and forgiving approach is required – one that more easily helps the patient bypass their DMN responses. *Psychedelic assisted therapy* under the guidance of a trained professional is rapidly emerging as the most effective method for treating most cases of PTSD. With a therapist trained in both methods, this and deep bodywork, the two can be combined for even greater success.

PSYCHEDELICS AND BRAIN CHEMISTRY

Our brain is a very complex processor that supports many different pathways, but it is also naturally programmed to only use a limited number of these for efficiency and survival. Our brains process chemical, magnetic, electrical, and other types of *information*; some of its processes and pathways are clearly understood, but many are not. If we change the chemistry, conductivity, connectivity, or *electrical field* of our brain then different types of processes and reactions can more easily unfold. Scientists have been actively probing and exploring how the controlled use of electrical, chemical, and magnetic stimulations causes the brain to see, feel, and respond, and through this type of bio-physiological research, they are now on the cusp of offering several new technologies. We are nearing a level of development at which, with new artificial implants, the deaf or the blind will be able to have their first experience hearing or seeing, and those with amputated limbs will soon be able to fully control their new artificial limbs with only their thoughts.

Occasionally we have experiences that fall outside of what is considered "normal" or routine brain functioning, and these offer new insights and awarenesses, opening us to different ways of *being*, or seeing life. These openings or shifts of consciousness can be triggered by healing, fasting, sensory deprivation, severe stress, exhaustion, dream states, meditation, illness, or the ingestion of *psychedelics*.

Psychoactive drugs that alter brain chemistry have been used throughout human history, but it wasn't until the twentieth century that we began to understand how they change the brain and nervous system functioning. One particular class of compounds, *psychedelics*, are particularly compatible with the process of exploring our individual and collective makeup. Several of these function well in this capacity because, while they may differ in their exact chemistry and just how they activate different parts of the brain, they all seem to have the ability to inhibit the brain's native tendency to automatically repeat everyday patterns (*Default Mode Network*). This interference with the DMN, along with some stimulation of new pathways and areas of the brain, allows users to revisit, rethink, and more easily analyze issues from new, different, and fresh perspectives. They are more easily able to revisit traumatic events in their lives without triggering their automatic unconscious response, or if they do trigger them, examine their reactions from a more distant, yet conscious perspective.

Psychedelics also help to break old patterns by facilitating the communication of important nonverbal *information* and meaning that normally only resides beyond the veil of our *conceptual horizon*. This *information* is always communicated, but our brains have been automatically filtering it out (again the DMN), so it rarely reaches our conscious awareness. This property makes them important "medicines" for psychotherapy and today, because of rapidly changing cultural and social circumstances, they are finally being seriously re-evaluated.

PSYCHEDELICS IN PTSD THERAPY

Throughout the 1950s, long before the hippies and the trippy days of Timothy Leary, The Grateful Dead, and The Merry Prankster's Acid Tests, many psychologists and psychiatrists around the world were aggressively exploring what was then a brand-new class of drugs, one which seemed to offer unprecedented success for treating trauma and addiction. *Psychedelics* were new, exciting, and quickly becoming a game-changer for the mental health field. Professional therapists were discovering that through *psychedelic-assisted therapy,* they could quickly unravel several very destructive mental disorders or conditions that had previously resisted treatment. Largely because of their availability in pharmaceutical quality, the primary *psychedelics* explored were *LSD,* derived from the Ergot fungus, and *psilocybin,* extracted from mushrooms. *Mescaline,* from the Peyote cactus, was less available but also used, and some studies were conducted using *DMT,* which can be found throughout the natural world in many plants and animals. Later, an entirely new class of drug, *MDMA,* proved to be as effective, if not better, for this type of assisted therapy. In the 1980s, pure *MDMA* began to be mixed with other drugs to create a popular and extremely abused recreational cocktail called "ecstasy." While early researchers using *psychedelic-assisted therapy* were consistently seeing success and cure rates many times greater than any other available methods, it was recognized from the beginning that it was not the drug alone that was responsible for the results. Instead, it was understood that, when guided by a knowledgeable professional, the traditional therapeutic environment could be greatly enhanced by these new medications.

There were, however, several major obstacles preventing the integration of *psychedelics* into standard therapeutic practices. Like many pharmaceuticals, these were extremely powerful medicines and subject to misuse. Because they demonstrated the ability to "cure" a condition in one or two small doses, the major pharmaceutical companies had little or no economic motivation to test, finance, or explore this promising new class of drugs. Most importantly, by the mid-1960s, the political and social situation of the times set a unique scene where these were destined to become powerful and effective political tools. Despite the desperate outcry from the many therapists and researchers who had been successfully exploring these medicines, *psychedelics* were politically "weaponized" and classified by the FDA under President Nixon as "Class One" pharmaceuticals, meaning that they were listed as having "no potential value as a medicine" and possession automatically led to long-term incarceration. Research on these promising pharmaceuticals quickly ended, except for the few, limited studies that the military supported, which explored their potential as a weapon for war or espionage.

For the next forty years, only a small handful of committed, private advocates continued the research, laboring secretly, with little or no funding as they also lobbied to convince the government to reopen the research for this important and promising class of medications. One particularly committed individual, Rick Doblin, started and maintained an organization named the Multidisciplinary Association of Psychedelic Studies (MAPS), which has tirelessly pushed this important cause for over three decades. Several wonderful books chronicle this history and the struggle. *Acid Test*, first published in 2014, is an extensive history of early research by the well-respected *Washington Post* journalist Tom Shroder.

As the political usefulness of the law subsided and the problems of over-incarceration and PTSD grew, a few new trials were approved by the FDA. These trials also faced unique and difficult obstacles, including funding and the ongoing problem of obtaining pure pharmaceutical-quality samples at a *time* when no one was legally allowed to manufacture these compounds. As pharmaceutical companies are quick to point out, trials are complex and expensive. Because there was no economic motivation for pharmaceutical companies to fund them, these new trials had to rely entirely on private donations. There was also an enormous *information* gap because so many research records had been hidden, destroyed, or lost, and generations of new doctors and therapists had been trained without any knowledge or positive *information* regarding this important class of pharmaceuticals. However, two recent events have dramatically altered this situation for the better.

First, the US government was finally forced to acknowledge that PTSD is a very real condition that now affects hundreds of thousands of veterans. Presently there are no other treatments that are nearly as effective as *psychedelic assisted therapy* for treating PTSD. The current psychotropic drug treatments (antipsychotics, anti-depressants, anti-obsessive agents, antianxiety agents, mood stabilizers, stimulants, and anti-panic agents) are not only less effective, but they tend to require long-term use, are very expensive, and come with many unfortunate side-effects including unexpected and volatile feelings of suicide and the outward expression of rage. The deep pain many veterans experience has profound and obvious psychological roots; yet powerful, superficially acting, and very addictive pain killers are being routinely prescribed. ***It is ironic and somewhat tragic to realize that the economic and social incentives for new research into*** *psychedelics were only re-stimulated because the conventional and legal treatment for the great number of veterans with PTSD proved ineffective and became too expensive.*

The second breakthrough changing this situation is unfolding as I write. The brilliantly researched, very informative, and entertaining book, *How to Change Your Mind,* about the history of *psychedelic* research and therapy by the best-selling author Michael Pollan was recently released amid great public fanfare. Within just a few short months of its publication, many well-positioned minds had been changed, and its general popularity has substantially shifted the direction of our collective cultural attitude about these very promising treatments.

One interesting discovery about this exciting treatment method is the common thread in the way that all the *psychedelic* medicines work. They do their work by depressing the activity of our *Default Mode Network*; the brain mechanism that shapes and replays our automatic learned responses. This brain mechanism exists for good reason; it permits the efficient, quick response often necessary for self-protection because it automates and makes every-day repetitive tasks much more efficient. However, when the patterning becomes "burned-in" through repeated extremely stressful situations or severe one-*time* trauma, it can then become so automatic and powerful that normal rational thinking and analysis are impossible. This entire class of medicines effectively slows or shuts down the DMN so our experiences, including difficult and triggering ones, can again register as "new" in much the same way they did when we were young and innocent children. This allows the patient to look at and analyze life events in a completely different way. Because these medications give patients the fresh ability to consider and consciously register a more impersonal and abstract viewpoint, with the added guidance of a skilled therapist, brand-new and life-changing perspectives can be more easily seen and realized. With these medicines, a trained therapist, and a well-planned setting, those affected can now re-explore their lives and the events that led to their condition without the burnt-in PTSD response being automatically triggered. Difficult events can then be seen anew and analyzed from a somewhat removed vantage point; sufferers are granted a fresh look at traumatic events through a process that, in some ways, is similar to rebooting a computer.

Psychedelics are obviously valuable tools for mental health when their use is integrated with trained professional therapy. Unlike many other pharmaceutical options, *they do not require long-term use or lead to addiction. In lab tests, when rats have the free choice of food or an addictive drug, like cocaine, they will choose the drug over food almost every time; but in the case of psychedelics, they will push the lever to dispense it only that one first time.*

PART FOUR – DIMENSIONAL HAZE

UNDERSTANDING THE VEIL

Throughout this book I discuss some of the better-understood reasons why so much of the architecture of our universe is invisible to our eyes and senses; why its deepest truths seem to be hidden behind the invisible and mysterious veil that is our *conceptual horizon*: the outer limit of the way our minds and senses work. To communicate these ideas I must rely on words and rational logic. Humans think and communicate through words that are then shaped into larger concepts, but at some point, we discover that our most unusual experiences can no longer be communicated in this familiar way. Words are no longer useful for sharing some of our unusual experiences; we discover that we are not able to describe that which is not describable.

However, words are still useful for this inner journey. While words may not work for describing our most profound experiences of reality, they can guide us in our discovery of what reality is not, and they can also help us to break free from our limiting shackles. This is the purpose of this section where I discuss our interesting relationship to extra-dimensional *space*, the complex filtering of our *information*, and *time* being far different than we usually imagine. Some aspects of these discussions may seem a little more technical than the rest of the book, so if you find any part of this discussion too obtuse or difficult, please feel free to jump to the next section.

PLATO'S CAVE

Plato gave us one of the earliest recorded discussions about the idea that human awareness represents only a small and often confusing vista into a much larger and unfathomable existence. His famous allegory "The Cave" brilliantly demonstrates how we always lack *information* about the next, bigger perspective, and, therefore, our existing world view will always be limited, or even mistaken. "The Cave" describes the perspective of a man who has spent his entire life as a prisoner in a cave, chained so that he was unable to move, always facing the same wall inside his prison world, the only world he has ever known. His only understanding of anything beyond is through the flickering of shadows that he observes on the surface of the cave wall that he faces, strange shadows cast from the unseen fire and movement of guards behind him. These shadows represent his entire universe, as this is the world that he was born into and the only world he understands. The shadows are his entire reality for they are all that he ever sees, interprets and interacts with; they are his only experience of life, making them his reality.

The prison life that is ongoing behind him within the cave is beyond his awareness. He hears sounds and sees shadows but never can understand much more about this hidden cave world. Not only is he unable to see the fire, guards, and objects that are casting these shadows, but he is also completely unaware of the rich life, full of light, bodies, and interaction, that unfolds outside of his darkened cave. The vibrant sunlit world that exists outside his cave is completely unimaginable, for he lacks even the most basic conceptual framework for interpreting the relatively rich and textural experience that we consider to be our everyday "reality."

In Plato's story, this prisoner is eventually freed and wanders into the bright sun, which immediately blinds him. The sudden shift of brightness and all that he sees are far too much for his restricted senses; he quickly becomes completely overwhelmed. After a period of adjustment, he begins to see a little better and attempts to interpret a small part of this new and wondrous world. After more adjustment *time*, he returns to the cave and attempts to relate the "good news" to other prisoners, but once back he finds that he lacks the "words" to even begin to explain what he saw. Of course, the other isolated and restricted prisoners are unprepared for this unexpected and radical vision and simply assume that he has lost his mind.

Plato argues that in our lives we define and limit our understanding of the universe in exactly the same way as these chained cave prisoners. From our history of exploration – scientific, spiritual, and experiential – it has become increasingly clear that we, too, have been living our lives chained within a cave formed by our culture and sensory interpretations. Today, as this new physics opens doors to previously unseen vistas and our conceptual chains have become partially loosened, we are better able to imagine new and very different ways of seeing and understanding our world, and through these openings, we are discovering new ways to live. We are being shown, and invited to enjoy, a much more responsive and interconnected universe, one that was previously impossible for us to even imagine.

For each of us, this change of awareness takes *time*; but once this new vision is fully internalized, it resonates powerfully within and this draws us further into its profound illumination; our fears dissipate and any desire to turn back quickly evaporates. Because *time*, as we will further discuss, is not what we once thought, and the universe is actually "rigged" in our favor, we always will have exactly the right amount of *time* needed to make any necessary adjustments.

FLATLAND: A NOVEL FROM 1884

INTRODUCTION

One of the most important foundational ideas, germane to all of my books is that this amazing *space* that we occupy always involves more *dimensions* than just our familiar three, and not only do these extra dimensions actually exist, but they fully interact with our lives. Mathematicians and physicists commonly work with dimensions, using calculations that describe these different *spaces*, but, for the rest of us, it might be helpful to learn a little about *dimensions* and just what they represent. Since art often predates science, I will use a Victorian-era (1884) novel, *Flatland*, to introduce this important concept. This novel may have been the first *time* that the general public was introduced to the radical concept of a fourth dimension.

Twenty years after this novel was published, a young Albert Einstein would redefine the world forever by publishing the theory that solidified the science and math of four-dimensional *spacetime*. It took many years to prove, but eventually, his two *theories of relativity*, with all their impossible-seeming implications, were proven to be far more accurate than Newton's previous long-standing deterministic description. Today, as scientists collect data from the many new tools for exploration, measurement, and experiment, his theories continue to hold up flawlessly.

It is not possible for us to visualize this type of expanded *space* because, as three-dimensional specialists, we do not have the senses, tools, or conceptual abilities to see or understand four-dimensional *space*, at least in any conventional sense that can be built on our experiences within a three-dimensional *space*. Human *beings* are simply not designed (at least yet) to be fully functional in this dimensionally expanded environment. That this extent of existence remains mostly hidden from us is not a problem that needs to be overcome, for it is no different than our not being able to hear that ultra-high pitched dog whistle. However, while four-dimensional *spacetime* remains completely impossible to experience through our senses, or "understand" through our three-dimensional logic, it still has become fully integrated into our mathematics and all of our latest technologies. As our technology expands, we will likely discover and develop new devices and technologies that reveal more and more of this hidden world. Beyond just these four *dimensions*, some current *quantum* theories propose the existence of as many as eleven *dimensions.* This expansion of *space* is completely unknowable because it lies so far beyond anything that the human mind could ever imagine. It sits firmly within the realm of gods!

However, what we do have right now is a detailed understanding of one, two, and three-dimensional *space*, the familiar spatial dimensions that we live with every day. Using this experiential knowledge, we can then observe various relationships between these known dimensions and note how things change as we move from one type of *space* to the other. The changes that we observe through these transitions can help us understand, and infer more about, the nature of further expanded dimensional *spaces.* The changes that occur as we shift from one-dimensional *space* to two-dimensional *space*, and from two-dimensional *space* to three-dimensional *space* are completely familiar and comprehensible. After we fully analyze and understand the nature of these transitions, we can *extrapolate* (using similar prior knowledge to predict) to gain a sense of what types of things and relationships we might expect from the next transition: the shift from three dimensions to four. *By going "back" to two-dimensional space and then looking at how things change as we expand to three-dimensional space, we can gain important insights into the probable nature of the shift from three-dimensional space to four-dimensional space.*

This is exactly what E. A. Abbott did 130 years ago in his ground-breaking 1884 novel, *Flatland—A Romance in Many Dimensions.* First, he imagined and carefully described what life would be like in a world of only two dimensions; then, once the reader was fully immersed, living in this limited and restricted world, he introduced a three-dimensional visitor: a simple sphere. Using this device, he effectively describes how the newly introduced third-dimension might be experienced for the first *time* and the types of physical and psychological difficulties that would be encountered by the two-dimensional inhabitants of his Flatland.

FLATTER THINKING

The residents of Flatland exist on a tabletop-like surface, where every person and object is completely flattened like the page of this book. Because theirs is a two-dimensional world, they are "allowed" only two directions (*coordinates*) for movement. Inhabitants can move forward or backward, left or right, or in some combination of these; **a *Flatlander cannot look or travel up or down because these directions do not exist in their two-dimensional world. Actually, they cannot even conceive of an "up" or a "down." From the Flatlander's perspective, our third dimension does not even exist; it lies*** *beyond their senses and outside their ability to comprehend. To Flatlanders, their relationship to the third dimension is just like ours is to the fourth dimension: completely invisible and incomprehensible.*

Everything within Flatland is understood from the perspective of this flat tabletop. A Flatlander sees only the parts of objects that are in direct contact with the table. Any object or part that is above, or below, the surface of the tabletop will not be visible or discernible. *If a coffee cup is sitting on the table, all that can be seen is the ring formed when the cup bottom touches the table. From the Flatlander's perspective, all parts of the coffee cup above the tabletop, including the coffee, do not even exist because they lie outside of their "conceptual horizon." Therefore, the vast majority of any three-dimensional object will remain invisible to them; all they can see and experience is where the three-dimensional object touches (intersects) their flat universe.*

THE PLOT

When I first read this novel, as a nineteen-year-old, I experienced a dramatic and unexpected shift in my awareness as I first became aware of, and then subsequently became mesmerized by, what I now call the "dimensional problem." While *Flatland* predated Einstein's theory of *special relativity*, it did not predate the mathematics that originally inspired Einstein's physics. Fifty years before Einstein even presented his first *theory of relativity*, its foundational mathematics was already in place, having been birthed and developed by Bernhard Riemann, James Maxwell, Marcel Grossman, and several others.

Abbott penned *Flatland* to make two very specific but different points. On one front, it was designed as a clever and cutting criticism of the Victorian social caste system, but it was also written to introduce the public to the new idea of extra dimensions. *Abbott's book does an excellent job of helping readers understand the conceptual and social challenges of a society that is meeting a new spatial paradigm for the first time – this is exactly what we are experiencing today.*

The author does this by conducting readers into his well-crafted, two-dimensional, Victorian universe. He slowly and carefully brings us into what becomes a very believable two-dimensional world, one that is complete, understandable, and comes with a full cast of relatable characters, but characters who are leading only two-dimensional lives. In Flatland, as well as Victorian England, there existed a strong social caste system built upon birthright of the inhabitants. Then in England, it was race and family that divided people into socioeconomic classes. In "Flatland" inhabitants are similarly divided, but this *time* it is their shapes that create the divisions. In his story, all the women are portrayed as lines, the most simple and "lowly" shape, yet a shape that also introduces a special danger: the threat of their extremely sharp point, the dangerous weapon that all women possess. The male warrior class is represented by triangles because, while they also have sharp points, their shape makes their points much stronger. Squares and other polygons, whose increased number of sides also corresponds to an increase in rank and status, are the business and ruling-class males. At the top of the caste system are the male religious leaders, represented by circles having no sides or points.

Abbott successfully creates and describes the lives, limitations, interactions, and experiences of all these characters by crafting a storyline that unfolds entirely within their two-dimensional world. He does this so well that when I was first reading *Flatland*, as I identified with characters, I had the very tactile and visceral experience of feeling "trapped" by the unexpected constraints within this dimensionally reduced version of the universe.

Once Abbott's two-dimensional perspective has been fully internalized by readers, he then introduces a simple three-dimensional object; a sphere visits Flatland. This sphere, a "messenger from another dimension," arrived intending to introduce the concept of three-dimensions to one of Flatland's citizens, the book's main character. Through this character's two-dimensional perspective, the reader then experiences this extra-dimensional object, the sphere, through a Flatlander's eyes. This well-described perspective helps us to understand the "dimensional problem" and how it might directly apply to our three-dimensional lives. Later in the book, we witness the lead character's impossible struggle to explain to other Flatlanders his encounter with this three-dimensional *being*, not unlike the dilemma of the prisoner from Plato's *Cave*. Somewhat predictably, as a result of religious persecution and public confusion, our lead character winds up in an insane asylum. Abbott's story is a very clever device for helping us better understand just how difficult, or even impossible, it would be to describe a four-dimensional existence from our limited, three-dimensional perspective.

A SIMPLE EXPERIMENT FOR HOME

To get a better feel for a Flatlander's two-dimensional worldview, sit at your kitchen table, place a toothpick on the table, and then lower your eye to the table edge so that you are looking at the toothpick from the height of a tiny insect walking on the table—even try to imagine that you are that very small insect. Close one eye and then slowly spin the stick and observe. What you see only looks like a line that gets longer and then shorter as you turn the stick. Next, from a piece of cardboard, cut a circle, a triangle, and a square, all about the same size, and place them flat on the table. Close one eye and lower your open eye so you can see only the thin edges of these objects, and observe. Because we only see the closest edges of each object, the circle, triangle, and square should all look about the same; they all appear as a line.

This effect is enhanced because we are using only one eye and therefore we also lack *depth perception*. While this loss of *depth perception* is not a requirement for a two-dimensional worldview, it does add another useful experimental condition; by sacrificing our *depth perception*, we can better understand what it means to have additional limits on our senses – such as our inability to directly see x-rays or hear dog whistles. Losing *depth perception* helps to illustrate how critical our senses are for the perception and interpretation of our worldview, and how a restricted sense can add complexity and misunderstanding. ***Because we lack certain senses and abilities, we still miss many things that actually exist in our three-dimensional world, even when they are right in front of us.***

LIFE IN FLATLAND

Since Flatlanders also lack *depth perception*, circles, squares, and triangles all look the same. As we observed in our tabletop experiment, unless they are rotating or the sides have different illumination (shadows), all these different shapes just look like lines. Flatlanders can only recognize the different shapes by touching or feeling the sharpness of their points and, therefore, they sometimes describe different shapes as "feeling" different. As we discovered in our table-top experiment, when triangles turn, they become a little shorter, then longer, then shorter again, and so on. Rotating squares look much the same as triangles, except their changes in length would be more frequent for the same number of turns. When circles turn, they don't change length; circles always look the same no matter how they rotate.

Because the Flatlander's senses and mindset are restricted, they have no understanding of "above." They cannot observe these simple shapes from above and simply count the sides, as we can. To do this would require the understanding and use of an extra dimension and, in their case, this needed dimension would be the third dimension; a dimension that is not a part of their direct awareness. Therefore, when Flatlanders explain the difference between triangles and squares they might, instead, introduce the idea of *time* into the description. For example, Flatlanders might say that a spinning square changes its size more "frequently" than a triangle, while priests (circles) don't change over *time*. From our broader, three-dimensional perspective, we can easily see that it is only our view of the shape that changes and *time* really has nothing to do with these observed differences. *Time* has become a part of their description only because of their limited two-dimensional worldview and their lack of depth perception: their *viewport* and a sense. Descriptions formed from limited perspectives can quickly become very complex and confusing, especially when *time* is introduced.

This is an example of where our *dimensional limitations* lead to what I am labeling the "dimensional problem." **The "dimensional problem" is when things appear to behave in complex, weird, or strange ways only because we are trying to understand or explain them from a perspective of fewer dimensions than the actual space they occupy.**

In Flatland, residents describe the differences between shapes by talking about how objects **feel** and how things change with **time**. "*Time*" and "feeling" are two complicated ways Flatlanders describe qualities, that for us are easily understood through simple differences in shape. If they could just add just one more dimension to their perspective, they would easily see that these different classes of citizens are all distinct shapes: triangles, squares, rectangles, and circles. They only use *time* or feeling to describe these objects because of their built-in limitations.

A SPHERE VISITS FLATLAND

Once Abbott fully establishes the nature of ordinary two-dimensional life in Flatland, the three-dimensional sphere makes its appearance. Imagine yourself living in Flatland as the sphere comes to visit. The sphere arrives at your tabletop world by moving unseen through the surrounding three-dimensional *space* and then, suddenly, the ball lands on your breakfast-table world. When it first touches the table only the bottom tip of the sphere can be seen from your insect eye-level tabletop view. Two-dimensional *beings* can't see or experience the third dimension, so they have no sense or idea of anything up or down, and this means that they can't see any of the parts of the ball that are above that single point on the very bottom of the ball that touches the table. The inhabitants of this two-dimensional table would not see or sense the rest of the ball until these parts reach and pass through the plane of their flat world. At this very first moment of contact, observers in this flat world only see a dot; they cannot look up and see the rest of the ball that is sitting above the tabletop. At this moment, from their perspective, their entire experience of the sphere is that single point where the sphere touches the tabletop.

To better understand their view of this "first contact," set a ball on your table and look for yourself. Then imagine this sphere is a special magic ball, one that can also pass right through the table. If this is too difficult to imagine, visualize a ball that is slowly being dipped into the surface of a flat pond, and you are a water-strider living on the surface of that pond. As the ball passes through the plane of the tabletop (or the top surface of the water), that initial point of contact (dot) grows to become a circle. As the sphere or ball continues to move, the Flatlander will see the circle growing bigger as a thicker part of the ball is now passing through the tabletop. Once the equator or midpoint of the ball passes through the tabletop, the circle will then begin to get smaller. At this point, from the Flatlander's perspective, they see a "line" that starts getting shorter, until, once again, it becomes only a small point or dot; this dot is the last thing they see at the very last moment of contact. Then, in the very next instant, the dot disappears entirely from Flatland, having passed beyond the surface plane of the tabletop (or pond surface). From the Flatlander's two-dimensional perspective, the sphere has just mysteriously disappeared. To a three-dimensional viewer, it is still there, but it has simply moved beyond the flat plane of Flatland.

OURS IS A FLATLANDER'S EXPERIENCE

From the Flatlander's perspective, this "alien" visitor to Flatland displayed extremely surprising and mysterious qualities; they had no good methods, or tools, for explaining any of its strange seeming behavior. In their futile attempts, the Flatlanders relied upon the concept of *time* and said things like, *"This line appeared from nowhere and then grew to a large size in one minute, and then shrank just as quickly before completely disappearing."* This strange and curious object was seen as fully *paranormal*, breaking all their "normal" rules. It came from nowhere and disappeared just as mysteriously, and while it was visible, it appeared to them as a circle that continuously changed in size. What priest could appear from nowhere, grow, shrink, and then completely disappear? What happened to it? How did it become invisible? Nothing in their world could explain such unexpected and mysterious behavior. They might even conclude that only a very powerful god could do something like that! *However, we can see that these are just overly "complex" observations that miss the much simpler truth about the actual shape of a sphere and the true nature of three-dimensions, a truth that can be instantly and easily understood from a space that includes just one more dimension.*

From our three-dimensional perspective, there is no weirdness or mystery; the physics and geometry of a sphere intersecting a flat surface are fully understandable. We naturally think and experience life from a three-dimensional perspective, so all of this falls well within our *conceptual horizon.* From our perspective this "strange-acting object" was not odd or paranormal at all; its appearance can be easily explained. However, seeing and understanding the actual geometry requires the ability to think in three dimensions; an impossibility for two-dimensional Flatlanders. *Our native ability to perceive in three dimensions is a type, or level, of "awareness" that is beyond the Flatlander's ability.* Our everyday perception lies beyond the limits of their nervous system, sensory organs, and their brain's ability to form these mental concepts; it lies beyond their *conceptual horizon.*

This story illustrates how, within our three-dimensional world, phenomena that appear mysterious to us, "feel" strange or seem to be changing in very odd ways over time, would become simple to visualize and understand if we were able to understand and fully employ one additional dimension. This is a key concept to grasp, so I will repeat it in a slightly different way: *strange or unexplained phenomena, and even our understanding of the way things "age" or change with time, are nothing more than our very limited way of trying to explain and understand the geometry of extra-dimensional space.*

This exercise demonstrates some of the difficulties we can expect within our three-dimensional reference frame when we are interacting with four-dimensional "things." We only observe and experience odd-looking pieces, artifacts, or partial views of these objects. There is no way to understand a four-dimensional object from the confines of our three-dimensional perspective. Actually, as illustrated in *Flatland* and our own lives, we are not even experiencing the full range or extent of all aspects of three-dimensional objects. Like the residents of Flatland, who lack *depth perception*, we are missing the sensory systems to perceive many things that exist within our own three-dimensional *space*; high and low-*frequency* colors and sounds are just the most known and obvious.

If Flatlanders could suddenly understand three dimensions the same way we do, then at that moment, their entire view of the universe would dramatically change. Of their many potential discoveries, two are of special importance for understanding the ideas of this book. *First, they would discover that they did not understand the full extent of many of the objects that are in their own two-dimensional realm. They would realize that even two-*

dimensional objects simply cannot be fully understood or described using only their two-dimensional concepts and mindset. For example, even though the entire triangle is within their world, they never actually "see" the full shape of a triangle; they only observe a line that changes length.

Flatland mathematicians could speculate why this was so, but their theories would probably look extremely messy and complicated. Objects can only be described easily and accurately when we also understand the *space* that contains them. **Three dimensions are required to fully describe and understand two-dimensional objects or two-dimensional space.** This means that Flatlanders can never even fully experience the entire extent of their two-dimensional world and its shapes. Because Flatlanders have such a restricted understanding of their *space*, when they attempt to describe three-dimensional objects or *space*, they automatically introduce a great amount of complexity, mystery, and confusion. This is precisely what happens when we try to describe four-dimensional *space* or "objects."

At the end of Abbott's novel, the reader is asked to imagine how difficult it would be for three-dimensional humans to understand and describe a world of four or more dimensions. **Through this cross-dimensional comparison, we realize that, from the perspective of someone living a two-dimensional existence, three-dimensional objects, even simple ones, appear as very complex and involve inexplicable phenomena.** Again, this appearance of extreme complexity when describing an aspect of existence with only the limited tools of fewer dimensions is what I refer to as the "dimensional problem." Because of this quality, any description of four- (or more) dimensional "objects" or *space* will always appear to be very complicated from our three-dimensional perspective.

The obvious implication is that, as three-dimensional beings, we cannot fully see or understand all aspects of even our three-dimensional shapes. How then could we even hope to begin to understand or explain four-dimensional realms and "things?" Because of this "dimensional problem," all of our attempts to describe four-dimensional space or "objects" will become very complex and confusing.

Like Flatlanders, we also only interact with that limited part of existence that is accessible from our dimensional realm. If Flatlanders were suddenly able to perceive three-dimensional *space*, a second discovery would be about the role of *time* in their "universe." **In dimensionally restricted Flatland, phenomena involving "time" offer clues or "windows" into the structure**

of the next dimension! We, in our *space*, also describe many phenomena using *time,* but are we only trying to describe a geometric quality that simply lies beyond our ability to see or understand? It is likely that for us, *time* provides a similar "window," but one that only can begin to describe our true relationship with the next dimension: the fourth dimension.

I have been speaking mostly about "objects," but all of this also applies equally to people, ideas, *energy*, and everything that is part of our three-dimensional reference frame. **Behavior and phenomena, which seem complex or strange from our limited perspective, become clear and even simple once they are observed from an expanded, dimensional reference frame.** As discussed later in the science addendum, this is exactly the type of relationship Einstein illuminated in his *theory of general relativity*, where he demonstrated that *gravity* was not some mysteriously behaving force; instead, it is caused by a simple change of shape or deformation that invisibly occurs in four-dimensional *spacetime*.

I use the term *artifacts* to describe our three-dimensional misinterpretations of extradimensional things and ideas. An instructive look at the way we make maps can help us to better understand *artifacts*. However, before diving into mapmaking, I must first discuss two very important relevant concepts, *viewports,* and *projections,* and then introduce a bit more about *time*.

VIEWPORTS

I often refer to *viewports*, a useful term and wonderful metaphor for describing our personal experience of the universe. I borrowed the term from an architectural software program that I rely upon daily; in this software, *viewport* refers to a special presentation "window," used to hide or filter certain elements of a drawing, making the remaining visible parts easier to understand. Our limited senses and inability to see beyond the veil of extra-dimensional *space* functions much like this *viewport*, allowing us to focus on three-dimensional living by hiding *information* that is not directly useful for this purpose.

Today's architectural drawings, constructed on computers, include all three-dimensional *parameters*, along with an enormous amount of additional *information* about specifications, mechanical systems, and structure. These drawings are virtual models of the building that include wiring, equipment, appliances, trim, and everything else found in a modern building. However, if we always displayed everything in the drawing, it would be very confusing because there is simply too much *information*; the amount of *information* would be so overwhelming that important parts for each step would be obscured and lost because of all the other *information*. **Viewports are used to simplify and show only the necessary information for each viewer, tradesperson, or step of the process.** The *viewport* chosen and printed for each trade is a filter, removing any unnecessary *information*.

In this software program, *viewports* can be easily adjusted and changed. They can be set so that only two dimensions are visible; they can hide unnecessary notations; or they can be easily adjusted to remove all utilities (plumbing, electric, HVAC, security, etc.) and show only the structural walls. This reduction of *information* ensures a simple and direct presentation of only the most critical *information*. The other *information* is not gone or lost; it is simply no longer visible. **The extra information is still present, but it is now invisible because it has been hidden beyond the viewer's senses.** I can freely adjust the software settings to control what my clients, suppliers, inspectors, and builders each get to see.

Viewport is a perfect metaphor for describing our three-dimensional experience within a multidimensional universe. Each of us has only an individual and limited *viewport* into the much fuller universe. Incapable of understanding and processing the total amount of *information* within the universe, we are happily restricted to our three dimensions and only a small obscured window into the fourth, which we call *time*. The *parameters* (or settings) of our *viewport* determine our experience of the universe. **To help us function efficiently and effectively in our three-dimensional world, our viewport intentionally and automatically filters out all of the noncritical or extraneous information.**

The outer limits of the physical size of our *viewport*, our visible universe, are presently limited to about 15 billion *light-years* in any direction. This is as far as we can presently observe with our astronomical tools, and, not coincidentally, this is also the distance that light has traveled since the cosmological event named the *Big Bang*. Our *viewport* is also further restricted in size to things that are within the range of our senses and our tools. If we were much smaller or larger, our direct observations would be very different. Additionally, our *viewport* is shaped by the other ways our brains filter and process information. Built into our nervous systems are filters, such as *cognitive dissonance,* and the limitations of our five recognized senses. These factors work together to limit the quantity and quality of *information* that we consciously process. These are the "settings" that contribute to the form and shape of our "point of view," and many of them are internally "adjustable."

Of course, any human *viewport* has both a conscious and a *subconscious* component. Just how big and important is the role of our *subconscious* to the whole of our expressed *being*? One way of measuring its effect is to look at how much relative *information* our conscious and unconscious minds can process in a set amount of *time*. According to one recent scientific study, our conscious minds can process between twenty and forty bits of *information* in one second. In that same second, our *subconscious* minds will process almost 20,000,000 bits of *information*. Since the *subconscious* can process up to a million times more *information*, all of it hidden from our conscious minds, we are all essentially on autopilot 100 percent of the *time* without even realizing it.

The *subconscious* is dominant, fast, and powerful, but because its world is normally hidden from our conscious minds, its activity is not easy to self-monitor or access. **Due to the preset and functional parameters of our three-dimensional viewport, we are rarely aware of our subconscious. This means that the dominant parts of our viewport are effectively hidden from our everyday consciousness.**

We can define our "viewport" as the "window" through which we on Earth consciously experience our universe. The precise extents of personal *viewports* will be slightly different for every individual, but generally, all human *viewports* will fall within a range of common and expected experiences. Einstein once described the nature of our *viewport* this way: "*Nature shows us only the tail of the lion, but I do not doubt that the lion belongs to it even though he cannot at once reveal himself because of his enormous size.*"[1] The famous parable of the five blind men describing an elephant – one at the trunk, one touching the tail, one at the foot, one at the belly and one on his back – also playfully describes this idea of an individual, but limited, *viewport*.

PROJECTIONS

INTRODUCTION

When the three-dimensional sphere passed through Flatland, it was experienced by Flatlanders as a two-dimensional circle that changed diameter as *time* passed. These dimensional translations can be predicted and

---

[1] As quoted by Abraham Pais in *Subtle is the Lord: The Science and Life of Albert Einstein* (1982).

described through a common mathematical relationship called *projection.* This concept is important for many of the ideas within this book, and it can be easily understood without any math. Everyday examples of *projections* include the *images* we see in movies, mirrors, maps, photographs, and shadows. All of these are secondary *images*; what we experience is not the actual object; instead, we see or interact with an *image* of the original object. With movies, we even use the term *projection* to describe the process of displaying it on the screen. Actors play their roles in three dimensions, but their performance is reduced to the two dimensions of the screen. (3D movies trick our brains by slightly shifting the position of the *image* that we see in each eye to add back a partial, but sometimes effective, approximation of the third-dimension.)

ARTIFACTS

In Flatland, the strange visitor was a simple, three-dimensional object: a sphere. Because a Flatlander's *viewport* was limited to only two dimensions, the sphere appeared only as a circle that changed size with *time*. Circles are the *images* of three-dimensional spheres that can be cut, cast, or *projected* onto a two-dimensional plane such as Flatland. (If an orange is cut with a knife, the visible cut edge forms a circle illustrating where the flat plane of the two-dimensional knife is intersecting the three-dimensional spherical orange.) The Flatlanders see a circle that changes size with *time*, but this circle is only an *artifact,* a confused misinterpretation caused by the *dimensional problem* of trying to describe something using fewer dimensions than the *space* that it occupies.

If we stand outside in the sun and look at our shadow, we are looking at an *artifact:* the *projection* of our three-dimensional body onto the flat, two-dimensional plane of the ground. This *projection* or shadow is not the object; it is a reduced or *flattened image* of the object that has been transferred, translated, or *mapped* onto another system of *coordinates.* In this case, this "other system" is the flat plane of the ground. The elongated, shortened, or odd shape of our shadow is not the actual shape of our body; it is only another *artifact* created by the *projection* (or *image*) onto a system of fewer dimensions (see *dimensional reduction*).

If we were to shrink to the size of bugs, living entirely on the flat surface of the Earth, unable to look up or down like the inhabitants of Flatland, much of our awareness of people, trees, and objects would be limited to encountering a small part of their flickering and changing shadows. From this bug's perspective, imagine how very difficult and mysterious the understanding of humans would be. How could this bug even begin to describe us from their confusing encounter with just our shadows? Our entire existence is actually filled with our interacting with similar types of dimensionally reduced *projections* (shadows), while mistakenly thinking we are interacting with full and real objects.

Plato's allegory, "The Cave," illustrates that our understanding of our world is built upon our belief that these shadow illusions, *artifacts,* or *projections* are real. The chained prisoner was aware of nothing but what he could see from his restricted *viewport:* the misunderstood *artifacts* cast upon his cave wall. Like all of us, he assumed what he saw and experienced was real and therefore, reality. Plato argued that the human condition is forever bound to the impressions that are received through the senses, and if we could miraculously escape our bondage, we would discover a much greater world, but also one that we would not easily or initially understand.

On Earth, we see and understand everything through our three-dimensional *viewport,* even though we now know that the universe is built upon at least four (and likely many more) dimensions. **What we observe and experience on Earth is only the very limited projection of the much fuller, multidimensional experience onto this limited three-dimensional space: our known universe. We experience only the shadows that intersect, interact, or map directly onto our three-dimensional world—a greatly reduced "image" of what actually exists. This "reductive filtering" of our experience is not to be viewed as a problem because it has an important purpose in that it allows us to function well and efficiently in our three-dimensional part of the multiverse. However, just our being aware that this is always happening will naturally expand our viewport.**

Throughout my writing, I will use the terms *projection, shadow, image, illusion,* and *dreamscape* to describe this phenomenon. *Projection* and *image* are the mathematical terms, while *shadow* more accurately describes our sensory awareness. *Dreamscape* and *illusion* more closely relate to the psychological aspects. Because of our *viewport,* we don't experience the world as "it is" (multidimensional), instead we experience it as "we are" (three-dimensional).

Physicists have recently been referring to the idea of a *hologram* to describe the *image* quality of our three-dimensional physical world. This idea is more than just an excellent metaphor, for it accurately models several aspects of our deeper existence. Later, I discuss *holograms* and related topics in more depth. We will also learn how *holograms* might be even more interesting because of their role in the way *information* may be stored throughout our universe.

DIMENSIONAL REDUCTION — DIMENSIONAL FILTERING

As we explore these ideas, we always need to remember the defining principle — **whenever we interact with or view something in fewer dimensions than its actual full geometry, what we encounter is only a "slice" of the actual object, idea, or experience.** We are never capable of visualizing or fully understanding any multidimensional object, experience, or concept because our input and processing ability is always limited by the extents of our three-dimensional *viewport*. Of course, this filtering is useful and important, for it is exactly what defines and shapes life as we know it. Our three-dimensional brains are specifically built to interpret these *projections* cast into our world from the greater existence beyond. **Our brain's interpretation is then based upon the information gathered from those parts of existence that we are capable of sensing and experiencing.** Since there are more than the familiar three dimensions involved, what we directly sense and interpret is then the result of sequential *projections:* the *projections* of the *projections*. Shadows are cast from the shadows of the shadows, etc.; shadows are sequentially cast from the original multidimensional source, and then, in turn, re-cast through each dimension. As these *images* pass through each dimensional realm, they are re-shaped, again and again, before they eventually intersect our three-dimensional world. What we see and encounter will not even resemble the real original "thing" because, through this process of multiple-level filtering, the original *information* is distilled, translated, and distorted many times over. Our *viewport* then allows us to experience the understandable "shadows" that eventually reach us.

"Dimensional reduction" is my descriptive name for this reductive process, involving multiple levels of projections. This "dimensional filter" is the primary reason for the appearance of mystery in our lives. This process shapes the veil that occasionally clears or aligns to reveal small glimpses of truth about the deeper nature of the universe. Dimensional reduction plays many tricks on our sensory systems, especially when it involves our ideas about *time* and *space*. Einstein realized that the *force* that we call *gravity* was only our limited three-dimensional experience of something geometric that was occurring in four-dimensional *space/time*. The *force* that we experience is only an *artifact*; our experience of *gravity* is just a *dimensionally reduced artifact* whose deeper source is the curvature or deformation of *spacetime*.

A better understanding of *dimensional reduction* and *dimensional filtering* will help the communication of many of these ideas. The simple and common task of making a map can help us better understand what these represent. (*Note:* These same two terms also have older but very specific technical or mathematical definitions that are quite different from this new use.)

MAPMAKING

INTRODUCTION

Mapmaking involves taking a three-dimensional object, such as the surface of our spherical Earth, and translating or *projecting* its *information* so that it fits within the flat, two-dimensional representation that we recognize as a map. Traditional mapmaking is an example of using mathematical *projection* to our advantage. It also provides a very clear demonstration of *dimensional reduction.* When things that belong in three-dimensions are presented in only two dimensions, *dimensional reduction* occurs, and this always causes a profound shift or loss of *information.* To accomplish this mathematically, we use one of many common and well-understood mapping *algorithms* (mathematical formulas). Since a map must be able to be displayed on a flat piece of paper (or, today, a smartphone screen), we must somehow remove the third dimension, and we do this by essentially flattening the Earth's curve—we change the *viewport.* The larger the piece of the Earth that is mapped, the more that must be flattened, thus the greater the *information* loss and distortion.

In other words, when we use a map, we are viewing an object that originally was three-dimensional: the curved earth, through only two dimensions: the flat paper. In the simplest terms, we do this by "squashing" the curve of the Earth to make it flat; there are many ways that this can be done mathematically, but no matter which method we use, as we "flatten" the globe into a map we must hide or distort *information* about the third dimension.

Visualize hammering a round globe until it becomes a flat map; after such a brutal process we might expect to have some problems accurately reading the map. While this method is particularly crude, there still is no fully accurate way to visually express the entire spherical Earth on a flat piece of paper; some *information* will always be lost or distorted. *Information* about the third dimension first needs to be *flattened,* a form of filtering, before it can be presented on flat paper, the two-dimensional *viewport* that we find useful for maps.

Fortunately, we have much better tools than a sledgehammer for this job. One of the most common methods for making a world map is the *Mercator Projection,* and the most common version of this is shown in the illustration below. This method uses a particular type of mathematical *algorithm* that results in a map that is quite accurate near the equator. However, as you move to the poles, the landmasses and distances get increasingly distorted and they become larger. We have all seen these types of maps where Canada and Greenland seem to be enormous landmasses, but this distortion of size is only an *artifact* from this particular mathematical process of *dimensional reduction.*

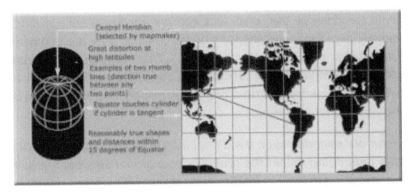

In a Mercator Projection, each rectangle represents an equal area between the lines of latitude and longitude. The areas near the equator are the most accurate. There are other versions of the Mercator that create different areas of distortion.

Hundreds of different types of popular *map projections* can be constructed using different *algorithms.* While all *projections,* or two-dimensional maps, of our three-dimensional world, are inherently inaccurate, each is useful for different purposes. Inaccuracies will always arise as *information* is lost or altered through the elimination of one or more dimensions.

OUR WORLD VIEW IS SIMILAR TO A MAP PROJECTION

We natively understand the three familiar dimensions, yet Einstein proved that there is at least one more dimension to our existence. What if this process of *projection* occurs seven or eight times in succession with each *time* causing more *dimensional reduction*? Imagine what compounded inaccuracies or *artifacts* could appear with eleven dimensions involved. This is exactly what occurs as *information* that originates in eleven-dimensional *space* is *filtered* or *projected* down through multiple levels until it eventually lands on our three-dimensional universe. What we ultimately experience: our reality, is many times removed (flattened) from the original *information.* If our universe has eleven dimensions as many predict, the *information* that we finally perceive will be only the shadow of a shadow of a shadow – eight times (eleven minus three) in succession. Our experience will always be radically filtered, distilled, and far different from the original source. *We can think of our entire universe as a three-dimensional "map" of information originating in a space of many more dimensions, which then is reduced, crammed, and filtered so it fits on three dimensions.*

Our conceptual minds understand the geometry of our *viewport*, one that is limited to three dimensions. Just as with my architectural software program, all of the original *information* is still available, but we are unable to view it directly: it is hidden by the constraints and *parameters* of our *viewport*. We only directly experience the shadows or *images* after they are geometrically *projected* onto our *dimensional space*. *This dramatic dimensional reduction is completely for our benefit, for it prevents us from being overwhelmed or distracted from our real purpose, which is to gain direct three-dimensional experience.*

Whatever our experience is, it will not be an accurate description of what is really happening within the greater dimensional spaces. We have no tools or experience to even begin to understand the deeper nature of our multidimensional universe, so, for now, we must understand our limits, and just let this mystery be.

DIFFERENT REALITIES

Because we add our personal *individual filters*, we each unconsciously modify the general cultural *dimensional viewport* to create our own specific *personal viewport*. Throughout our lives and the resultant personal growth, every one of us is constantly tweaking the *algorithm* that defines our personal *viewport;* therefore, *we each see our lives only through our individual and unique set of filters.*

The assumption that we all share the same viewport or worldview has been a common and historical source of many of mankind's problems and conflicts. Two different people can witness the same event, say a car accident, and their reports will often read as if they had witnessed two entirely different events. This is not just a memory problem, and it also is not that one view is right and the other view is wrong. Since we each unconsciously frame our *viewport* in a slightly different and personal way, we all perceive and cognize *information* uniquely. *Each, and every, one of us fully inhabits a unique universe; some of it is defined by the three-dimensional realm and some of it described by our particular sub-*

culture. However, the final layers of our personal and unique viewport are always self-constructed.

Many of the perceived "problems" in our world exist simply because we see and judge others based on our own *viewport*. If we allowed more "room" for the unique framing of *viewports* by different individuals and cultures, the blame and divisive idea of "right and wrong" could begin to evaporate. *It is, therefore, important to realize that every one of us is experiencing a different reality through a slightly adjusted and uniquely personal projection. We all inhabit slightly different realities, so our experiences are always ours, and ours alone.*

VIEWPORTS SHIFT OVER TIME

Our individual, group, and cultural *viewports* constantly shift, change shape and size, and evolve. When we observe our old paradigms changing around concepts such as "flat Earth," apartheid, women's rights, sexual identity, or civil rights, what we are actually experiencing is a gradual shifting of our collective *viewport*. However, because of the diversity within and the massive size and momentum of these sub-cultures, this adjustment usually unfolds through a very slow and difficult process; significant shifts in our culture often seem painfully slow. *Fortunately, waiting for others is never necessary; individuals can shift their personal viewports much more rapidly (even instantly) than our entire culture can shift.* Even more significantly, once the individual changes, from their new perspective in their newly expanded *viewport*, many of those "others" will also appear to be changed as well.

"Wait!" the careful reader might say. "You already said that others will not change spontaneously because change requires inner work, so why is it that these "others" appear to have changed?" This is an important and paradigm-shattering question that involves two related principles about the way our universe works through its built-in architecture of freedom: the existence of *parallel universes* and the *holographic principle* of the outer world expression always being a reflection of the inner world. These important topics are discussed in other sections of this book, but both are greatly expanded upon in the science appendix that follows.

TIME

THE COSMIC TRICKSTER

We often use our human concept of *time* to help describe how things appear to us. The sphere that visited Flatland was observed as a circle that changed size with *time*. *Time* is the wild card, the joker in our game of life. It may seem very important for our three-dimensional worldview, but really, *time* is just a misunderstood and orphaned trickster.

We often perceive *time* as our enemy: the constantly marching, relentlessly advancing quantity that defines our limited stay here on Earth. We imagine that if we could only stop the advance of *time*, we would not age or die: we could become immortal. Most of us are engaged in a constant battle with *time*.

Digging deeper, we discover that there is nothing that needs to be changed about *time* except our attitude. The physics of the last century demonstrates that *time* does not always "march on" as an absolute. Instead, in some yet to be understood way, within each of us we unconsciously create a sense of marching and ordered *time* to facilitate our three-dimensional living. We are starting to realize that instead of our living in *time*, *time* lives within us. *Time* and mind are completely interrelated. In the words of Einstein, "*Time* is a stubbornly persistent illusion." In the words of Eckhart Tolle, "Time and mind are inseparable. To identify with the mind is to be trapped in time." *Time,* as we are just beginning to understand, is one of our mind's most-misunderstood tools for living our three-dimensional lives.

Spiritual leaders and scientists are both in agreement that our perception of *time* is an illusion. Physicists and mathematicians think of *time* as being similar to another dimension. In mathematics, we might find ourselves referencing the future and the past in much the same way that we now look to the East or West. A physicist might say ***"The present moment informs the "past" and the "future" equally, through waves of information that travel in all directions."*** Our more human perception of "*time*" is as if we are stuck on a freeway heading north, not realizing that we have the ability to slow the car, or even turn around and head south. We seem to only understand *time* from the perspective of being in a car that is always traveling at 70 miles per hour in the northbound lane, a very limited perspective.

If we were traveling in a multidimensional hovercraft, we could travel in new and "different" directions to leave our realm of *time* completely in much the same way as flying liberates us from our restricted, two-dimensional, ground-level contact with the Earth's surface or the sphere was able to leave *Flatland*. To relate to *time* differently, we must shift our entire "time-based" paradigm. ***The very first step of this process is to better understand how and why our current sense of marching time limits our vision. As we shed our*** *old, restrictive ideas about time, we create room for the emergence and growth of a new expanded vision.*

PAST, PRESENT AND FUTURE: A CONTINUUM

A simple demonstration can help us visualize *time* differently. Imagine that we are on a rapidly moving train. Looking out the side window, we see a landscape as it passes. Having just finished our lunch and used the last paper towel, we notice its cardboard tube begging to be used in some creative way. Playfully, we put the tube up to our eye and look out of the side window with only that one eye open. Pinning the tube against the window, we now see only the small portion of the view that happens to be framed by the tube.

At any given moment, we can think of the landscape that we see through this tube as the "present." We cannot see the "future" landscape until it pops into our tubular view, but at every moment a new view becomes our present. We remember, but we can no longer see, those scenes that have already moved through our field of vision; those remembered scenes become our "past." With some imaginative childlike play, this tubular view can become entire our "world," as we discover that we can make reasonable guesses about the "future" scenes based on our "present" landscape and our memories of the past. The train moves forward, bringing new parts of the landscape into our view, as each next "view" becomes our next "present moment." *Time* marches forward relentlessly, and if we forget that we are on a train, this linear and one-directional "passage of *time*" can become our world and our reality until, playfully, one of our kids knocks the cardboard tube away from our eye. *Suddenly, right in front of our eyes, is the entire "past," "present" and "future"—all at once and fully interconnected!* The once divided and ordered "past," "present," and "future" now forms a continuous landscape, one that we recognized can be traversed in any direction, once we step off our train.

Time is a function of the way our brains divide, organize, catalog, and store an otherwise *infinite* continuum of *information* and experiences. *Time is our loyal assistant, which prevents everything from happening at once and completely overpowering our brain's limited processor.* It is nature's way of intentionally restricting our *viewport* within a much wider universe: a necessary constraint that is necessary to shape our three-dimensional lives. *Time* is the ordering system that allows us to form and organize our conceptual thoughts. *We are not built to handle everything unfolding, or existing at once, or more accurately, outside of time. For information to be useful, it must be processed by our brains in smaller increments. We stack and order these increments by creating the organizational systems we think of as "time" and "space." Time is just one way our brain organizes information—it is part of our brain's filing system.*

PART FIVE – BROADER DISCUSSION

INTRODUCTION

This section discusses several sub-topics that are more controversial, abstract, or philosophical. Here I will leave some questions hanging, unanswered, introducing ideas that require further thought. Where reasonable and helpful, I have made these discussions self-contained by repeating parts of earlier discussions, and therefore readers should expect some repetition of earlier ideas and *information*. Input or feedback is always enjoyed and appreciated.

BOTH SEPARATION AND UNITY CONSCIOUSNESS

I believe that our purpose is to grow more "whole" through our experiences in life so that we open to the greater universe that can only reveal itself through the illumination of our inner shadows. For this journey to have meaning, we must better understand the nature of our separation, and this can only be accomplished through a deeply personal exploration of life. Our existence in both separation and oneness must be integrated into our process because, in our physical universe, they always must exist together as complementary partners that together build our wholeness. As yin requires yang to exist, we are always both; so if either is held in isolation, it leaves us fractionated and incomplete.

Separation consciousness and *unity consciousness* are two descriptive names often assigned to this division of awareness – at least until the *time* when we will no longer feel the necessity to name and divide "oneness." Growth occurs every *time* we let go of resistance. Instead of our reflexive "NO," if we fully embrace everything that life presents with an expansive inhale of "YES," we continue to grow and become more whole. After thoroughly exploring and testing *separation consciousness,* we fully understand that our loving acceptance and embracement of all parts of our separation is just as critical as opening to our deep interconnection.

I have been using the word "illusion" when referencing reality, yet the physical world we each experience daily always appears to be very real and concrete. This is because, from our perspective, everything that we physically experience *is* real. The entire physical universe is absolutely real for us, yet we still live an *illusionary* existence. **The "illusion" is only our belief that these very real physical lives, bodies, and dramas are who we actually** *are. We have fallen into the illusionary trap of believing that these things define and limit who and what we are. However, deeper awareness reveals that our physical lives never define or limit "who we are." Our physical lives are only one experience in time and space that our greater being is temporarily exploring. We are spiritual beings having this very real-feeling physical experience, but we are not just this physical experience; we also simultaneously exist and flow within much deeper realms.*

TWO TYPES OF PEOPLE

Because of the way life's pendulum swings, at any single moment in *time*, any given individual could be experiencing either more connection or separation. Before these two polar aspects of the whole *being* can be integrated to dissolve the paradox, they typically appear, divided in form, and often expressed through two distinctly different types of people. Some individuals believe and live as if they are fully separate and competing, and some recognize their connections and inter-dependence and live their lives accordingly. Of course these are the extreme polar views and almost all of us fall somewhere in-between, including those that already recognize that we are simultaneously both. It is interesting to note that throughout most of recent history there seems to have been an overall balance in the numbers and influence of these two types of people. This makes sense because neither is right or wrong and both are just positions for exploring a part of the story. One clear way that we see these differences expressed is through our politics. In the United States, this division can be noticed in the platforms and actions of our two primary political parties. Over the years, these two "halves" of the political whole have been represented by an almost equal number of voters with only the small fluctuations around the middle affecting the swing of the political pendulum.

VICTIM OR PERPETRATOR – TWO SIDES, SAME COIN

I recently discovered an unexplored cache of judgments that I had formed about others. (No matter how many rocks are overturned and revealed, there always seems to be more waiting to be found.) The key to this discovery was a recovered memory of the drowning of my best friend when I was only eight years old. For almost sixty years, I had "blamed" our neighborhood "bully" for the "murder" of this childhood friend. This same "bully" physically beat me almost every day after that "murder;" a trauma-inducing pattern that continued for several years until we moved from that neighborhood. Years later when I learned that the bully died in an armed robbery, I quickly and righteously adopted the attitudinal belief that "he got his just rewards." In my mind it was "case closed," and there it remained, hidden until a few months ago when I was unexpectedly inspired to reexamine the entire incident through a different *viewport*: that of the perpetrator's perspective and feelings. It was a surprise to be reminded how much things can change with a single, and simple, change in perspective, and how many tricks our old memories can conceal.

My sister has a great mind for recounting historical details. She remembered being told that from her kitchen window, my mother saw the "bully" and my friend walking towards the bay at the end of our road. Later, when the "bully" returned alone, he looked nervous and quickly ran to the woods behind my house. My worried mother ran after him, and once found by my mother, he "confessed" that my friend had fallen into the water and that is why he became frightened. After discovering the drowned child, my mother quickly decided that this "bully" had pushed my friend in, and soon the entire neighborhood, looking for a rational reason to explain their loss, adopted and solidified my mother's perspective of blaming this boy. That young "bully" was only a small, frightened, eight-year-old child himself, and must have felt horrible and very isolated, regardless of his culpability.

My healing revelation came to me through a short dream involving a "movie" that might have taken place within the child bully's mind; in the dream, I became that boy. In this "dream movie," my playful boyish prank of pushing a friend into the water quickly turned into shock, surprise, and then panic as I, now the perpetrator, discovered that his young friend, unexpectedly, couldn't swim. The setting was particularly difficult and dangerous because of exceptionally tall bulkheads and lack of ladders in the area that made any rescue attempts difficult or impossible. Later, his anger at being labeled a "murderer" focused on my mom and was expressed through the only effective outlet he had at his disposal: repeatedly beating up her son. As the "movie" cycled forward through *time*, it became clear that being labeled this way by the entire community at such a young age became a self-fulfilling prophecy, which eventually contributed towards propelling him into a future that included armed robbery.

My new and sudden breakthrough was initiated by being able to better recognize the bully's perspective and seeing that it could have been any of us pushing this friend into the water; all the boys in that neighborhood played with each other in that rough and careless way. I saw that my own unexamined and definitive judgments likely contributed to the "bully's" ill-fated and shortened life, and this also meant that I, too, was a perpetrator.

This was a monumental realization for many reasons. This dream initiated a strange period in my waking life where I found myself in a series of dramatically staged situations that forced me to have very direct interactions with over a dozen people about whom, not coincidently, I still unknowingly held judgment. The process lasted for a month and there were days when this odd experience unfolded multiple times. Before this process, if I had been asked about "harboring any old judgments" I would have scanned, and replied, "None that I can think of," and yet this single clear revelation opened my memory to an entire hidden treasure chest that was full of them. The path to personal freedom is a journey that never ends.

RECOGNIZING BOTH VICTIM AND PERPETRATOR

As we begin to explore our personal history of trauma, most of us will begin our journeys by recalling the times that we felt victimized and remembering the wrongs that "others did to us." This is a very natural and expected reaction; most of us begin this part of our journey by discovering and exploring our *victim consciousness*. At some point we encounter the obvious: if we are "all things" and we seek "all possible experiences," then at this new level of consciousness we are not only victims, but we are also the perpetrators. To understand any *dualistic* situation fully, we have to be able to identify with and understand all aspects of the polarity; the perspectives of both the victim and the perpetrator must be recognized equally and together, along with everything in-between.

With this type of simultaneous exploration of all sides of an issue, there is an interesting and unexpected side benefit for our healing. When we work through only one side of an issue, we are addressing only that one side of any embodied tensions. Because we are only addressing part of the *energy*, physical releases can then create an unbalanced state in the body until there is more release or enough *time* to adjust, compensate, and re-balance. However, if we address and can recognize the entire issue as a unified whole, together with all its conflicting parts, then our releases tend to unfold and release in a more balanced and symmetrical way.

At a deeper level, we realize that taking the longer and more difficult path by only exploring one side of a polar division is never actually a "problem" because we are still gaining experience; there is never a need to rush. *Since all perspectives are necessary for wholeness, simultaneously exploring and embracing all sides of a new awareness, while not "better," can allow for a smoother, more balanced and more joyful process.*

WAR AND OUR CULTURE

Beyond the fundamental existential types of trauma that involve health, life, and death, lie the myriad of varied, energetic, and creative ways that our human culture creates and compounds additional layers of harmful trauma. One of the most significant human contributions to deep trauma is our deep, disruptive, continuous, and *infinitely* damaging "culture of war." From the ruins of war emerge every type of human traumatic experience imaginable, including extreme subjugation, sexual abuse, and physical torture. All of this is created, promoted, and enacted entirely by, and upon, ourselves; nothing outside of our minds and actions is ever involved, and much of it has even been codified into our laws.

Robert Ardrey, the playwright turned anthropologist, convincingly argues in his two books, *The Territorial Imperative* and *African Genesis,* that our brains and bodies developed their present form because of the evolutionary imperative to become better at war. We evolved stronger bodies to wield more-damaging weapons, but what is most important is that we developed more intelligent brains largely to design more powerful weapons and then the genes of the most successful victors survived, thrived, and multiplied. Through genetic adaption and *pre-adaption*, mankind's genetic makeup has, therefore, been continuously shaped to favor those who are most successful at war. War has been such an important part of our evolution that it can be found throughout our DNA.

Our entire recorded human history is been largely defined by our history of war. For thousands of years, mankind has been repeating this same damaging approach to resolve its conflicts. Fueled by our predominant belief that we are separate, competitive *beings*, conflicts inevitably arise. We first try to resolve our conflicts by attempting to "convince" the "others" that we are right; and when this does not work, we become increasingly aggressive until, at the level of larger groups, we eventually go to war to subjugate and often kill all those whose opinions differ, perpetuating this same traumatic pattern that has shaped our entire history. Many of the world's leading religions glorify and celebrate the idea of a "noble war." Our deepest political, religious, and socio-economic problems are the direct result of our reliance on this brutal, one-sided, ineffective, and unbalanced method for solving our problems. War has been, and is still, central to our cultural paradigm for many lesser reasons, but ultimately all these involve a complete misunderstanding of our deep interconnectedness. Driving this misunderstanding is our evolutionary competitive nature, which our culture keeps alive through its constant reinforcement.

This universal confusion about the nature of our *separation* has created an environment in which brutal warfare is viewed as a rational and even normal way of engaging and solving problems. Our economic power and political systems are, and have long been, completely intertwined with the enormous and financially very rewarding business of making war. Individuals and nations have historically used war as a quick path to enrichment, often hiding this purpose under some manufactured, but very motivating and easily manipulated, guise involving nationalistic or religious persecution and conflict. Our world has experienced many wars that were advertised to be the final war: "the end of all war," but these too just added more layers of suffering, resistance, and deeper conflict. We are feel it is an advancement to have codified "legal" international rules of warfare defining the correct ways to kill other people. How do we consider this civilized? This fact alone demonstrates that our prevailing culture is conflicted, unbalanced, and incorporates little understanding of our deep and profound connectedness. *Our cultural normalization of war is one of the clearest indicators that our worldview is far out of balance.*

From all corners of our world, this pattern of endless war continues to generate a universal, very loud, and pained human cry. From mothers, fathers, and all those who understand the importance of love and connection can be heard the pained cry: *"There must be a better way!" Of course, we know deep down inside that there is another way, but making this change seems to be impossible for humans. This is because our existing culture and belief system does not, and cannot, prepare us for the actual type of change that is required.* To change we must first understand the deep and hidden reason for this pattern that continues to create such endless and deep human suffering. *Our problem is the mistaken way we are taught to view the "self." Our culture has taught us to see ourselves as alone and competing with "others" for our very survival. It is a cultural view based mostly on struggle and fight so that is exactly what it manifests.*

FIGHTING BEGETS ONLY MORE FIGHTING

The world's dominant cultures have all been built upon and continue to encourage the splitting and judging of *dualistic* pairs such as "right and wrong," "ours and theirs," and "good and bad." We are repeatedly being encouraged to engage in some form of struggle or battle involving these concepts; we constantly struggle with each other; we fight other teams, other countries, other corporations, and thankfully, at least for today, we only fight other worlds in our movies. We fight against fighting, struggle to end the struggle, resist those who resist, punish those who punish others and never see the ironic and self-perpetuating nature of these culturally normalized reactions and responses. Some of us are motivated by power, some by money, some by fear, and many by all these. When and where we are not involved in outright war, we are still maneuvering, provoking, strategizing, and engaged in economic battles with each other. We are taught, trained, and genetically bred to be better fighters: to be more aggressive, to push harder and longer, to be more deceptive, and to use more destructive weapons. When empowerment and enrichment are the primary goals, then the deeper questions: What is really being won and what price is being paid? are usually ignored or spun. Human death and suffering are coldly quantified to dollar amounts and labeled *collateral damage,* and then justified as a "fair price" to be paid for whatever has been presented as "the prize." Politically, and in our iconic stories, war is usually portrayed as a classic struggle between the forces of "good" and "evil," with the home team always being understood to be the "good" or righteous side.

Within the vision of an *interconnected holographic universe*, this concept of a "good" or "righteous" war to solve a disagreement can be instantly seen as false narrative, designed to manipulate primal human nature: a type of propaganda only made possible by the widespread blindness to our interconnectedness. Since all warring factions throughout history have been convinced that their perspective was the righteous one, wars can never resolve our differences.

Yet struggle and fight define our individuality, our corporations, and our nations. We entirely miss that even when we "win," we also must "lose," because ultimately we are only fighting these battles with the divided parts of our greater selves. Our culture teaches us to explore and define ourselves through this process involving *dualistic* struggle, but because of the very nature of opposites, our lessons only fan the flames for more struggle in the future. ***Opposing forces feed off of each other, for both sides only exist and grow in the presence of each other. Both "sides" are empowered by this opposition and neither "side" would exist long in isolation without the other's energetic resistance.*** For any army to be financed, built, and maintained, the appearance of threat from an enemy is a political necessity.

Without an enemy, or at least the threat of one, armies would have no reason to exist. The perception of a common enemy is often what determines why, how, and when nation-states are formed, shaped, and defined politically. Throughout history, ambitious political leaders have created, perpetuated, and exploited this idea of "enemies" to enable them to expand armies and increase their power.

As long as we continue to be convinced that it is possible to benefit by fighting against "others," there will be continuous war and no war will ever be final. "The war to end all wars" is an illogical and impossible concept for many reasons. One of the most egregious is the fundamentally flawed concept that a solution can be achieved through one side's superior physical force and violence, where what is actually required is a more holistic balance. Until we each realize that it is conceptually impossible to "fight our way into peace," war will remain an active part of our lives.

Ultimately, after a deeper and more honest examination, we discover that we are only waging war externally because we have been continuously at war, within. If we each have difficulty accepting the conflict embodied within, how could the collective soul be any different? Because the appearance of the outer world is only the reflection of our collective energetic inner world, the critical work that must be done first is for each of us to free ourselves from this ancient, cyclic pattern. ***We cannot and will not end the war without until we end the war within!***

WAR AND DISEASE—FIGHTING OURSELVES

In our contemporary world, we are not only witnessing more pressing external problems, such as more damaging wars and environmental destruction, but we are also witnessing a dramatic increase in a particular type of disease: very mysterious auto-immune diseases, allergies, and a host of new psychological conditions, all of which can be understood to be outer manifestations of our confused and divided inner nature engaged in a battle with itself. The appearance and increase of these types of problems in the physical realm are a clear and powerful message warning of our increasingly confused inner condition. It is being revealed so that we can more clearly see and understand what is going on deeper within.

The outside state of world affairs, our physical health, and our inner struggle with separation will always look much the same because the inner and outer always reflect each other – this is a fundamental quality of all holographic relationships. If, on the outside, our extreme division

and separation appear as war, great human suffering, and the destruction of our environment, and within our physical bodies these same divisive energies are being expressed as an increase in diseases that parallel this same type of struggle, (allergies, autoimmune disorders, and psychotic breaks), we can be certain that the gentle ringing phase of our wake-up call is over. We are now being slapped into attention with this late and heavy-handed warning to look more carefully at the disorder within. What we witness individually and collectively will always reflect the inner vibrational landscape; as within, without.

The chaotic and destructive aspects of the world, which we think of as external or "outside," are just the secondary projections of the perfect "mirror" that has been gifted to us so that we can fully witness, experience, and better understand our inner division, disorder, and judgment. War, hunger, and environmental destruction will eventually be understood as the direct external expression of our collective inner division and conflict, while autoimmune disorders, allergies, psychological disorders, and other physical problems will become recognized as the individual manifestations of our long unaddressed separation, judgment, and ongoing battle within our psyche.

Autoimmune diseases manifest when the body becomes confused and starts to attack itself. Through a holographic awareness, war can be understood as an ultimate autoimmune disorder: the outer expression at a global scale of our unresolved and often hidden inner conflict. To end the war without, we must end the war within. The outer world's healing can only follow the healing of the individual.

We imagine that we are at war with "others," but in the deepest truth, we are only at war with ourselves. With an awareness of our interconnection, war can be understood as the deadliest, most widespread, and most far-reaching autoimmune disease. This also means that our planet's healing, or whole-ing, can't be, and won't be, accomplished through only external means such as laws, financial aid, and regulations. The healing of the "outer" world can only begin and unfold within each individual.

BEYOND WAR

Eckhart Tolle says, *"Just as you cannot fight the darkness, so you cannot fight the unconsciousness. If you try to do so the polar opposites will become strengthened and more deeply entrenched. You will become identified with one of the polarities, you will create an 'enemy' and so be drawn into unconsciousness yourself. Raise awareness by disseminating information, or at the most, practice passive resistance. But make sure that you carry no resistance within, no hatred, no negativity. 'Love your enemies,' said Jesus which, of course, means have no enemies."*

War does teach us many things about ourselves; we learn about hate, trust, greed, anger, fear, terror, the heroic, the tragic, and love. Driven by our desire to develop more powerful weapons, we have also discovered amazing new technologies. Through warfare we learn about our awesome human power to both create and destroy; as our awareness slowly grows, we eventually learn how to illuminate this part of our darkness and then expand this illumination into a type of consciousness that rests comfortably in the "oneness" that lies beyond conflict.

Once we expand our awareness about the nature of opposites and master the practice of becoming passive and loving witnesses, our old habitual struggles seem to fall away. With practice, we gradually expand and grow that part of us that carries no resistance within: that part of us that can fully experience the ever-present *flow* of *Love*. When we do this with our full *being* and clear intention, over *time*, our vision of the world gradually becomes lighter, yet simultaneously more meaningful and joyful. From this new clear and more balanced place, the outer physical world then appears to respond by mysteriously changing, adjusting, and shifting for the better. The outside world has not changed, for all possibilities already exist, at least as potential expressions. What changes are the always existing parts that we now resonate with and therefore engage. There is very great transformational power in this mysterious, yet very real, process.

The only way to end the constant war outside is to clear the internal darkness, embrace our entire being, and calm the struggle within. The journey towards a more peaceful world begins and ends with our learning how to let the "light of consciousness" illuminate our inner shadows.

HIGHER VIBRATIONAL CONSCIOUSNESS?

Some contemporary philosophies promise things such as "one day, our *vibrations* will increase in *frequency* to the point where much of humanity will shift into the higher *vibrational* mode of pure goodness." A related belief promises that "one day a "critical mass" of higher consciousness human *souls* will shift the planet's *vibration*, and this will instantly and automatically change the thoughts and actions of others so that everyone on our planet will be aligned and operate at this new level." Proponents often cite, but misunderstand, the Japanese WW II "Hundredth Monkey" experiment as "proof" for the validity of their claim.

While this idea of an Earth that can function with only what we generally consider to be "good" will and "good" vibrations, will appeal to almost everyone's senses, it is not a reasonable, valid, or possible outcome for many reasons. The *dualistic* universe simply does not function this way: instead, it always seeks a balance in all things. Physical life also requires that all change originates from within, not without, because the physical world is only a reflection of our inner state; the external world is only the secondary expression of our internal state, so its appearance will not change until the source within changes.

An unbalanced condition of just "good," or "higher," without all the other parts of a complete *dualistic system,* would not be sustainable for a physical universe such as ours. Lacking necessary balance and contrast, such a one-sided system would be radically lopsided and begin oscillating wildly. Because everything in the physical universe cycles and seeks balance, this unbalanced condition would begin to wildly shift back and forth until a more balanced and stable state was eventually reached. As the planet and universe re-adjusted itself to find balance, the swing of this pendulum would plunge us into equally long periods where our Earth was dominated by what many would consider to be very "bad," or negative, influences. While very natural, this desire for our physical universe to maintain a state of imbalance in this fundamental way is the only a product of confusion about the true nature of existence.

As cultures shift and change, the average centering point for the "goodness-badness" dynamic will have a natural ebb and flow but there will always be the drive to achieve a balance around the center. This is what defines and maintains the basic *dualistic* structure of our physical realm. ***By necessity, our physical universe must always allow for all possible expressions, the "good" and the "bad" alike.***

Our Universe can Change in an Instant

This new understanding of our universe brings an *infinite* supply of "good" news. Because *duality* must be built from a full range of possibilities and opposing *dualities* interact in ways that we find unpredictable, our universe is revealing itself to be a much more interesting, fluid, and *magical* place then we could ever have imagined. ***There is even better news. Since the outer*** *world is a reflection of the inner, we don't ever need to wait for "others" to change themselves for our world to completely change. Once we change within, from our viewport the appearance of our "outer" world also simultaneously changes. As within, without. It is not the external physical world that changes. Instead, as we shift our resonance, the aspects and qualities of the outer world that we interact with the most will be different. While it seems like the outside world has changed, it is only our perception of the world that is different.*

CHAOS THEORY MISSES THE DEEPER POINT

Because of the instantaneous and complete intercommunication of everything in existence, there is no such thing as an accident in our universe; everything is always interrelated and causal. Because our experience of the world is entirely the result of existent energies and their interaction, nothing can ever be accidental or random.

Chaos theory recognizes what is called *emergent properties*; but this just means that the theory recognizes that small changes early on in a process can build to have a dramatic impact on the final results: a characteristic some have named the "butterfly effect." Because humans don't have the capabilities to understand these many complex relationships, *chaos theory* then concludes that the universe must be random. We are once again looking through our old narrow *viewport* and as Einstein reminded us, we can't solve problems using the same type of thinking that created the problems.

WANTING TO FIX THE WORLD

While the embrace of everything in existence is an important cornerstone in the foundation of the deep healing process, it also is the most difficult idea for compassionate individuals to understand or accept. Our natural reflex is to want to fix the 'things" that we think are broken and push away that which we judge to be unpleasant, painful, hurtful, or horrible. A deeper exploration eventually reveals that the human condition does not need fixing, but reaching and abiding in this type of deep knowing requires time and experience. Life, just as it shows up, is the perfect gristmill for the evolution of human consciousness. Through it, we learn how to trust and follow our hearts.

If an individual feels called to devote their life to projects, organizations, or acts that directly aid "others" through their heart-felt desire to relieve human suffering (at least for the short term), then of course, this is exactly what they should be doing. Through these interactions and expressions of compassion, those giving and those receiving will all have wonderful new experiences that will stimulate and expand their evolutionary growth. However, if the action is taken with an expectation of changing or "fixing" some part of the human condition, then this "fixer" will also be engaged in a learning process about the nature of the human condition and our expectations. (See "The Nature of Our Expectations")

The deepest healing asks that we not only overcome our natural reflex to push away, reject, or try to "fix" what we consider to be the broken or unpleasant parts, but that we also take this process one bold step further. Our wholeness is asking us to recognize the important contributing role of these negative aspects for the expansion of collective consciousness. This initiates a process that eventually leads to the release of all judgment and serves to help re-integrate lost parts of our being. Everything in existence is a part of us, as we are also a part of everything. As within, without; there can be no other path to wholeness.

This is an extremely difficult principle to understand and fully accept, so it is one that will also take *time* to integrate. As always, it is best to process radical change bit by bit. Each challenging principle must be fully understood, and embraced internally, and only then does the next level of healing become possible.

ASCENSION

The need to facilitate our physical "ascension" from this world is another old but currently very popular belief that is now being recycled through new methods, teachings, and workshops. Major religions have always promoted the idea of just enduring this difficult physical expression of life so that, later, we can be rewarded with a free trip to some form of better paradise. There are even some confused adherents that are hoping for Armageddon to arrive sooner, believing their "rapture" to some higher "holy" place will be facilitated by the burnt ashes of this world; a few go so far as to engage in actions that they hope will speed up this process. Others believe that killing those that they deem "unholy" will elevate their status, while most adherents are just quietly praying to be transported to a different type of place, one that lies beyond what they see as a "horrible, limiting, and troubling world." While having this feeling of wanting to get out of this crazy world is very common, today this escapist view is, once again, being repackaged and marketed as a new age "technology" that promises to facilitate or hasten our "ascension." Offered as an expensive workshop, "ascension technology" is just the latest example of this long misunderstanding, but one that has recently been successfully commercialized.

While this book discusses the existence of other or parallel worlds that we can become free to enter and enjoy, there are enormous differences between this awareness of the new *architecture of freedom* and these escapist beliefs. We can't visit imagined worlds by simply leaving others behind; new worlds are only available to us through our embrace of everything in existence, including our current physical world, exactly as it appears. While not as harmful as blaming, judging, hating, and murdering others in our quest to reach "heaven," escape is never a pathway to expansion; our love and embrace of the world just as it appears to us, is.

The *holographic* nature of our universe (as within, without) means that these other worlds and places can only be accessed by first honoring and embracing every aspect of life here on Earth. If we wish to access other worlds, the required journey is always initiated from within, not without. To access these other worlds, we must first practice expanding our awareness to be resonant with the deeper nature of our existing world. We must fully understand that the world that we are experiencing, right now, is being created solely by the quality of the *energy* that is, right now, "vibrating within."

Miracles are only accessed through a path that includes a deep love and appreciation of life, exactly as it appears. All that is sought is only found right here, right now; this also means that there is nowhere to go and nothing to leave. **Our world will appear to change only when we expand our beingness by illuminating more of the shadows within. Because the appearance of our outer world is only the reflection of the world within, each of us is the only one that can change our world. Nothing could be simpler. Nothing could be more perfect.**

NON-DUALISM

Non-dualism is an ancient spiritual philosophy, inherited largely from Eastern traditions, which describes an obtainable state of consciousness where there is complete awareness that the separation of things and *time* is illusionary, while the eternal nature of the *soul* is not. *Non-dual awareness* often first appears in a given individual's life through spontaneous, but fleeting, moments of insight. These brief encounters with this much deeper state of reality may seem disconcerting, disturbing, or even psychotic if the recipient has not been properly prepared, but over *time* and with training, this state of *being* will become more available and recognized as fully transformational. The ultimate, but elusive, state of being that some believe can be reached through this type of *non-dual awareness* has been called *enlightenment*.

Some religions and spiritual philosophies encourage adherents to completely reject our *dualistic* world and instead only seek that which is *non-dual* or permanent in nature. Followers are encouraged to abandon or avoid everything involving our physical, or *dualistic,* world, and instead immerse themselves in more "pure" thoughts centered on *non-dualism* or God. As observed through thousands of years in monasteries and retreat centers, this can be a useful practice for gaining experience, but ultimately it must be recognized for what it is: only a temporary retreat from the foundational *dualistic* nature of our physical lives. These extreme versions of *non-dualism* are inherently unbalanced because they reject life's most valuable gift: the rich experiences available for building deeper self-knowledge through the living of a full life. By encouraging followers to avoid and dismiss worldly experiences, extreme *non-dualism* misses this primary purpose of life in this realm, which is to gain this direct experiential knowledge of, and for, our greater *being*.

Some *non-dual* traditions even go so far as to view *dualistic* thinking as "bad" and *non-dual* thinking as "good." This is an amusing, but interesting, irony, since this type of labeling or judgment of "good" and "bad" is about as *dualistic* as one's thinking can be. While these extreme ascetic practices can be very useful for various stages of spiritual development, they can also lead us to miss the greatest invitation and opportunity of life. Through its diverse tapestry, woven and shaped from *infinitely* rich and meaningful *contrast*, life is inviting us to fully explore and appreciate the world exactly as it is being presented to and for us in each, and every, moment.

Our lives are lived most beautifully when a deep awareness of non-dualism can be held close to our hearts while we are simultaneously living full lives in our dualistic universe. This is the type of awareness described in my first definition of *dualism*, at the beginning of part three. There exists a peaceful and permanent state of awareness where both views can be held together in harmony without creating the *paradoxical* struggles that lead to more *cognitive dissonance* within. Today, assisted by the exponential rise in opportunities for broad experience and enhanced intercommunication, a particularly large number of individuals have found, and can rest peacefully within, this expanded state of *being* while also living full and busy Western lives. They are found everywhere in our world, completely invisible to those still wearing the blinders of *identification* and *attachment*.

CONCEPT OF "SIN"

The Christian concept of "Sin" has a very different meaning at the level of the *multiverse*. Our Christian-shaped culture, defines "sin" as a violation of divine law, sometimes with the additional clarification of sin being an act that harms others. Because there are actually no "others" within the "oneness" of our fully connected *multiverse*, from this perspective, "sin" can be understood, instead, as any act or thought that is not aligned with our own best self-interest. ***Sin is, therefore, only the natural and inevitable result of not knowing, trusting, or following the deeper truth of our interconnectivity and "oneness."*** When witnessed from this deeper level of *being*, sin can be understood as nothing more than an unconscious form of careless "self-harm," not different in intent or meaning from when an unaware child walks out into a busy street. Because no "others" are ever actually involved, even the most heinous appearing acts against "others" ultimately can be understood to be only a form of self-abuse. We "sin" because we live in fear, not understanding, or fully trusting, our interconnectedness, our eternal *beingness*, and the deeper nature of life. ***Our "sins," are the result of nothing more than living in fear and still believing that life happens to us, rather than for us.***

Because we are all just different parts or perspectives of one single *being,* at some subconscious level of *awareness,* we all sense that any act directed against "others" is, ultimately, only an act against ourselves. The "sinner" is simply not yet conscious of this. One reason that we may feel guilt, shame, and pain in association with our "sins" is because, at a deeper level, we intuitively understand that our sins are really just acts against ourselves: self-abuse, and with these actions we have just created even more suffering and difficulty for our *Self*.

As we mature and expand to consciously recognize the deeper truth of our "oneness," what we once labeled as "sin" can be, instead, understood as little more than an "inefficient" expenditure of *energy* and *time*, the normal and expected result of taking a slightly less than direct route while exploring the *infinite* possibilities of existence. As we grow, evolve and connect through our experiences, we begin to see how nothing real is ever really harmed or lost through our "sins"; the only important thing happening is the expansion of awareness from each *present moment* experience. ***The most egregious "sins" can then be seen as little more than temporary wrong turns and*** *inefficiencies during our inevitable and collective journey towards freedom.* Because we learn something new, and therefore grow, from all of our experiences, the process of life is always working perfectly towards its express purpose of expanding our consciousness.

Everything that we label as "sin" is really little more than an unconscious, inefficient expenditure of *energy* when viewed from this deeper and *timeless* perspective. Ultimately, "sins" are not even "wrong" turns, because every turn that is possible must be explored and experienced for *Being* to become fully self-aware and compassionate; all "wrong" turns contribute to our self-knowledge. It is the nature of life to fill every void and explore every crack or opening in existence, and a critical part of our purpose is to contribute to this process by "leaving no stone unturned". We may occasionally become lost in the exploration of our separation, but even these misadventures add to our breadth of experience. *Since everything is intimately connected, everything is also completely self-correcting as every natural system will return to equilibrium over time. Therefore, there can be no "sins," mistakes, or wrong turns within existence; Being is witnessing, learning, and growing more compassionate with every type of experience, including our "sins."*

LIFE IS PERFECT – GET OUT OF OUR OWN WAY

How are we each able to recognize the perfect outcome, adventure, activity, challenge, event, or person for us? *What is best for each of us is really very easy to recognize, because it always is whatever is showing up in our lives at any given moment.*

There is a special type of wisdom that arrives when we are finally able to integrate two lesser awarenesses. The first is when we fully know that we always had, and still have, the ability to choose any path that we desire; we fully understand our awesome ability to create. *The second is when we simultaneously realize that the most beautiful creations are those which occur when we finally let go of our desire to control the outcome.* These combined awarenesses work together to form a new and beautiful threshold of understanding; one which is inevitably discovered at some moment within every individual's journey towards personal freedom. *We are the creators of everything, but our egos are always inserting themselves into the creative process.* They are busy sabotaging, controlling and limiting our *infinite* possibilities. *Our self-created egos, by the very nature of their limited vision, lack the wisdom and depth of vision to ever understand*

what is best for us. Instead of allowing the universe's perfect wisdom to unfold, they choose to manifest through their perspectives, which are full of their own fear-based egoic desires. This is an automatic and habitual practice that certainly adds a lot of diversity and color to our lives, but it also leads to a great amount of human suffering.

As many sages have reminded us (including Ram Das in his last book, *Polishing the Mirror*), life does not happen *to* you, it happens *for* you. When we allow life to *flow* naturally, free of "mind-made" obstacles, our lives unfold in the most unexpected, yet perfect, ways. At those times when our lives are not *flowing*, and difficulty after difficulty appear, we need to remind ourselves that these "problems" are just being presented to help wake us up; they represent the best possible experiences for expansion and growth at that particular point in our journey.

As we open to let the light of consciousness in, we illuminate all that we choose to focus upon. If we choose to focus our attention on drama and intrigue, that will be what we see in the world because that is what becomes most illuminated. If we focus on "they" and how badly "they" are treating "us," then this focus, and the inherent separation it defines, is exactly what becomes illuminated and more real. On the other hand, if we illuminate what we love and what brings us the most joy, then these outcomes are reinforced to become a greater part of our lives. **This is an extremely simple but infinitely powerful principle. "Anything, and everything, that we focus our attention upon, becomes a more substantial part of our reality." The phrase, "Where our attention goes, energy flows," has been repeated by teachers around the world. We must be careful and aware of where we focus our attention; it is a very powerful life-shaper.**

QUANTUM MYSTERY NOT EXPLAINABLE

Today it is very common to hear public speakers and authors referencing "quantum healing," "quantum energy," "quantum shifting," and similar references. The term "quantum" has almost become synonymous with the magician's hat, from which anything can be pulled. What all of these speakers are really saying is that their argument, technique, or method is utilizing newly discovered relationships, forces, and *energy* forms that are real and repeatable, but also, at least presently, unexplainable. **No person— absolutely no one—understands or can use words to explain the real meaning of the experimental results that have consistently appeared throughout the history of quantum physics. Its deepest truths reside in**

a type of space that lies far outside of our thinking, words, and concepts. However, while it may be impossible to understand, it is very possible and extremely valuable to learn how to better work with, instead of against, this much different reality.

QUANTUM SPACE AND QUANTUM FIELD

> "Out beyond ideas of wrong-doing and right-doing
> there is a field. I'll meet you there." *Rumi*

One of the most recent sub-theories of *quantum physics*: *quantum gravity,* is a well-known recent attempt by physicists at combining *relativity* and *quantum physics*. The unification of these two important theories requires finding a way to combine *gravity* with the other three known forces: *strong* and *weak nuclear forces* and *electromagnetic force*. *String theory* is an earlier sub-theory that also shows promise for this type of unification. Both of these sub-theories are looking deeply into the *quantum* mystery to determine its meaning and they are both attempting to describe the smallest things that make up all of existence.

While both seek to describe the same deeper truths, they approach the puzzle from quite different *viewports*; they are approaching the problem from entirely different types of *dimensional spaces*. *String theory* theorizes that these smallest bits of matter are tiny *vibrating* objects, string-like or membrane-like, which exist within ten or eleven dimensions. *Quantum gravity* is examining Einstein's *spacetime* (four-dimensional *space*) and proposing that it exists as a result of a *quantum field* or *quantized matter field*, which when disturbed, produces discrete packets or *quanta* that can act as tiny *particles*.

Quantum theory began in 1900 when Max Plank unexpectedly realized that *electromagnetic radiation* was *quantized* or clumped into well-defined packets. Through experiments, he discovered that his measured results only made sense if *energy* was released in small discrete amounts, called *quanta,* which always contained an amount of *energy* that was inversely proportional to the radiation's *wavelength*. The fixed proportional constant that he measured one hundred and twenty years ago still stands today as a firm foundational principle of *quantum theory*. Early *quantum* experiments that observed and measured *photons* determined that these packets of *energy*, or *quanta,* were like *particles* because, when these *quanta* strike *electrons*, they knock them into motion like billiard balls, causing the *photoelectric effect* that is now so fundamental to the solar electric industry.

However, at the same *time*, other experiments confirmed that these same *particles* called *photons* also acted like *waves*. This confounded scientists, for how could this *photon* be both a *particle* and a *wave*. Eventually, physicists realized that this *wave-particle duality* could be explained by *field theory*. They concluded that these *waves* were simply disturbances in the already discovered *electromagnetic field,* which only later appear as *particles* when they finally interact with something.

Then physicists discovered that *electrons*, a component of all *matter*, also showed the same behavior. Over *time* they also found that *protons, neutrons*, and even *atoms* and *molecules*, some quite big, also behaved in these same strange ways. Suddenly, these "strange-behaving things" were no longer just limited to the *particles* associated with the long-recognized *electromagnetic field,* meaning there was now a need to expand the earlier *field* theory.

Many physicists now believe that a greater and more fundamental *field,* the *quantum field,* sits at the root of all physical existence. When it becomes excited, moves, or *deforms,* it then creates all the *particles* that shape our universe. ***This new understanding means that everything we perceive as physical matter is only a secondary result, or image, created by this vibrating quantum field.***

This *quantum field* fills the entire known universe but still acts like a single entity, communicating with all its parts and extents instantly, as it produces all the separated *particles* that we then experience as our reality. This *field* occupies a *space* as large or spread out as all of existence – a pretty impossible conclusion in a *time*-ordered universe of immense or *infinite* size, but much more "reasonable" in a universe built upon a *timeless oneness*.

Both of the major and well-tested theories of modern physics: *quantum physics* and *relativity,* are more accurate descriptions of our flexible interactive reality than the rigid and fixed deterministic view that evolved from Newton's description. However, because both are trying to describe "the indescribable" from the very limited perspective of our rational thinking processes, they appear to be very different, and sometimes even paradoxical, descriptions. They do meet and join, but only outside of our understanding, beyond our *conceptual horizon,* in a *field* of greater *oneness* that scientists are naming the *quantum field*.

EARLIEST RECOGNITION OF OUR "CONNECTION"

One of the most important ideas for initiating deep change is the recognition of our profound connection to everything and everyone. The only lasting way to understand this type of "oneness" is through direct experience. The more *whole* we become, the more we can experience our connectedness.

By definition, "everything," must include all that we judge to be good, bad, holy, or evil: the entire enchilada that is life! The most profound and deepest type of healing is always built upon rediscovering and reclaiming our wholeness—all of it. When any parts of our *being* are hidden, ignored, banished, incomplete, or missing, we suffer. Healing can only unfold once we recognize, embrace, and fully integrate this life-changing awareness.

The recorded history of our human awareness about this deep interconnection between everything dates back to the scribed "Hermetic wisdom" of ancient Egypt. This ancient philosophy was founded upon a recognition of an "all" or "one" that exists beyond the material cosmos: a deep unity through which the entire universe participates. Legend proclaims that these principles were first written upon an Emerald tablet and then passed down to *Hermes Trismegistus* (Thoth/Hermes) approximately 5,000 years ago. One translation of its most basic principle beautifully parallels the *holographic principle* of modern physics: "That which is Below corresponds to that which is Above and that which is Above corresponds to that which is Below, to accomplish the miracle of the one thing."

Today, in the light of the advances of our latest physics, these ancient principles resonate more than ever. More than 100 years ago, at the same *time* that *quantum physics* and *relativity* were being "born," this ancient wisdom was again translated and reinterpreted into the seven principles of the *Kybalion* as part of the *New Thought* movement. These seven revised Hermetic principles can be summarized as follows: Everything is (1) mental (thought creates), (2) *holographic* (as within, without), (3) cyclic, (4) polar (dual nature), (5) vibrating, (6) causal (cause and effect), and (7) gender-based (again polar). These ancient principles, not coincidentally, parallel many of the twelve arbitrarily divided principles described earlier in this book.

Throughout history, humans have wondered whether our lives are restricted to just the "three score and ten" years of a single lifetime, or if the human *being* has a greater connection to an *infinite* existence that lies beyond the limits of this physical world? Since the times of Pythagoras (2500 years ago), there has been a lively documented debate about whether the *soul* is a temporary creation of the body and mind, or does it have an existence beyond? Does the *soul* perish with the body, or does it continue after the body dies in some form? This question has always been the most profound and frequently discussed topic about our existence. Do we continue our existence beyond this earthly form?

Most world religions were formed upon these Platonic and Hermetic ideas, yet throughout our history, instead of the different faiths focusing on what is common and unites us, powerful, competing, and self-serving interests have often used these same religions to intentionally create more confusion and division. To manipulate large masses of people for their purposes, political and economic special interests have weaponized faith by exaggerating and highlighting our perceived differences, instead of focusing on our common roots, creating great conflict instead of bridging divides.

Despite special-interest influencing and manipulating the world's major religions, these important connecting principles still find room for expression. Ideas that are similar to these are easily found within the earliest discovered Christian records such as the gnostic scriptures of Nag Hammadi. This idea of our deep interconnectedness is also central to much of the recent *gnostic Christian* revival. References to our connectedness and the illusionary nature of life are found within the more scholarly and esoteric writings from every major faith, as well as Buddha's direct teachings. The Hindu faith is clear about the illusionary nature of the world (Maya), how life is like a "play" (Lila), the appearance of separation as just a temporary phase, and our deeper oneness. The Sufi Muslim Poetry and the ancient Chinese philosophy of the Tao read almost like they are direct translations of modern *quantum* philosophers. Buried deep within every world religion are the same profound insights that currently are being illuminated by our newest physics. The language and words are different, but not the ideas. These are ancient ideas that periodically resurfaced to gradually prepare us for the impossible-seeming implications of our deep, eternal, and profound interconnectedness.

We live in a fortunate *time* because, today, with the discoveries of twentieth-century physics and the internet, there are many options for fresh insight. Until this recent revolution in our sciences, outside of religion, there were few places to turn for guidance or answers to these, our most profound questions. Today there is a steady stream of new experimental data that is revealing new, but very strange-seeming, possible arrangements for our universe, all of which signal the existence of interconnections that were once unimaginable.

Relativity and *quantum physics* have opened previously unseen doorways, revealing fresh vistas of these ancient, but *timeless*, ideas and worlds. This extent has always existed beyond what we could imagine with our human minds, but now our technical skills and computer sciences have illuminated new ways to grasp this reality. These discoveries have shattered our conventional beliefs about *time*, *space*, and separation, raising new perspectives like, "if *time* is not absolute or marching as we thought, then what is so absolute and final about the *time* that we die?" This new shifting awareness forever changes the trajectory of our thinking and acting and, therefore, our entire lives.

ENTER "PARALLEL UNIVERSES"

"When you are transformed, your whole world is transformed, because the world is only a reflection." *Eckhart Tolle*

From our perspective, one that is locked into time and space, something unexpected occurs as we gradually shift our core resonance; as we slowly illuminate our shadow world, the externals of the world around us also appear to change. Other players in our lives also seem to have quietly shifted as events unfold in completely unexpected and surprising ways. What is happening? Why did all these "others" appear to change just as we did?

One very important and interesting interpretation of the *quantum* mysteries is Hugh Everett's 1957 *Many Worlds* Interpretation (*MWI*), which proposes that when we make any decision or take an action, the universe, itself, actually splits and completely divides. (This and other interpretations of *quantum physics* are discussed in the science addendum.) The *MW* interpretation means that all possible outcomes from every possible decision (all paths taken and not taken) manifest as new worlds and universes that are equally real; an entirely new full universe is created for every choice, but these now-divided universes also have a common history up to their *time* of division. The *Many-Worlds* interpretation proposes that all of these alternate histories and futures are equally "real" and that each represents an actual "world" or "universe" that manifests somewhere. By extension this also means that there is an *infinite* number of parallel universes; and everything that could have possibly happened in our past, but from our perspective did not, has actually played out in some universe that exists right along with the one that we are directly experiencing.

This is wild and impossible stuff for our rational minds to believe but if, as modern *quantum gravity* theory proposes, there are no actual things in the universe, and instead there is only the *quantum field* that produces *particles* whenever it becomes "excited" or "disturbed," then this wild *MW* interpretation suddenly becomes much easier to visualize. An *infinite* number of *parallel universes* existing in physical form side-by-side seems a bit crowded and hard for our rational minds to digest; but an *infinite* number of equally possible worlds held in a state of potential expression by the *quantum field* is much more comprehensible, especially after witnessing the potential of this through digital *virtual reality (VR)*.

If it is our attention and awareness that excites the quantum field, then what happens when we simply change our awareness? Through this changed awareness, we would excite the quantum field differently and have instant access to an entirely new world. We can understand this type of radical "world" change much more easily, now that we have seen how quickly "worlds" can change within *virtual reality*. Because we have been enculturated to imagine our "real" world as solid and enduring, we are hard-pressed to believe such a change is possible; but if we are playing or working with computer-generated virtual reality, we are quite able to accept a rapidly shifting external world. *What if our "real" world works more like VR?*

With this new metaphor, we can now see that a shift to an entirely different "world" can be as simple as simply changing a few pieces of basic *information* – the *parameters*. This new perspective helps us to more easily cognize something that is happening to our physical world, not only occasionally, but continuously. Our *viewport* which is controlled by our *core vibrational resonance (CVR)* is what determines many of our universe-shaping *parameters*. By learning how to shift our *CVR,* we are also changing these *parameters* and this causes the *projected* physical world to instantly respond; our experience of the physical world changes as our *CVR* shifts. To be able to *flow* within all of these new and constantly changing worlds requires that each of us becomes fully open and aligned with our *CVR*.

The surprise is that once individuals fully integrate this awareness, they often lose interest in their own "creating" or "manifesting" because, simultaneously, it becomes clear that the deep wisdom of the universe is already creating the best possible outcomes. The biblical expression "thy will be done" suddenly takes on a new level of meaning. *It is important to remind ourselves that the availability of information in a form that can be comprehended by humans is not a requirement for something to be true. That said, being able to visualize this impossible-seeming process through a similar phenomenon, like VR, certainly can facilitate our integration.*

Because of the way our brain works (cognitive dissonance, smoothing, etc.), from the perspectives of most people, little will change when we make these shifts to *parallel universes.* Our world will still appear relatively unchanged and we will have no sense that we are constantly jumping (or *quantum leaping*) into different universes or parallel realities. This is much the same as the way we don't experience the wild "loop the loops" of our planet through *space.* In terms that a *quantum physicist* might use, "With our intention and attention, the many possibilities for expression (*superposition*) *collapse* into a single measurable outcome." When we make this shift to a new parallel reality, the players or characters often will not change, but because we have shifted our *viewport*, the way everyone intercommunicates and relates becomes completely altered. While these differences can be very dramatic, from the perspective of the participants, the transition usually seems fluid and continuous. This effect is similar to the way travelers going the speed of light also notice no changes.

Another way to visualize the *Many Worlds* interpretation is to think of every person "already" existing in an *infinite* number of different, but already existing, worlds. In each of these similar worlds, the dominant *viewport* will be only slightly different because these "worlds" were identical until the last decision split them. The traits of each person expressed within each parallel world are adjusted because each individual's *core vibrational resonance* (*viewport*) also changes to match that particular *parallel world.* For some reason not yet understood, our awareness is only centered in one of these worlds. What if our awareness (determined by our CVR) can just hop around between worlds?

When we make changes in our lives, our *core vibrational resonance* adjusts and our awareness shifts. Our entire experience automatically adjusts in *time* and *space* to whatever already-existing parallel universe *resonates* best with our new *core vibrational resonance.* Because our experience is the result of a *holographic* and fully interconnected *projection*, most of the other "players" from our "life drama" will also have *doppelgängers* in that new universe, but their interactions will reflect the changed *resonant* tone of this new universe. From our perspective, because we are not aware that we shifted worlds, it appears that these "players" have just changed their attitude. Since all parts of the universe are interconnected (*enfoldment* and *entanglement*), our new *resonance* instantly links and connects us to all other universes and "players" that are *resonating* in the same way as we now *resonate*. Such a model also explains the frequent occurrence of phenomena such as highly improbable *synchronicities*.

One more way to imagine how this type of shift is possible is to understand that in this new parallel universe, we automatically vibrate with those already existing aspects of any "others" that are of similar *resonance*. (See *resonance* in the science section of the Appendix.) None of these attempts to use words can accurately describe the *magic* that is happening because here we are probing far into the realm of the indescribable.

This idea of *resonance* might seem similar to the popular concept of "being your higher *self*," because both ideas recognize that there are many (*infinite*) possible (or potential) vibrational expressions of our *being*. However, we must keep in mind that all of the possible forms of *self*, including what we might define as "higher" or "lower," are required for completing our wholeness. This means that no one *vibrational* version of *self* is better, more important, or preferred because all versions are critical, needed, and contributing parts of "oneness." To aspire to vibrate at only the level of our "higher self" misses this most important guiding principle of *whole-ing*. It is our "higher self" that is active when all aspects are *vibrating* together in an orchestrated harmony of wholeness. Our "higher selves" leave no parts behind.

Once again, we need to constantly remind ourselves to embrace the perfection that presents itself in every moment. Each moment, just as it is presented, is the perfect physical expression of our *core vibrational resonance.* Each moment is only being presented to us as "reality" so that we may better witness ourselves and illuminate our shadows. The entire "show" is for our benefit; it is through this clear and illuminating presentation of our inner world as our external physical life that we can fully witness, and finally "know," ourselves." **Becoming aware that our external life is shaped through constant adjustments to our personal viewport is an enormous step towards personal freedom.**

Anais Nin famously said; "We do not see the world as it is, we see it as we are." As we change ourselves, we also simultaneously change the world we live in. **This can also be viewed through the holographic principle: because of the holographic organization of our universe, any change within will also be reflected without.**

INFINITE IDENTICAL WORLDS

There is a particularly crystalizing moment in each personal process of opening to this amazing architecture of freedom when we finally understand that from our perspective there exist *infinite* worlds exactly like ours, and even identical selves on each identical world. When one of these identical selves makes a different thought, movement, or decision, their worlds split or diverge, never to be identical again after these different paths are chosen. Within an *infinite* and *timeless* universe, there will be, by definition, an *infinite* number of these other identical worlds, (To better understand this seemingly bold leap, please read the science appendix.) For every world that appears, an identical *doppelgänger* makes a different choice but, with that different choice, we are no longer aware of this divergent path because our awareness only follows one path through this *infinity* of choices. Each *time* we make a choice, our world shifts; but somehow our consciousness weaves all these changing worlds together into the fluid continuum that we each understand as our personal "life story."

This is a question that I have long been curious about: "Why does my consciousness follow only one path?" I imagine that from all my other *doppelgängers'* self-perspectives, it might feel much the same, so why is this particular thread or path the one that I "remember" as mine? One possible explanation is that all these "paths" only exist as "potentials" (superpositions) until our awareness chooses a path which then becomes conscious, by collapsing the wave of potential expressions. Is this the mechanism that allows each individual to create his or her reality in every moment?

VIEWPORTS IN THE WEB OF POSSIBILITIES

Human *beings* are not built to grasp the complete multidimensional picture of "all that is," because that was never our purpose; our purpose is to gain experience for a collective experiential "knowing." However, we are capable of recognizing some of the important implications of this magnificent vision and then adjusting and living our lives accordingly. ***Knowing that the universe is only a reflection of our inner being and that it always communicates and responds to everything that is felt, thought, or done completely changes our perspective about how we might best move through our lives.*** Many have had occasional hints, revelations, sensations, or fleeting glimpses of the world that lies beyond our normal *viewport;* these are just invitations to focus within, instead of without. If we are open to this type of subtle inner guidance, we begin making the lifestyle changes that will ultimately result in our ability to inhabit a different, larger, and more flowing landscape within the "Web of *Infinite* Possibilities."

In the twelve principles listed near the beginning of this book, my description of the Web includes the word "possibility," which refers to a specific and important idea from *quantum physics*. One hundred years of physics research have determined that, at the *particle* or material level, there is a quantifiable *probability* for every possibility that is *permitted*. As we learn more it seems more likely that *permitted* may mean something as broad as "imaginable" and if this is the case, this then means that "anything that can be imagined is possible." We are all built from these same *particles*, so this also applies to each of us—nothing is fixed, or a given, for we are all blessed with *infinite* possibilities, including that of being able to entirely alter the appearance of the world that we experience. This is wild stuff, but if we follow the physics, logic, and examine our actual experiences, this idea starts to make a new but wonderful kind of sense.

Once our individual viewport shifts, even slightly, if we pay particularly close attention, we will notice that the entire "outside" world has also appeared to change. When this happens, the "outside" world has not changed at all; what has changed is our position or perspective within its continuous Web of Possibilities. We have only changed our *viewport* (a *parameter*) and this makes for an entirely new type of experience. Through a shift of *resonance* that alters the size and extent of our personal *viewport*, we expand our *awareness* to include new, more, and completely different locations throughout the Web.

From our new expanded *viewport,* many things will persist unchanged. Nothing real is ever excluded or lost in this process because we grow by expanding our view or perspective; we only increase our "bandwidth" and add to the extent of our *beingness*. ***Nothing real is lost; all that is ever discarded are our self-limiting ideas.*** If the shift is small, most of our old friends remain, our home is usually unchanged, and we may still work for the same company. We might need to be unusually attentive to even notice the subtle changes, but as we observe the details of our life more carefully we discover that enough has changed to effectively alter our relationship with the outside world! Maybe our smile is met with a warm smile from a once-distant co-worker, our boss notices our extra work on a particular project, the dog's barking didn't keep us up last night, the sore wrist disappeared or that tired feeling is suddenly gone, replaced by the renewed excitement around some new endeavor or relationship that unexpectedly came our way.

These are small and ordinary changes: common everyday events that can easily be overlooked. Incremental, small shifts like these are more effective for healthy change because most of us would have difficulty adjusting to larger and more dramatic shifts, even those that we consider to be extremely positive. *Over time, the accumulation of many of these tiny shifts results in a very different personal adventure.* The deepening that unfolds by dropping resistance is a process; small changes continue to occur unnoticed, moment by moment until, one day, we wake up and realize that we are living a completely different life, one that is much more harmonious, joyful, and more like the life that we imagined was possible. Major shifts in awareness often sneak upon us. This is the common observation of many individuals who have committed to this lifelong process of growth, change, and opening. "Shift happens" naturally and smoothly when we become fully open to this new way of *being*.

All shifts of our position in the Web are the direct result of a change in the way we vibrate at the core of our being. The way we *vibrate* at our core determines the size and extent of our *viewport*. The common result of this type of growth is that we have a richer experience and relate to more of the "outside" world. Life usually becomes much more vivid and interesting.

INFINITE UNIVERSE

INDIVIDUALS A AND B HAVE
NO COMMON EXPERIENCE

INFINITE UNIVERSE

A EXPANDS AND NOW
THEY "CONNECT"

Individual "A" has shifted and expanded his or her awareness. He or she now shares common ground with "B." Where they once were unaware of each other and had very separate lives, they now experience connection. The "outside" world did not change, nor did individual B. We only need to change ourselves to change our interaction with "others."

Understanding quantum mechanics or relativity is not a requirement for creating or experiencing this type of personal shift. We cannot accomplish this change with our minds; while knowledge can be a helpful tool for guiding or focusing, shift only unfolds at the deepest levels of our *being* when our *core vibrational resonance* (*CVR*) changes. From our perspective, the extensive *Web of Infinite Possibilities* exists now and forever, but at its root, it lies completely outside of *time*. Because of our special relationship to *time*, this Web appears to us to be dynamic, growing, ever-changing, and evolving. Through practice and evolution, we will eventually discover that we can move consciously around the Web, and we will ultimately discover that we have the freedom to journey wherever we choose. For the present, however, we are only being asked to appreciate and build upon the simple experience of witnessing our small individual shifts of perspective.

QUANTUM WEIRDNESS – MY PERSPECTIVE

Quantum physics has given us an entirely new way of looking at and describing physical interactions at the level of small, subatomic *particles*. In the appendix, and above, I discuss the *Many Worlds* interpretation, in which all possible decisions, actions or new ideas produce outcomes that can be viewed as actual "realities" that co-exist in *parallel universes,* unfolding somewhere beyond our senses and awareness. I noted that there is always some *probability* that anything "allowed" could happen; and how, with the right *energetic* conditions, when two *masses* meet, rather than collide, *particles* could potentially pass through the vast *spaces* that exist between *particles*. In the *Many Worlds* view, even extremely improbable outcomes like this would occur in an *infinite* number of *parallel universes*. If we could learn how to shift our awareness to one of those other universes, could we then walk through walls, manifest objects such as bread or flowers in our hands, or even "walk on water?" *Miracles* like these helped to ensure the reputations of several famous religious figures, yet it seems that this ability might be intrinsic to human nature, its access requiring nothing more than a slightly different approach or mindset. *We need to seriously consider the possibility that it is only our deeply ingrained patterns, conceptual ideas, and conditioning about the solidity and fixed nature of our universe, as well as ingrained ideas about the controlling influence of "outside" circumstances or "others," that keeps us from participating in these different types of experiences. Ultimately, our "reality" is determined by what we actually believe is possible in the foundational depths of our being.*

As the reader knows by now, my personal belief about the meaning of all this, while not unique, is still different from that of most mainstream or "New Age" writers. I have the clear sense that our inability to see a greater truth is mostly because of our brain's three-dimensional focus; our brain needs to divide oneness into individual things and to *time*-order the unfolding of reality. This causes it to divide, organize, and *time*-order all the *information* that it receives, creating a sense of *time* and *space* that inhibits our awareness of other possibilities, especially that of a *timeless* "oneness." *Space* and *time*, by their very nature, imply a separation between things and events. To demonstrate, when we observe instantaneous *non-local* behavior between *paired particles*, it is not because some *infinitely* fast *information* pulse or wave is connecting the two *particles*, as the physicist David Bohm theorized. Rather, it is much more likely that this instantaneous transfer occurs because they are not two different *particles* after all; both *particles* are unified at some deeper level of existence. No matter how far apart the separate pieces appear from our perspective, a singular oneness exists at the core of existence. For another example, in that famous double-slit experiment, the fact that we are surprised when *electrons* or *photons* do not behave differently when their rate of release is slowed down to where they can no longer "interfere" with each other, is only because we always see our world through our "lens of *time*." If we clearly understood that *time* is just an organizing concept for our brains, then we might not be quite as surprised by these results.

Our perception of "oddness" in *quantum experimental results* is largely due to our attempting to view and explain existence rationally on our terms, using fewer dimensions than the true *space* that these *particles* occupy. The difference between the old black-and-white movies and the three-dimensional version of *avatar* only hints at the kind of spatial transformation that we will discover as we unlock this secret. ***The real meaning of non-local behavior is that everything in the universe is always completely and deeply connected. The appearance of separation and time passing is only the result of the way our three-dimensional brains need to separate, organize, and view information.***

Within our thinking, we can only explain multidimensional phenomena using terms and ideas solidly grounded in our current paradigm. For example, when we try to imagine what it would be like to occupy other dimensions, we still talk or think about *time*, distance, and shape. Our conceptual brains and our language alone cannot walk us out beyond these old ideas. To better understand multidimensional *space*, we need direct experience, and this requires *time* for our gradual evolution and expansion.

Fortunately, there is no need to search far and wide to find answers because we each hold within all that is needed; as J. Krishnamurti reminds us, "You are the World." Each of us can and must do this on our own; the entire universe and all its secrets are always, and only, found within. **We never have to look towards others for answers to our questions about the true nature of life, for at this level of truth, there exist no others.**

Evolution is a process; it does not happen all at once, at least within our world of *time*. A practical first step is relaxing our ties to, and need for, our old concepts; this practice makes room for many new ways of experiencing life. To understand why this is important, it helps to watch and observe how our old ideas act like anchors and hold us firmly within our old *viewport*. *Freedom comes with discovering just how and when to let go of these now-obsolete concepts that once served us so well. Freedom already exists within, but it can only surface through the discovery and release of our restricting ideas. Discovering and freeing our true self is never about needing different or more information.*

THE "MOVIE OF OUR INNER LIFE"

Our primary mission is to "know thyself." Life unfolds the way it does only so that we can better see and understand all the invisible and hidden aspects of our energetic selves. All of our adventures, all that we might consider being "good" and "bad," are manifested on the big screen that we call "life" through an elaborate presentation of visual and tactile sensation – all for the benefit of our individual and collective *being*. All of our hidden "shadows" eventually become expressed through this, the ultimate "movie" of our inner life.

Most of us reflexively blame our unhappy experiences on "others" or "outside" circumstances. While this is natural and completely understandable, this type of deflection is always an indication of a profound misunderstanding: it completely misses the most important gift of life. *Life's greatest gift is the clear physical expression of our unresolved energies through the perfectly choreographed "movie" of our lives. It is only presented this way for our individual benefit, so that we may more clearly see and recognize the truth of our being.*

Rather than understanding this and welcoming life exactly as it is presented in each moment, we usually struggle to change the way it appears. We try to control outcomes to fill some imagined need, while simultaneously blaming others or circumstances for our problems. Most of us fail to recognize that this dramatic "show" is for our benefit, so we prefer, instead, the illusion that our ego is constantly creating. Since most of us are convinced that this physical life and body are our "reality," it is quite natural to want to control the outcome.

When observed from a broader perspective, our typical egoic response is not unlike attempting to change the plot of a movie that we are watching. However, with movies, we can easily recognize it as a story; one that excites and stimulates us, but one that is clearly "not us." We are usually able to experience the movie as entertainment or *information* while remaining open to what the movie is attempting to communicate. Almost universally, we believe that the "movie" of our inner life is completely different. Having become convinced that our life movie is somehow more "real," we try to control it through our actions. This typical habitual response to control completely misses the critical central point of our lives: the clear physical expression of our shadows to create opportunities for deeper self-discovery.

LIFE ON THE "BIG SCREEN"

At a certain point in each of our journeys through this world of separation, we slowly begin to recognize, understand, and embrace a deeper understanding of our true nature. Within each of our journeys, we gradually realize that we are not our *ego*-constructed *image* of *self*, nor are we these individual physical bodies. *Eventually, we "know" that this appearance of our physical separation is just the secondary image produced by the underlying energies and there also are no "others" to "do" anything to us; we understand that thinking otherwise is to believe and be invested in the illusion of separation.* We "know" that everything that we experience as physical life is just a reflection or expression of the existing but hidden energetics of our collective *being*. *Only then, do we fully recognize this physical life as the immense gift that it truly is.*

At this point in our journey, we fully understand that all those who we once experienced as "others" are only shadowed aspects of our own deeper being that are only appearing in their separate roles to help express our deepest truths, and that this all is manifesting only for our collective benefit. With this deeper understanding, instead of blame, we experience only gratitude for how life is always aiding and supporting us through its continuous revelations. *From this new level of awareness we can recognize and forgive all trespasses, not through the egoic and calculated act of "forgiving another," but through recognizing that no "others" even exist, and nothing actually "happened;" it was only a scene from a movie. As we realize that we only need to forgive ourselves*

for not understanding this, we then move forward by no longer resisting life's revelations. Once we learn how to honor and appreciate all that is being revealed as just *information* about our *being*, the inner experience of life completely changes. A new level of forgiveness now becomes possible, and this changes everything.

THE NATURE OF LEARNING

Within the evolution of our species, through every level of our evolving form – cellular, amebic, protozoan, amphibian, and reptilian – as a central nervous system developed, most of its processing was devoted to basic biological functions and the protection of the physical body. A very complex sensory system evolved that allowed our ancestral humans to continuously scan the environment for *information* as they mapped obstacles and searched for signs of food or danger. All of this was automatic. Only recently, in our long evolutionary history, did our unique capacity to think, reason, and rationally decide become available.

Throughout this amazing evolutionary process, we acquired, modified, and then even lost many different sensory abilities. Other animals can hear and sense many things to which we are completely oblivious. For example, we lack the ability of dogs and rodents to hear extremely high frequencies, we don't have the nighttime infra-red or low light vision of many nocturnal and deep-sea creatures, and we still don't understand how many species navigate when they make those impossible-seeming, yet perfectly choreographed, annual trips required for migration or nesting.

However, because of our reasoning minds, we have compensated for and surpassed many of these missing senses by adding new scientific technologies that rely on *radio waves*, *microwaves*, and *x-rays*: capabilities that were completely unseen and unknown only one hundred and fifty years ago. Our unique set of human senses and abilities defines our role in the physical world; other species have very different receptors and neurological systems allowing them to sense, compare, and measure other, and often very different, aspects of the same "external" environment.

Yet, even with all of our expanded technical capabilities, humans only interact with a very small part of the physical world. Our senses only register the most important *information* required for biological function and survival, while our instrumentation only adds a little more. We know that there are many types of *information* that we don't directly sense (*energy* forms and physical expressions), but because we also have not yet developed the necessary technologies, these invisible forms of *energy* can't be monitored or measured. Beyond all these, there still is the greatest part of all: all of that which we are completely blind to even its existence: those vast extents of our *multiverse* that lie beyond our *conceptual horizon*.

As we expanded our exploration of the cosmos, scientists made an unexpected discovery. Not only can we not explain the vast majority of the phenomena that we already observe and sense, but also that part which we "know to be unknown" only seems to be growing. We recognize that there is an enormous amount of phenomena in just the observable portion of our universe that we still do not understand, but, beyond this, we are seeing many clues to a far greater, unseen, or "not yet seen," existence; hints of a type of cosmos that we have not even begun to cognize. The meaning, extent, and importance of this *infinite* unknown is a mystery whose understanding is so far beyond all of our current abilities, instruments, and senses that we don't even know what we don't know. Our current human brains were not built for this understanding; they are completely incapable of "understanding" this vast unknown.

One of the most revealing discoveries of *quantum physics* is that, for all of us, our expectations and participation seem to create and define our personal experiences. As discussed, psychologists have long understood that we learn by building upon what we already know or understand; our previous experiences become the foundation for all new learning. Because we learn through contrast and comparison, all of our journeys must begin from some known place. We build our perceived world, block by block, upon that which we already think that we understand.

However, when building our worldview, if the basic foundation of understanding has voids or weak spots, our entire construction, no matter how elaborate, will continue to be flawed and remain unstable. Because of this, the most revealing journeys towards truth can only be initiated by first discovering and then releasing or unlearning enough of what we "already think we know" to allow for completely new beginnings. It is critical to understand the role of our previous "learning" and how it determines our perception of the next experience and, then, train ourselves to move beyond it. This is an enormous challenge since we all have been deeply programmed for this efficiency and we have few tools for envisioning beyond our built-in *conceptual horizon*.

Art is a helpful tool for this process because it can provide hints of what may exist beyond our day to day experiential life. Because of our long tradition (habit?) of relying upon the written word, one of our most easily interpreted, and therefore effective, art forms is literature. *Flatland,* and *Plato's Cave,* both discussed earlier, beautifully illustrate how our inherited and learned concepts function to cage and limit our awareness. Poetry can add new or different meanings to words, and when most effective, communicates while completely leaving rational meaning behind. Other art forms, those not relying on words, bridge the gap by entirely bypassing language and reason, which have long been two of our strongest shackles.

THE HUMAN BODY – A COLONY?

What exactly is a human *being*? Where do we hold our thoughts and "awareness?" Where is our identity contained within this collection of cells that we once considered to be our sovereign body? While we often feel that we are alone and isolated, we have evolved on this Earth, not as a single entity but, rather, as an interactive community of many different parts and interdependent organisms. Looking out to our external environment, we understand that we could not exist without air, water, the earth, or the sun, so why do we even imagine ourselves as being separate from these *elements*? Inside and on the skin of our physical bodies are found human cells, bacteria, viruses, fungi, plankton, planaria (simple worms), prions, insects, and other forms of life that have yet to be discovered.

When we count the cells that make our "body," we discover something very interesting, unexpected, and difficult to reconcile with our traditional ideas about independent selves. Studies have indicated that as few as one out of ten cells that live within this contained form that we consider to be our body are, in fact, human. Most cells of our body are the other small micro-organisms listed above. If we consider just the basic cell count, each of us is technically a colony that seems to be only about ten percent human. One recent study found the number of human cells in our bodies was greater and closer to thirty to forty percent, meaning that our non-human cells still make up about sixty to seventy percent of the total cell count. ***Either way, the majority of cells contained within our bodies is non-human.***

For millennia, all these elements and organisms interacted and communicated together to eventually evolve into the collective organism that we now separate and label as "human." We did not evolve in isolation, instead, we co-evolved with all these elements and living organisms. Because of this co-evolution, our lives are fully interconnected with these "others" through a wide variety of complex and completely misunderstood mechanisms.

The significance of this type of co-evolution is just now being recognized and explored. As we dive deeper into understanding our co-evolution with these other organisms, we discover a well-developed system of interdependence. *Research is now revealing that not only do most of these co-evolved cells and organisms not harm us but, instead, they assist us in many ways. In many cases, we fully depend on them for the very chemical and biological processes we need to live healthy lives.* This will seem a bit weird, even uncomfortable for many of us, but even at the basic level of the physical, what we think of as "us" has never been just a single *being. We are, instead, a thriving, enormous, complex, and very inter-dependent colony constructed from a majority of non-human organisms and cells functioning in what we think of as an "external" environment that is also fully intertwined with our being.*

Chemists and biologists are presently discovering that many of these cells and organisms are vital to our well-being: they produce many of the bio-chemicals we need for health and function and assist in critical biological processes. In a healthy and balanced body, the vast majority of these "outside" cells are not trying to invade or attack. Instead, because they co-evolved with our bodies over many generations, most of these cells work in unison for the benefit of the entire colony. Of course, just like with people, there will always be some microorganisms looking for a "free ride," (parasites and saprophytes) but a healthy body naturally keeps these in check, sometimes employing "other" microorganisms for that very job. We have long been aware of medical problems caused by environmental degradation, but today doctors are just beginning to realize that some of today's most epidemic physical illnesses may be related to the loss of these internal helper colonies through the overuse of antibiotics, pollution, processed foods, and our over-sterile lifestyles. Recognizing that some of these problems are caused by severely depleted colonies of helper organisms, we have been trying to replace these by taking *probiotics,* and even more recently doctors have introduced an "exciting" new technology that has been proclaimed to be the latest in advanced procedures: fecal transplants.

Symbiotic relationships have been known to be an important part of ecosystems throughout the world and some of the best-known examples involve larger mammals. It has long been understood that some mammals (cows for example) are entirely dependent upon bacteria in their rumen to complete their digestion of food, yet it was only very recently realized that we too are highly dependent on many of these smaller organisms for our health. Studies of preserved ancient humans indicate that our bowels were once home to hundreds of types of beneficial bacteria. Today, largely as a result of antibiotic overuse and our overly sterile living conditions, our guts typically support, at the most, only a few dozen different types of bacteria. New research now indicates that this changing bacterial composition of our stomachs and intestines may be related to the increase of issues such as Parkinson's, Crohn's disease, and dementia. Since these are new discoveries, little is understood, but it is rapidly becoming clear that when the size, diversity, and health of the human colony suffers, our overall health suffers. Scientists are just at the beginning of their exploration of the many unknown secrets of our "human" colony. *Because our human cells co-evolved with these "other" organisms; our health suffers when our helper organisms are harmed by the habits of our disconnected modern lifestyle.*

Where, within this thriving colony, does the identity that we consider to be the "I" reside—in which cells? Are our memories, fears, and joys resting in these non-human cells as well? In which cells does consciousness reside? From just this biological perspective, it is clear that we are more communal and interdependent than that fully individual separate *being* that we once imagined. Even at the most basic biological level, we are constantly interacting with and completely dependent on what we have always thought of as "others." *Before we even begin including extra-dimensionality and expanded consciousness in our discussions, the concept of existing as a separate and independent self is already discovered to be a misunderstanding.* Without air, earth, or sun we don't exist: they are also integrated parts of us, and when we dive in fully, we soon discover that there is nothing in existence that is separate from us. All that we once considered separate can ultimately be understood as just another extension of our *being*. *Our physical body can be understood as another metaphor for the holistic universe; we are a part of everything, as everything is a part of us.*

THE COLLECTIVE

Over the last one hundred years, we have learned that, at the level of the *quantum*, all behavior is *probabilistic;* this means that while anything can happen, some occurrences are much more likely—they have a higher probability of being realized. *Quantum superposition* is the term physicists use to describe this "pre-physical" state of multiple possibilities that exists before any actual expression. What causes one possibility to be expressed over another is not yet understood, but our engagement through personal attention, observation, interaction, or participation seems to be a very important factor. Physical manifestation and human consciousness appear to be tied in ways that we have yet to understand.

At the macroscopic level, where our minds and bodies engage, the world seems *deterministic*, fixed, solid, and surprisingly enduring. The world we directly experience is the result of all these *infinite* potential possibilities intersecting and interacting with the *energy* of our collective mindset. Collectively, we determine (vibrationally) which possibility is to be expressed – the possibilities with lesser probability still exist, but they are completely hidden or buried beyond the range of our collective *viewport*. **Our solid-feeling "reality" is the resultant product of our majority mindset. Three-dimensional life is the direct result of our "collective" majority – a true democracy; but as we all know, "democracy" is always messy, confusing, and leaves many with a feeling of discontent.**

Fortunately, as with any system of government, local action is where all real change begins. Our own "local" mindset or *viewport* has a much greater influence on our day-to-day existence than other more distant components. The most effective way to influence the greater collective mindset is through the expansion of our "local" mindset. This is one reason why group experiences such as workshops, meditations, sporting events, or political rallies can be so powerful; the local mindset that they shape can be very *coherent*, palpable, and easily felt by participants. While it can be very difficult to stimulate and maintain *coherence* in larger groups such as nations, we each have much more personal influence on local groups. Of course, we have the most influence on that very special local group or colony that we think of as our "self."

For deeper healing, it is always necessary to honor and embrace all aspects of our greater selves, even those parts built from different-appearing DNA. **Expansion, growth, and healing always involve inclusion, not exclusion; and the understanding and integration of this principle are critical for living in freedom. All of existence must be equally welcome because even the strange, unexpected, unfamiliar, and sometimes scary microorganisms are important contributors.**

SCHOOLING FISH, FLOCKING BIRDS

When we carefully observe schools of fish, we witness a degree of intercommunication that often seems extraordinary or even miraculous from our perspective. Hundreds, even thousands, of fish are somehow able to choreograph their movements to dance in a unified, fluid, yet often complex, ballet. They don't just move in unison for protection and efficiency; schools of fish will dance and play. This playful dance can be witnessed by anyone who scuba dives in clear waters amongst large schools of reef fish. Birds, many types of herding animals, and bees behave similarly. This ability to move together, anticipating, and instantly communicating in highly choreographed "organized" group movement, is clear evidence for the existence of forms of intercommunication that lie beyond those already known or understood.

Schooling, flocking, swarming, and herding behaviors demonstrate that a large number of what we consider to be separate "individuals" can be more tuned to interacting as a single, *coherent* organism. Many other species seem to have developed similar neurological system capabilities that lie beyond our current or "normal" human abilities. Do we have these same systems in place? Have we just forgotten how to activate them, or have we yet to learn how? Will some of these heightened senses always remain outside of our *collective human viewport*?

CONNECTION AND "KEEPING SECRETS"

Our *viewport* defines the limits of this slice of existence that we can see and consciously participate within. Because most of it unfolds outside of our conscious awareness, deeper aspects of our *being* must be communicating and interacting on a much grander scale than what we typically imagine. Below the surface of our conscious minds, we are fully *entangled* with the entire universe at all *dimensional* levels. We fully interact and communicate with the entire *multiverse* in every moment of our lives without ever being aware of these invisible connections.

This also means that nothing we do ever goes unnoticed by the universe; meaning there is no such thing as a secret. Everything that happens, including all that we do, think and feel, creates ripples of *energy* that instantly inform all the other parts of the universe. ***In a holographic universe, there is nothing that ever can be withheld or hidden; the universe is always fully informed.*** Later we will discuss why the opposing directions, "out-beyond" and "deeper-within," become equivalent ideas once they are explored and understood from a more expansive and *holographic viewport*.

ILLUSIONARY NATURE OF TIME AND SEPARATION

Two fundamental parts of the three-dimensional illusion are extremely deceptive and, therefore, particularly difficult to break through. The first is the nature of the separation of *space* itself, where our "oneness" is always masquerading as a plethora of separated individuals, energies, materials, and things. The other is the appearance of incremental and passing *time,* also a form of separation. Seeming very "real," these are both just very powerful illusions that are necessary for our brain to function and handle its critical job of navigating in three dimensions while also protecting the body. These two persevering illusions, *physical separation* and *marching time,* work hand-in-hand to create an internal environment that allows us to process "*time* and *space*" in smaller bites that are delivered in a certain order so that we can have and benefit from, experiences. Through the illusion of "stretching" *time* and *space,* we are given a unique and wonderful opportunity to evolve, grow, expand, and explore *self* deeply.

In the deep interconnected "oneness" which eternally exists beyond our familiar structural veil of "*time* and *space,*" all events unfold interactively in what might be described as a single "moment" or "event" that transcends *time, space,* and our ability to understand. Marching *time,* distance, divided objects, and separate-appearing *beings* only form as a result of our neurological system's need to subdivide this "oneness" into discrete manageable increments. *Time* appears subdivided so that we have the "*time* and mental *space*" to effectively process *information* within the operational limitations of our three-dimensional brains. If everything happened without this apparent separation of "*time* and *space*" (which for us would most closely mean at once), the *local processor:* our brain, would become so overloaded that we would be left in a perpetual state of confusion – forever doomed by the biological equivalent of our computer's "spinning wheel of death."

These two illusions mesh together in our minds to create the single greatest human misunderstanding: our deeply entrenched belief that "we are these bodies." Much of our culture teaches us that we are just biological organisms, destined for certain and final death. Many of us live our short lives hoping for some kind of reward, later, after a life of suffering. Others are thoroughly convinced that we will soon die, decompose, and be forever banished to a dark and cold void of nothingness or worse. Because this illusion of being tied to our body is so powerful and effective, we find it almost impossible to consider "real" any existence beyond the context of these physical bodies. Even Christianity's teachings about the resurrection of its most important religious figure are usually portrayed as his return to physical form.

This universal "misidentification" is the primary cause of our angst and primal fear, which directly leads to most of our reflexive and often destructive behaviors. The primary and very dysfunctional belief that blindly drives and confuses our entire culture is the deeply ingrained idea that our lives are limited to only the "three score and ten" years that we occupy these vulnerable and fragile bodies. How might our world appear if this belief changed?

However, the understanding and integration of such a new and different level of awareness can't be achieved through language, concepts, or words; it always requires direct personal experience. To gain this experience, we have been gifted with this amazing body and its special ability to divide, spread out in *space*, create, and spend *time* examining all of the different aspects of every possible situation. Each individual has their own set of distinct experiences but taken together over the entire history of mankind, we are collectively assembling a complete palate of possible human experience. Our experiences are never lost or forgotten; all of these human experiences are remembered and recorded *holographically*, in the depths of our collective human psyche: our collective *soul*. This *holographically* "remembered" record of experience can be accessed in special circumstances, and has been named by some, the "Akashic records."

Together, we are exploring the *infinite* nooks and crannies of conscious manifestation so that we can better "know" our Self. All human drama exists solely to create a "set" and series of "stories" designed to stimulate our interest and then motivate and guide us through all types of experiences so that more aspects of our *being* can be fully witnessed and expressed. Through all of our fascinating experiences, involving the constant unfolding of the infinite variety of possible human interactions, we individually and collectively grow, evolve, and expand to better "know" our collective *self.*

Within our three-dimensional expression, our individual experiences appear to be unfolding within *space* and *time*, a filtered and reduced *viewport* that allows our brains to process and make conceptual sense of what would otherwise be an overwhelming jumble of confusing *information*. Through this slowing, distancing, and *separation,* we are able to experience emotions and feelings repeatedly; they have the *time* and *space* to deepen and eventually become our deeper "knowing." Through lifetimes of collective experience, we then can witness and come to "know" the entire range of human expression. **Such depth of self-knowledge is the ultimate purpose of our physical and dualistic experience, yet simultaneously, this three-dimensional** *experience is also only one small aspect of our much deeper and fully interconnected multidimensional lives.*

THE VOID WITHOUT AND OUR EMPTINESS WITHIN

Many hold the paralyzing fear that nothing but void exists beyond our short lives, while simultaneously experiencing a profound unexplained emptiness within. These two energetic forms are completely related. As long as man has been pondering the meaning of life, there have been angst-ridden odes to the "dark and bidding void." From this fear-based sense of emptiness comes our ubiquitous, and sometimes overwhelming, desire to escape this feeling of angst by somehow filling this void. Not understanding the nature and purpose of this drive, we engage in all kinds of busy, distracting, and counterproductive activity in our constant attempts to quell this misunderstood fear.

Our initial attempts at "filling the void" are usually focused upon the idea that "more of something is needed": more sex, more money, more fame, or more fun; but at some point in our attempts, we usually discover that no amount of external activity ever satisfies this deep inner longing. This realization leaves us at a loss and then, unfortunately, our most common reaction is to resign ourselves to just the continuation of living with this lingering sense of incompletion; a choice that often manifests as severe depression or worse.

This difficult and long path is very typical and the natural result of a fundamental misunderstanding about the meaning and value of our persistent sense of "void." Also, of course, it is only a temporary detour; there are no "wrong" paths in life, for all journeys ultimately lead to our deeper understanding. This emptiness itself is just another of nature's great teachers; a mechanism that often initiates and motivates a lifelong spiritual search that ultimately fills the void by opening us to deeper meaning, purpose, joy, and love.

OUR CRAZY WORLD – FINDING PEACE WITHIN

While our world is absolutely amazing, in many ways, it is also quite extreme and difficult to navigate. Every day we witness the exponential population increase, the fast pace of life, the uneven resource and power distribution, the proliferation of high-tech weapons, climate changes bringing bigger and more dangerous storms along with the rising sea level, the worldwide production and distribution of unhealthy food, the release and spread of new diseases, and the pollution of our air, land, and water. All of this contributes to creating a multi-tiered dystopian inner mindset where any hope of creating sustained inner peace seems almost impossible.

However, all these pressing external "problems" also serve us by conspiring together to help guide us to the only place where we can create real change: inward. ***Unexpectedly, the world's chaos turns out to be the hidden blessing because this lack of good external options is what eventually steers us to look "within," the single direction where we can find the deep peace that we are all so desperately seeking.*** Beyond adding greatly to our experiences, the chaos of our modern world is a very powerful stimulant guiding us directly to the quiet place within.

THE NATURE OF OUR EXPECTATIONS

One common stumbling block interfering with our ability to "live in the now" is that we unconsciously bring our many personal and cultural expectations with us to every interaction. We do this automatically, without ever thinking about it, because we were trained in a culture of "doers," where we were taught to always have some personal agenda that we would like to see realized. While having a personal agenda is considered very "normal" and even healthy in the West, it fosters a state of *being* that does not allow for any *flow* or chance of experiencing the "now."

When we approach any interaction with a personal list of hopes, wants, and needs, we are also shaping the constraints and boundaries that serve to define the direction that this interaction can take. We are no longer "open" to all of the possibilities because we have either consciously or, more often, unconsciously defined and edited constraints that favor our predetermined objectives. ***Any personal agenda unknowingly collapses the potential for expression and limits the possible outcomes to what has been already experienced, known, or imagined.*** When we bring our agendas to our interactions, we completely sacrifice our ability to step outside our personal constraints and abide in the natural creative *flow* of life.

Preferences are very different. Most people have preferences, but these only become damaging expectations that block the *flow* when we become too *attached* to some particular outcome. The key defining difference between agenda and preferences is the degree of our *attachment* to some predetermined outcome: our expectations. Our ability to consciously and quickly drop our expectations is the acquired skill that makes all the difference.

EMBODYING THE ILLUSION

Our persistent experience is that these bodies and our physical universe are very real, but this is based entirely on *information* that has been gathered by our senses and then processed by our minds. It is only a product of *mind*. When we identify with our bodies and this perceived "realness," many of our experiences can be quite terrifying and leave us feeling fearful, wounded, and more closed down. However, once we fully understand the illusionary nature of our physical "reality," we can more easily abide in a different level of *being*, one where we no longer fear the unknown because we know that our deeper *being* can't be harmed by these experiences. They are, instead, recognized as wonderful new opportunities for expanding our awareness even when they appear as great difficulties.

As discussed earlier, one way to better understand the illusionary nature of our reality is to imagine our physical lives as being similar to a three-dimensional interactive "movie,' complete with fully integrated sensations and feelings. This more impersonal view of a "movie" experience allows us to better feel and engage a wide variety of different experiences, so that we may observe, open to, and come to "know" all that we are. Through this externalized impersonal "movie," the serial experiences of life can unfold interactively on the fly to shape a causally integrated "story" that teaches and *enlightens* both our individual and collective perspectives. Through these "stories," we individually and collectively learn about our *being*, compassion, and eventually the "oneness" of a deeper kind of love.

With recent advances in physics and technology, there are now many new tools that function even better than the old "movie" metaphor to help us understand the nature of this important sensory illusion. ***Two of these new tools, virtual reality and holography, provide extraordinary clues for understanding the true nature of our reality.*** When we sample the latest in *VR*, we are easily surprised by the convincing and perception-fooling power of these computer-generated *images*. As outside observers, we might laugh as we watch others wearing VR goggles duck and flail while responding to their new *virtual world*. Once we put the mask on, our perspective quickly changes because our bodies and minds begin to automatically respond to these computer-generated *images* as if they were "real." A well-produced *VR* experience is visceral and stimulates dramatic physical reactions in the user as any previous line demarcating *VR* and "reality" quickly disappears.

Many of us have also walked around life-like *holograms* on display in museums or art shows to observe how a well-produced *hologram* looks fully three-dimensional. While we know that its appearance of dimensionality is only an "illusion," it is easy to imagine how we might be even more fooled by combining *VR* technology with *holography.* What if we add electrical stimulation to parts of the brain that control smell, taste, and/or touch? At some point, although this is only a computer-driven *image,* it will begin to completely fool our neurological system and rational brains; our bodies will interpret and respond to these precise signals as if they were actually from our physical reality.

Eventually, through technological advancement, our *VR and holographic images* will become so dense, life-like, and *information*-filled that they will produce a computer-generated experience of realness that will be difficult or even impossible to distinguish from our other "real" lives. Our bodies are extremely complex *transducers,* but eventually, when our sensory system is stimulated and supported by enough *virtual information,* we will reach a point where our system will no longer be able to distinguish the "real" from the *virtual.*

Einstein reminded us that *time* is only an "extremely persistent *illusion,"* and more and more modern *quantum physicists* are now saying that the physical world is only some kind of secondary *image* related to the *energetic* stimulation of the more fundamental *quantum field.* **It is becoming clearer and clearer that our entire world, which seems so real and solid to us, is more like some kind of secondary projected VR image. However, and this is critical and fundamental, because it is only a projected image, its entire physical expression can also change in an instant.**

Like with any movie, *images,* by themselves, are completely neutral; any meaning applied to their manifestation is completely defined by the personal internal interpretation of the viewer. The entirety of the "realness" of any of these *images* results from a carefully orchestrated series of biochemical and bioelectrical signals. What these experiences represent is only the result of all this raw *information* being interpreted by our minds. Eventually, we understand that our entire experience of life is only what is formed and interpreted within our brains. **This means that our entire experience of reality is completely internal, regardless of what we believe exists externally. Ultimately, it is only our interpretation of these images that makes them seem so real.**

All of our personal experiences are the result of our inner processing of information that has been projected onto spacetime, the three-

dimensional screen for displaying our great "movie" called life. We eventually recognize that everything that is "time-dependent" or "spatially located" is just a part of this information model and therefore not "real" in the deterministic way we once imagined. Our persistent and absolute sense of a structural physical reality is ultimately an extremely refined and complex sensory illusion that is not dependent on the existence of any "real" externals, for it is, instead, what creates the illusion of them.

Eastern faiths have always reminded us, "That which is impermanent is not real, only that which does not change with *time* can be considered to be real." Ultimately our human journey is about the expansion of our *collective consciousness,* and from our *spacetime* bound perspective, it might appear that our expansion, evolution, and growth only can unfold over *time.* **However, because our movement through time is just another illusion formed by our minds within the greater illusion of our separation, the deeper truth is that no time is passing and nothing is ever really "happening" in the physical world of time and space.** All that we think of as "real" is only an *image* created as a dramatic physical metaphor to clearly and accurately express the invisible underlying *energetics.*

One main purpose of life is to make these invisible energies visible. This manifestation of the invisible is a fantastic "show" and it is all for our benefit; here our internal energies can be seen and fully explored on the "screen of life" so that we all can better know the deeper nature of our true being. It has only been spread out through time and space so that we can better see and understand, piece by piece, and moment by moment.

HOLOGRAPHIC UNIVERSE

> "The eye through which I see God is the same eye
> through which God sees me." *Meister Eckhart*

Throughout the known three-dimensional universe, extending from deep within the tiny *quarks* to out beyond the most enormous *galaxies,* a repetitive and consistent physical pattern seems to rule. Bigger "things" are constructed from mostly "empty" *space* with sparsely spaced, smaller, physical-appearing "things" that are all held together by various forms of *energy.* When we examine these smaller "things" more closely, they, in turn, reveal that they are made up of even smaller "things" that are distantly spaced and held together by *energy*. Each smaller piece is roughly modeled after the larger, and vice versa.

Quarks, *atoms*, *molecules*, *solar systems*, *galaxies*, and *universe* are names given to some of these different levels of structured physical systems. Every *time* that scientists have proclaimed that they have found the smallest component of matter, later they find an even smaller structural building block. We do not yet know if there are large or small limits to this pattern, but it is possible that as our technology evolves, our paradigms shift, and we begin to dream new possibilities; we will discover that this pattern continues, forever, in both directions.

These layers of repeating patterns evoke physicist David Bohm's groundbreaking *holographic paradigm*. His theory proposed that the universe (including our bodies and brains) is constructed in a way wherein every smaller part contains all the necessary and same *information* as all the larger parts. *Information* is nested so that every piece contains all the *information* necessary to recreate the whole.

More recently and specific to cosmology, this idea has been expanded into the *holographic principle*, which states that "all the *information* contained within a volume of dimensional *space* can be contained on the surface boundary of that region." Many physicists now believe that *black holes* might be extremely dense storage systems holding all the *information* about our universe. This means that the entire record of our physical existence might be fully mapped onto the surface of *black holes*.

HOLOGRAPHIC IMAGES SPAN DIMENSIONS

Holographic images are generated by splitting special light beams, formed of tightly focused and single-*frequency* (*coherent*) light generated by *Lasers*. One-half of the split beam, the *object beam*, is *projected* at the object that is being reproduced. This beam is then scattered, by reflection off of the object, towards a special photographic plate that is similar to a film negative. The other half of the split beam, the *reference beam*, is *projected* directly onto this same photographic plate. Once the two halves of the once single beam re-join, the differences in these two halves of the beam form an *interference pattern*, which is then permanently recorded onto the photographic plate. To our eyes, this recorded pattern on the *holographic plate* looks random and abstract, but when the same *frequency* of *Laser light* is *projected* through that plate, a life-like *holographic image* of the original three-dimensional object unexpectedly appears. Many of us are familiar with the visual and visceral power of these *images*; and until we try to touch a well-constructed *holographic image*, it does seem quite real.

An extremely interesting side property of this type of *holographic image projection* is that if we take the flat negative or *holographic plate*, break it into smaller pieces, and then *project* the *Laser* through a single broken fragment; we do not see only a part of the *image* as we would expect with a standard photographic negative. Instead, we see the entire *image*, although it may be less sharp-edged, less bright, or less dimensional. While it may be less detailed from particular angles, each broken piece still holds enough *information* about the whole to create a lower *resolution,* yet complete, *image*.

As scientists learn more about *holographic storage*, they are also simultaneously exploring another interesting aspect that involves the cross-dimensional transfer of *information*. While the *holographic plate* is only two-dimensional, it creates an *image* that has some of the qualities of a three-dimensional *image*. Such a well-constructed *holographic image* appears to be able to occupy, at least partially, three-dimensional visual *space*, even though its *information* is only stored on a two-dimensional surface*. **Holograms can, therefore, function as a type of bridge, or window, between two- and three-dimensional space.**

What else might it mean for our universe to be both *holographic* and multidimensional? Since two-dimensional *holographic* plates store all the *information* necessary to create three-dimensional *holographic images*, what if each lower dimension holds and stores all the *information* for the next "higher" dimensional *projection*. This would also mean that all the *information* stored in any dimension would be, in turn, contained, filtered, and then stored, again in the next "lower" dimension.

With this type of layered cross-dimensional storage, we quickly realize that even one-dimensional *space* will contain all of the necessary *information* to describe the entirety of existence. No *information* is ever "missing" from any dimensional realm, and this includes our three-dimensional world and universe. Instead, *information* is just dimensionally "flattened" or filtered, making it challenging for us to understand; in a sense, it is *encrypted*. While we directly experience only our small three-dimensional slice of a much greater cosmos, it still contains all the *information* for describing the entire *multiverse*. **In our multiverse, where all potential worlds exist together in a unified, multidimensional soup, so deeply "enfolded" that all possible outcomes are adjacent and instantly available, any and every piece of**

information from anywhere or anytime is directly and immediately available to any and every other part of this web-like system. This fully inter-connected web also acts as a *holographic* storage system; no matter where we look or how small the part, all the *information* about every possibility is always locally and immediately available. **This also means that everything within existence, everything that is, ever was, and ever will be, is already contained, right here, within our bodies! We are deep within our journey of self-discovery—one that is about opening, evolving, and learning how to freely access and use this infinite amount of information.**

I naturally find myself wondering about other possibilities. As discussed, some *cosmologists* have proposed that *information* describing our physical universe might be stored *holographically* on a particular type of "flat" surface that may be functioning like a *holographic plate*: the sidewall of *black holes*. If our three-dimensional universe is functioning as a *holographic plate,* is it then, in turn, *projecting* a four-dimensional *image*? As dimensional levels compound and the amount of raw *information* increases, do these *holographic images* begin to take on a more "solid" feel? Could it be that our solid-feeling, three-dimensional world, exactly as we experience it right now, only feels physical because it has been created and "solidified" through multiple levels of *holographic imaging*? Is the appearance of solidity in our world the result of multiple layers of *projected information* between dimensional levels? Could this then mean that our world, while appearing solid, might be only a very dense *holographic image?* Could this be the mechanism to explain how we can have such a real and solid-feeling experience within an *image*: our universe feels so solid only because of the interaction of many layers of *information* stored throughout the *multiverse*? Could this be the actual mechanism of the illusion or *Maya* from ancient Hindu texts?

Within each of our healing journeys, we eventually will discover the state of *being* where we embrace everything in existence with an awareness of balance and equal importance. At this point, another shift in perspective occurs as we begin to see the internal world being reflected without, and the outer world being described within. "In" and "out" then just become two directions for viewing or perceiving our *oneness*; as we open to all of existence, our entire *viewport* becomes fully *holographic*.

WE EXIST SIMULTANEOUSLY AT MANY LEVELS

Because of its *holographic* nature, the *information* describing our lives and world is repeated again and again throughout existence. Every individual has these *holographic* extensions of their *being* that are always interacting at multiple levels. We have aspects of ourselves reflected *infinite* times throughout the entire dimensional structure and their interactions are what shapes our existence. Within each *dimension,* there are also regions, or areas of increased *resonance,* aspects that interact most strongly with our *being.* While our conscious human *awareness* seems to be restricted to only one level of awareness, our fuller *being* is always connected to and expressed throughout, all the levels, or *dimensions,* of existence.

What this also means is that any two or more people can "meet" at many different places, or levels of *awareness*, depending on their interacting degree of openness and their *resonance*. Both individuals don't need to change for a personal relationship to completely shift, because if only one person changes, then the place of intersection will change in both extent and *frequency* of *resonance*, and this means that the entire nature and quality of the relationship will also change. We always meet and interact at multiple levels so if there are any changes at any level our relationships shift. Through these intertwined and complex *holographic* relationships, entire worlds can appear to change instantly.

OUR BRAIN, MIND, AND COMPUTERS

Our rapidly evolving age of computer technology is conveniently introducing devices and technologies that provide us with rich and surprising metaphors, which we can then use to model, understand, and better explain many aspects of this new vision. Before we had access to this technology, these ideas were much more difficult to understand and integrate. Processors, *RAM, LAN, hard disc, Internet, downloads, wireless,* and *the cloud* are all concepts from computer technology that can also be used to describe living systems and their interrelationships. Computer and internet technologies turn out to be surprisingly helpful for describing the nature of the individual *soul* and its temporary use of the human body form. They help to better illustrate ideas such as collective consciousness, remembering, thinking, being in the *flow*, synchronicities, ESP and déjà vu, and the difference between *Mind* and *mind*. They also can provide new ways for understanding birth and death, and they even give glimpses into our potential for developing future forms of life that humans might, one day, choose to inhabit themselves.

I was reading a recently published book by a critical-care physician who was discussing how patients had accurate memories of events that occurred while they technically had no measurable brain activity. He struggled as he tried to explain where these memories could have been recognized and stored while the brain was inactive. Today, with our growing familiarity with the "cloud," we have at least one useful model for easier visualization.

COMPUTER NETWORKS PROVIDE A MODEL

Computer networks are particularly helpful models for explaining otherwise difficult to understand relationships. One way to employ this model is to think of our brains and nervous systems as our local hard disk and processor, while full access to the more expansive *Mind* or "cloud" can be compared to a controlled and throttled *internet*. Here, the experience of "oneness" could be understood as being similar to what might occur once the high-speed fiber *internet* was suddenly no longer throttled or regulated.

Up until very recently, the idea of mankind functioning as an interconnected network of minds was usually dismissed as pure mysticism belonging to the realm of science fiction. Our dominant culture convinces us that all of our brains and bodies are independent and separate, that they only retrieve and process *information* belonging to each separate individual's private learning and experience – like the hard drives of computers before the days of the internet. Unexplained phenomena like ESP, past-life memories, channeling, precognition, and other experiences, which could not be understood through this deterministic model, were largely ignored or dismissed in a huge act of collective *cognitive dissonance.*

With the many realizations of our new physics and these new technologies, the assuredness of our old obsolete vision, one that was founded upon our separation, begins to fade as more and more of our ingrained concepts start to be questioned and found to be inaccurate. As dramatic new observations, such as *Non-local entanglement* open our thinking to other types of perception, we begin to recognize that our universe is much more mysterious, interactive, and interconnected, but in ways that previously seemed impossible to visualize or logically explain. *Space* and *time*, while very convenient forms for human organization, begin to be seen as only temporary and illusionary structures. As our minds adjust to this new perspective, a new type of flexibility is being formed within our awareness, bringing with it fresh ways of experiencing and understanding our universe.

Our ability to consciously retrieve and process non-local *information* varies from person to person, depending on each individual's experience and degree of openness. If we imagine an individual "mind/body" to be a "local computer" that has limited or filtered access to the full internet, the difference between "mind" and "Mind" can be seen as a continuum. Depending on the amount of "bandwidth" and filtering (Google fiber is used in US cities, but they still use highly filtered dial-up in North Korea), each individual's mind may be able to access more, or less, of the universal "Mind." ***However, unlike computer networks, no outside forces control the depth of our access, because it is only our deep and often rigid personal beliefs that form our filters and throttle our access. The amount of personal "bandwidth" is entirely regulated by our internal openness and availability.*** We, and we alone, operate the controls that determine the degree of our access. An individual's ability to connect with the greater Mind will increase as we open and expand our limits to become more conscious of "that which we truly are, always were, and forever will be."

Our Bodies Are Like Avatars

Another wonderful concept-defining technological metaphor that can be helpful for better understanding our relationship to our bodies comes from the world of virtual computer simulations: *avatars*, the representational identity of a particular participant or gamer. Normally, it seems that we are our bodies even though there are many clues that this is not an accurate assumption. Despite our best knowledge, this assumption still shapes a belief system that leads directly to most of our confusion and deepest pain.

A much more accurate model for a better understanding of our real relationship to our bodies is to imagine our bodies as temporary devices that are functioning in much the same way that we employ *avatars* in computer simulations. We move closer to *Truth* when we can relate to our physical form as nothing more than a handy "virtual body," or a temporary point of reference. When we can see our bodies as just these functional devices that are fully independent of the true source of our *being*, our most fundamental fears instantly vanish.

As computer simulations become more detailed and life-like, we will also find it easier to use this technology to model the deeper workings of our universe, and this alone will greatly facilitate our integration. Future developments in *Virtual Reality*, computer simulations, artificial intelligence, and biological computers will continue to add new levels of clarity to this model, initiating a deeper understanding of our relationship with our bodies, our cosmos, and even that "external" *parameter* which we have named God or some other equivalent. As we continue to expand our integration, we will also better locate and understand the deeper "Self": that which animates our *avatar*. When the once-obscure relationship between *self* and body becomes clearer, those parts of our *being* that transcend the physical will also become more visible to us.

The "illusion" of our confinement to the body is largely the result of our believing and thinking that this physical experience is our only "reality" so it must be what defines and describes us. Because of this "confusion within the illusion," we also believe that our entire experience of life is actually "happening" to us when, in deeper truth, nothing in our universe ever "happens" because what we are experiencing is only the physical expression of the hidden underlying energies – there is only the appearance that things happen. Just like with VR experiences, nothing "real" is actually "happening," as appearances can change instantly by simply changing the *parameters*. Once our old ingrained and programmed perspective of a fixed and solid world unfolding in linear *time* can be relaxed and released, our external world will also shift in an instant. Nothing is absolute, including our history; it resides only in our limited memory, built from our own uniquely filtered and shaped *viewport*. If anything is ever happening to "us," it is only "happening" to our temporary *avatar*, so that our *collective consciousness* can have another opportunity for a new experience. Through playing and engaging with our *avatars,* life is continuously learning about itself. We discover that our separation and aloneness are only scripted parts for our limited *avatar* roles, which we then can experience through this special interactive drama. The purpose of this drama is to add to our experience so that we may better understand our hidden underlying energetics and, through this process, become more compassionate and whole.

Life is not unlike watching a movie where we are also the actors, except this "movie" is a full 3-D version that allows for sensations, feelings, and a full menu of interactive choices. In other sections, I discuss how every human choice, and all of the *infinite* choices that conscious *beings* can make, must and will be fully explored somewhere in the *infinite multiverse*. Through an *infinite* number of these precisely interwoven "movies" and all the *avatars* (conscious perspectives) playing roles in the drama of life, the entire range of duality must, and will be, expressed. All of these *infinite* possibilities are then woven together in a unified and fully integrated web: the Web of *Infinite* Possibilities that is discussed thoroughly in my earlier books. Through our integrated collective experience, *being* is then able to explore and "know" the totality of life so that we, collectively, functioning as a single holistic being together in "oneness," may become more "experienced," compassionate, complete, and fully integrated. *In our lives, just like in a movie or video game, nothing ever really "happens."*

IMAGINE A GAME

Using another computer-related metaphor, we can also think of each of our separate selves as isolated personal computers building networks that then link with an infinite number of other computers. Imagine a computer-based virtual reality game that everyone on Earth plays. This game is built upon artificial intelligence so the program also learns and changes over *time*. It is a self-contained closed system. All programming comes from the participants themselves – there are no signals or information introduced from the "outside" that interact with this program. As the participants engage, they also constantly modify the underlying program; most of this programming is unintentional and automatic (unconscious), but it can also become fully conscious. This program is entirely built from instructions, both conscious and unconscious, that have been created by all of our collective feelings, emotions, and thoughts. Unknowingly, every player contributes as they join to produce the master code that runs the entire game. Because the program is constantly changing, the rules also change as the game progresses.

This game employs a very technologically advanced blend of *artificial intelligence (AI)* and *virtual reality (VR)*. Through the neurological sensors of smell, taste, touch, hearing, and seeing, it inputs enough *information* to stimulate our neurological systems to imagine and produce the real-seeming *holographic multi-dimensional image* that is our entire physical universe. This "game" also leaves the interpretation of these signals entirely up to every individual.

Continuing with this digital analogy, the most important factor for each *avatar's* experience is the default settings for its personal "filter," the settings that determine what *information* can be sent or received. This filter can't be adjusted or re-programmed by our minds: it is only be regulated and changed by our deepest held beliefs. If these default settings, controlled from our deepest recesses, include the "fear of death," then these are the signals that will pass through the "filter" and be amplified. If, instead, the setting is "knowing that we are not these bodies," then the signals will be very different: they will, instead, resonate more strongly with the *vibrations* of love and joy.

WE MAKE UP THE RULES AS WE GO

Because the "rules of the game" are being created "on the fly," anything and everything is possible. The number one rule of the game is "we are making up the rules together, so life will mostly appear to be whatever it is that we collectively agree upon." An important self-empowering corollary to this primary rule is "anything we focus our attention upon will grow in importance, power, and influence just because of our attention and participation." *Another way to say this is that "anything that we believe and focus our energy upon will become more solid and 'real' appearing."* The most profound realization is that we actually create and then manifest the entire universe through our beliefs and attention. *This, of course, also means we each have full responsibility for the way our lives appear, and for many of us, this can initially be an impossible mind-blowing and very difficult realization. Initially this new level of awareness might stimulate feelings of guilt or shame, but as we already understand these expressions are just the mark of ego trying to regain control. Eventually, as this idea is better integrated, it becomes a mind-changing, life-changing, and fully liberating awareness; as we change our hidden belief systems, the outer world will appear to mysteriously change for each of us.*

What we ultimately discover is that as we change the focus of our attention, the rigid way that our external physical world appears to each of us will also gradually and incrementally change. For example, if we become obsessed with the judgment of and the struggle between good and evil, our entire world is eventually seen as a manifestation of that judgment and struggle. Our attention creates our "world-view filter;" and over *time* this dictates how our world looks to us. This is just the *magical* illumination of cause-and-effect that life so generously provides to guide us towards a better understanding and integration of these deeper truths. Because this is an awareness that many have already realized, many are now making the conscious choice to focus more of their attention upon that which they love.

This realization also applies to the size and extent of the "we" envisioned by each of us. Our *self* can be expanded to include and encompass anything we choose to allow; there are no real limits, for we can be as expansive in our expression of "we-ness" as "we," alone or collectively, are prepared to be. The "we-ness" that we experience may only involve those of the same family, values, religion, or politics; or it can include all of humanity. Our "we-ness" can further expand to include all the birds and bees, rocks, and trees, and ultimately all of existence. As we each open to more of *being*, more of our "oneness" is experienced within.

However, we must always remind ourselves that we choose not with our minds but, instead, with our entire *vibrational being* or *core vibrational resonance (CVR)*. As we heal and expand our awareness, there is a parallel expansion of the once isolated "self" to include the larger collective "oneness" as our understanding of the "self" grows to include all of existence.

SPIRIT HAVING A PHYSICAL EXPERIENCE

At the same *time* that *being* is exploring and remaking the physical realm, it is also creating the rules. Each of our lives is a contributing part of this "game," a part that helps us to grow a better understanding of our relationship to *Being*. *Being* might also be using our current form to create new forms for it to, one day, utilize or occupy. As new forms are created, new choices and possibilities are presented and, through this evolving process, more options become available. The possible expressions for human life are always expanding, and our human role is pivotal for this process. As we progress in this "game," the day may arrive when our current physical forms (physical bodies) are no longer the only available or best forms for our *being* to utilize.

As I write, the intersection of our biological and digital worlds is rapidly expanding our expressive possibilities. Remote and artificial appendages, enhanced hearing, sight, and touch, data banks linked to our minds, internal *quantum* computing, and the promise of *quantum*-based teleportation are all new, but very real and evolving technologies. "Beam me up, Scotty" may soon become a normal and integrated part of our everyday lives. Released from its strict dependence on our physical bodies, *being may quite* naturally be attracted to these "man-made" forms simply because they offer many more new possibilities. **Nature likes new forms that facilitate different experiences, so it seems likely that at least part of our work on this plane is to design, build, and physically integrate new forms: better vehicles that may allow us to, one day, greatly extend our human capabilities.**

TIME – THE COSMIC TRICKSTER

Our human sense of *time* is always a critically important factor when describing how things appear to us. "The summers were cooler in the past." "Tom sure is looking older and heavier than he used to be." The sphere that visited Flatland was observed as a circle that changed size with *time*. *Time* is the wild card in our deck: the joker, often used to explain phenomena that otherwise would make no sense to us. We do this unconsciously without even thinking about the way all of our observations are tied to our inbuilt sense of one-directional marching *time*.

Time seems very important in our three-dimensional *worldview*, but it is only a terribly misunderstood and orphaned trickster that dropped in from another realm. We often perceive marching *time* as our enemy. We become anxious or upset by our consumption of what we think of as a limited resource that often cuts short our length of stay here on Earth, and determines what we can accomplish. We imagine that if we could only stop the advance of *time*, we would not age or die and, instead, we would become immortal.

Digging deeper, scientists are beginning to realize that our sense of *time* is related to our expectations, focus, and level of engagement in various activities. We all understand that subjective time is relative; when we are enjoying what we are doing, five hours can seem like five minutes. When I was a young boy, the last five minutes of every school day always seemed to drag on forever. Earlier I related how time stopped completely for me during an automobile accident. Last week while mid-air in a bicycle crash, I experienced time slowing to a crawl. When I was raising young children and just beginning my career, I never seemed to have enough *time*. These experiences, while still very personal, seem to touch upon more general objective realizations. I eventually discovered that with *time* (and all things), that with just a change of attitude, my entire experience of time changes. **Time does not "march on" or exist as an absolute. Instead, we create and employ the concept of time to facilitate, organize, and regulate the three-dimensional unfolding and expression of life.**

Time, as we experience it, is a necessary conceptual tool for living in our three-dimensional reference frame. *Time* and mind are completely interrelated. In the words of Einstein, "*Time* is a stubbornly persistent illusion," and according to Eckhart Tolle, "*Time* and mind are inseparable. To identify with the *mind* is to be trapped in *time*." Spiritual leaders and scientists agree that our experience of *time* is internal and subjective making our typical experience of it, entirely illusionary. In physics and math, we can easily treat *time* as we do other variables. In mathematics, if the variable of *time* is necessary at all, we might find ourselves referencing the future and the past in much the same way we might look to the East or West. A physicist might say: **"the present moment informs the "past" and the "future" equally through waves of information that travel in both directions."** Our more human perception of *time* is usually as if we are stuck on a freeway heading north, not realizing that we could slow the car, or even turn around and head to the south. Usually, we only understand *time* from the very limited perspective of a car that is always traveling at 70 miles-per-hour in the northbound lane. We cannot easily fathom the possibility of slowing down or stopping the march of *time* and making that U-turn to travel the other direction.

To understand *time* differently we must first shift paradigms. ***The very first step of this process is to better understand how and why our persistent sense of marching time limits our vision. We must first shed our old, restrictive ideas about time if we are to allow for the emergence and growth of a new and different vision.***

RESONANCE

INTRODUCTION

Resonance is an extremely important phenomenon in *wave physics*. It is also something that we can easily understand because we observe it every day in many ways. *Resonance* is defined as *"the intensification and enriching of vibratory phenomena by supplementary vibration."* *Vibration* produces *waves.* When two *waves* of the same type meet in the same medium (air, water, etc.), they interact (*interfere*), either destructively *(destructive interference)* or constructively *(constructive interference),* or more typically, in some combination of these two. We respond to and *resonate* with many types of *vibration.* Some types, such as *visible light,* we see, while other types, like *waves* in the ocean, we both see and feel. *Sound waves* are heard and often felt (throbbing bass) but cannot be directly seen. *Radio waves, microwaves, x-rays,* and *gamma-rays* can be recorded by our scientific instruments, but they remaining unseen, unheard, and mostly unfelt by our bodies.

Our everyday experience with sound provides easily understood examples. In music, certain *frequencies* or notes can combine *constructively* to create a more *resonant* whole—a fuller harmony. When two *waves* of the same or related *frequency* (pitch) vibrate *in-phase,* the *waves* will add together (*constructive interference*), and we say they *reinforce* each other. This new, combined wave will have an increased *amplitude,* and if these *waves* are *sound waves,* this means they will sound fuller and louder. If these same *waves* (*sound waves*) are *out-of-phase*, meaning that if the vibrational directions of the *waves* are opposing each other, then they will cancel each other, resulting in the full or partial destruction of these *vibratory waves.* This is known as *destructive interference,* the common vibratory phenomenon that noise-canceling headphones utilize.

The mixing of *constructive* and *destructive interference* helps form and shape the music, language, noise, and all of the everyday sounds that we all hear.

Feedback in a PA system is a dramatic example of extreme, uncontrolled *constructive interference (reinforcement)*. A beautifully sung harmony is partially the result of carefully controlled *reinforcement*, while the discordant *diminished seventh* chord is an intentional mix of *reinforcement* and *destructive interference.*

All physical things in our world possess one or more *natural resonant frequencies:* the *frequency* at which something will vibrate naturally and easily. When we strike a drum or crystal glass, the note that we hear is its *natural resonant frequency.* When a wave that is already oscillating at some multiple of the object's *natural resonant frequency,* reaches an object, this object will also start to *vibrate* at that same *frequency* or at some multiple of that *frequency.* **Resonance is the term used for describing this reinforcing relationship between two or more things that are vibrating together at the same or related frequencies.**

UNDERSTANDING RESONANCE

Humans have a deep, natural, and physical understanding of vibration. We have experience with *sound waves, electromagnetic waves (radio waves, microwaves, visible light, x-rays, gamma-rays,* etc.), ocean *waves* (including tsunamis), earthquakes, and many other types of *vibration,* all employing different mediums. Air is the medium for sound *waves,* ocean *waves* are generated in water, while earthquakes involve *vibration* of the soil and rocks of the earth; all these are easily and intuitively understood. However, the medium for *electromagnetic waves,* including light, is much harder for us to visualize or understand because it appears to our senses as empty *space.* Almost two-hundred years ago it was discovered that an invisible *electromagnetic field* is responsible for all *electromagnetic waves.* Today, *fields* continue to be a critical area of research in modern physics and as I write, new data on *gravitational waves* is quickly transforming our understanding of the universe.

Because we all have direct experience with *sound* and it is such a well understood and relatively simple type of *vibrational wave,* we will continue to use it for this deeper exploration of *resonance.* Musical instruments are always communicating with each other by exchanging *energy* and *information* in the form of *vibration.* Everything that we call music is simply the carefully controlled interaction of sonic *vibrations.*

Guitar strings provide a convenient way to begin our discussion of basic *sonic vibration* and *resonance.* Pluck the low E string of a guitar or bass and watch it *vibrate*; it *vibrates* at a *frequency* slow enough to see the string move back and forth. *Frequency* is the number of times that the string makes one complete back and forth trip in a single *second*. For example, the *frequency* of the A string on a bass guitar might be expressed as "110 vibrations per second"— also referred to as 110 Hz (Hertz), or sometimes 110 cycles per second (CPS).

In music, the term for *frequency* is *pitch* or *note*. If we hit two strings side by side and they both vibrate at the same *pitch*, say 440 Hz (modern "A" tuning pitch), we say that these strings are perfectly tuned. This causes *resonance* and the combined tone sounds much louder and fuller because it is *reinforced.* If one tone is 430 Hz and the other is 440 Hz, then when they vibrate together they will compete with each other and produce a third vibratory sound, called *interference* or *beats,* which represents the difference between these *frequencies.* In this particular example, the difference in frequencies is ten, which results in a ten-beat-per-second, audible vibration—a "wah-wah-wah" sound that most guitarists listen for and seek to eliminate when they tune their instruments. (Sometimes in music, this type of *interference* or discordance is desired.) Sound also *reinforces* when the *frequency* is halved or doubled; we call these special cases *octaves.* Lesser *resonances,* called *harmonics,* also occur at multiples and divisions of 3, 4, 5, 6, etc. A trained guitar player knows the exact places on a guitar string to "tap" and create these special-sounding *resonant harmonics.*

Sympathetic resonance occurs when a secondary object begins to *vibrate* on its own because it has the same *natural harmonic frequency* as the source *vibration.* For example, another musician's strings could start to spontaneously *vibrate sympathetically (*on their own without being touched by fingers) if they have the same or related natural *resonant frequency* as a string that was plucked on another instrument. Physical *vibrations* from the first string are then transmitted directly through the air to the strings of the second instrument, setting them into motion. These *secondary sympathetic vibrations* can then be heard and felt by the ears and bodies of the audience, creating a more complex orchestral sound or feel.

This effect can be intentional, as it is with well-tuned symphony orchestras, or it could be accidental and undesired. Controlled *sympathetic resonance* can be built into the instrument itself, as it is with the drone strings on a sitar, dulcimer, or other similar instruments. Carefully sized resonant chambers on many types of instruments are specifically designed to take advantage of *sympathetic resonance.* If a musician accidentally leaves an electric instrument near their amplifier and walks away, an electronic version of *sympathetic resonance* can build into a deafening sound called *feedback*. Sometimes, in rock and roll, punk, and contemporary electronic music, this *feedback* (amplified *sympathetic resonance)* is created intentionally as part of the music. Jimi Hendrix was famous for utilizing this principle.

In 1940, the wind blew at just the right speed to *vibrate* the Tacoma Narrows Bridge much like a guitar string (or like the reed of a saxophone), creating a deep howling sound and so much movement from *sympathetic resonance* that it caused the bridge to fly into pieces. Because this collapse happened to be filmed, today you can easily view it on the Internet. The Tacoma Narrows Bridge disaster taught engineers about the need for *damping* in bridge and building design; as a result, modern structures are designed to prevent this physically destructive type of *sympathetic resonance*.

Vibration and *resonance* are of fundamental importance to all physicists. *String theorists* propose that all matter begins with the *vibration* of tiny *strings* or string loops, and the way these *vibrate* determines which *particles* (quarks, *electrons*, *photons*, gluons, etc.) appear. *Quantum gravity* theorists believe that the *quantum field* becomes "excited' and this, in turn, causes *particles* (*matter*) to appear. Either way, *vibration* and *resonance* seem to be involved. Because we also are made from *matter*, our very existence is a sign that we are responding and *resonating* with some aspect of this deep, original, source *vibration*.

Our natural resonant frequency at any given moment is what determines just what part of this multiverse appears as our experience. As we evolve, we change, open, and expand our resonant frequencies, and this changes how we vibrate and just what we resonate with, which then alters how the physical manifestation of our world appears. As our patterns of deep internal vibration change, the world we each experience will also change. I have named this deep level of individual, but fully adjustable, resonance, the "core vibrational resonance" or CVR.

RESONANCE THROUGH MULTIPLE DIMENSIONS

Everything that we know and consider to be real in our world is formed and communicated through vibration, much of it originating in deeper and more fundamental levels of existence. Some physicists now believe that there exists a quantum field that when disturbed (vibration) creates all the particles and matter that make up our universe. This is where the original "symphony" of creation is played; its vibration then manifests everything that we experience. The particular forms that appear in our personal lives are the result of our being able to resonate with specific different parts of this infinite symphony. Our personal universe is created and experienced only through our core vibrational resonance, and as we open to more possibilities, our CVR expands to develop a wider range of possibilities for our lives. As our core vibrational resonance adjusts, we resonate with new parts of this symphony, and then, as we suddenly "hear" more, the appearance of our world instantly changes and also becomes more.

We can think of the universe as a giant vibrating musical instrument (*m-theory*). **Sympathetic resonance is producing harmonics that continuously flow and connect through multiple dimensions. This extraordinary but deeply hidden symphony is the way the universe creates and organizes itself.** At the very core of existence, throughout the *quantum field, vibrations* combine to form a "symphony" that is *infinitely* rich with *harmonics* and other *vibratory information.* This complex *vibratory information* creates and animates appropriate forms of energetic expression (*sympathetic resonance)* that then continue to *flow* through all dimensional layers. **As it reaches our three-dimensional realm, we perceive the resonant information from this deep-level "symphony" as the physical manifestation of our universe.**

Everything in existence is intimately and directly connected (*non-local, enfolded*), meaning that we are always communicating with the deeper levels of existence through our *vibrational* state of *being.* The *frequencies* and *harmonics* to which we "tune" our individual receivers determine the characteristics of the "song" we experience. **Through this type of deep-level vibration, we connect and communicate through the dimensional barriers with everything in creation. This multidimensional vibrational interconnection is what manifests and then connects everything, not just in our universe, but also in every possible parallel universe. How we each vibrate determines exactly which parts of all these parallel universes we resonate with, and this, in turn, instantly determines our experience and how "our" world appears to us. We come to "understand" other dimensions and other dimensions "learn" about us through this vibration—this continuous and fluid form of communication travels both, and all, ways.**

"Going with the flow" is simply allowing the natural resonate frequencies to vibrate us and move us along by fully utilizing

sympathetic resonance. As our instrument opens to more types of vibration, our location and reach in the universe will constantly shift, as we travel and dance within a universe that is always supporting us through natural sympathetic resonance. At this level, we are just "hitching a ride," letting the universe do the" heavy work" for us. Finding the "path of least resistance" involves learning how to expand our natural frequencies and then allow them to resonate most fully so that we can enjoy the energetic "help" of the vibrating universe.

WHAT IS FREQUENCY OUTSIDE OF TIME?

All of our *frequency* discussions have included, by definition, the human concept of *time. Frequency* and *wavelength* as we understand and define them, are three-dimensional concepts because we include *time* and length in their descriptions. Since *frequency* is measured in *cycles* or *vibrations per second, time* becomes a necessary part of our description of any *vibratory* phenomenon occurring in the physical realm. S*tring theory* predicts that the fundamental nature of the universe is *vibratory,* but its ten- or even eleven-dimensional *space* is almost surely not subject to our same concept of linear marching *time*. Within this type of *space, frequency,* and *vibration* must have very different meanings than they do in our three-dimensional world. At the very depths of existence, a place where there likely is no *time* and, therefore, no beginning or end (alpha or omega), there must exist a different *timeless* form of communication, one that somehow is related to what we understand as *vibration*.

What would *frequency* be without the measurement of *time*? Such questions highlight the limits of our three-dimensional thinking. We do not yet have the conceptual tools to comprehend *vibration* outside of *time*. However, we do have the ability to feel or sense what reaches us through *resonance* from these deeper levels without relying upon our normal conceptual thinking. *The development of this type of sensitivity to "presence" is an important step in unlocking our ability to experience and participate more fully in the deeper dimensionality of existence.*

So just what is *vibration* if it originates, "outside of *time*," and what changes as our awareness of *vibration* becomes liberated from the constraints of concepts and *time*? Like *Zen koans*, pondering these questions can help us open to new ways of *being* as we further explore and experience our deepest interconnections.

BIG BANG

The *Big Bang theory* was the first and most widely adopted cosmological theory about the creation of our universe that was derived directly from *relativity*. The theory was first introduced in the 1930s, but the name *Big Bang* was not used until some 20 years later. As the name implies, it proposes that our universe began as a very dense, hot, and microscopic *mass* that very rapidly expanded during a unique event that occurred about 15 billion years ago. The actual rate of expansion required to fit the most current data is so dramatic and rapid that this event makes *supernovas* look puny. A more recent version of the theory introduces two phases of *Big Bang expansion*: the initial moment of the actual "bang," and then secondary, but even more rapid *inflation*. Both phases are required to explain the even distribution of residual background *energy* that is spread throughout the universe.

There are several aspects of the *Big Bang theory* that do not personally ring true for me. The theory is built upon our old concept of linear one-directional *time*. A "beginning," with a single *Big Bang*, does not seem to follow the normal pattern of most natural processes; nature tends to repeat, reuse, and recycle. For my sensibilities, such a massive one-directional process seems quite unnatural. One idea proposed by some physicists is that the *Big Bang theory* caused ten-dimensional *space* to split into two smaller-dimensional realms, and the part we now occupy is a *space* that contains only the four dimensions that we call *spacetime*. The other part with six or more dimensions is said to have "curled up" into pieces that are so minuscule that we cannot see or measure them with any of our devices. Have these physicists also become so constrained by the old conceptual paradigm that they must still resort to describing invisible things as "very small"? When we combine all implications, it seems to me that we are trying far too hard to make the known data fit into a box constructed around our three-dimensional concepts.

To my thinking, the "Big Bang" is just another dramatic creation story that sounds strangely similar to those of some from the world's major religions. *It is my clear sense that, again, time and dimensional limitations are interfering with our ability to understand what is going on; a creation process unfolding in multiple dimensions cannot be explained using only our old concepts, thoughts, and language.*

If we could stand back and see all of existence from a broader, multidimensional perspective, this *Big Bang* "event" might not look much different than the top of a large cresting wave in the ocean. From this bigger picture, we might see the *Big Bang* as just another marker, a passing but likely extra dramatic event (possibly a tsunami) in the rhythmic wave of creation that energizes the fabric of the *quantum field* which then, in turn, manifests our universe. The *Big Bang* might look special and important from our reference point, but from a broader perspective it describes just one part of the *vibratory waves* of *information* that are forever shaping our universe – a symphony that has no beginning or end because it occurs outside of *time*. Ultimately, the *Big Bang* is just a passing reminder of the ever-changing rhythmic dance of the universe. ***Events like the Big Bang can be thought of as the bass notes and kick-drum beats of the infinitely repeating and rhythmic song of beingness; they help to pump the deeper rhythm of the vibratory cosmic groove.***

THE UNIVERSE SPLITS – THE MULTIVERSE

In the science appendix, I describe the nature of *infinity* and how in an *infinite space* that is also not bound by our sense of *time,* there would be entire universes, not just planets, identical to ours. ***Due to the very nature of infinity, in our infinite cosmos, an infinite number of planets, galaxies, and full universes, which are identical to ours, must also exist.*** No matter how impossible this seems, this is exactly what must happen in *infinite space,* both mathematically and in its physical expression. (If this seems impossible, please read the *infinity* section of the science appendix.)

With a deeper understanding of *infinity,* Hugh Everett's "Many Worlds Theory" can be understood in a broadened way. Viewed from our three-dimensional perspective, it appears that every *time* that we make a decision or change something, the entire universe is also changed; it splits off to create another possible path, making another possible future. Every possible variation of each decision for everyone creates a different possible outcome and, therefore, an entirely new world. At every point of change or decision, the entire known universe splits off anew and is expressed somewhere in existence, multiple times, as an *infinite* number of identically replicated universes. For example, imagine one leaf falls off a particular tree in one universe – at that instant the old universe has been changed and an entirely new universe is created. Every other possible variation – no leaves falling, two leaves falling, three leaves falling, etc. – is also expressed somewhere as an entirely new universe. This not only happens once but it occurs again and again an infinite number of times. This results in the existence of an *infinite* number of precise replicas of every possible variation of our universe. This is beyond "mind-boggling" for all of us.

As bizarre and impossible as this might seem, if our universe is infinite, then because of the very definition and meaning of infinity, this infinite collection of identical universes is also an absolute fact. To our three-dimensionally bound finite brains, this idea sounds crazy and completely insane, yet this is not just a theory! If existence is infinite and marching time is not a limiting factor, then there must be multiple entire universes exactly like ours, along with every other possible slight variation. These copies are also completely enfolded so that they are always in instantaneous and intimate communication with everything else in existence.

What if we were to unconsciously shift and slip between these very similar and adjacent universes all the time but rarely even notice this change? What if we could control, or steer, our journey from universe to universe? What if we could learn how to "steer" this shifting of universes to instantly change the appearance of our lives? The truth is that we have always had this ability, and part of why we are here is to discover how it works. We occupy an incomprehensible Web of *Infinite* Possibilities and our very specialized conceptual minds make this idea very difficult to comprehend, or even believe. It seems too fantastic, *and it is fantastic, but not in the way we might be thinking.* Our old concepts about numbers, distance, and *space* do not easily allow us to process and understand ideas that are without bounds or limits. *The real difficulty is not that space is so expansive, but that our minds and senses are not; they are designed, built, and programmed to work only within certain limits.* One day, as the paradigm shifts, it will only be these "old limits" that seem strange.

Extending these ideas further, there is a one-hundred percent probability that there are other exact "copies" of each of us, living our lives and making many of our same decisions. In many of these universes, choices are made that are ever-so-slightly different. Every time we make any decision, thought, or movement, somewhere other people just like us are making that same decision, thought, or movement; and simultaneously, others just like us are making different choices. At every moment our universe divides so that every possible personal choice is fully realized and expressed somewhere in our infinite web of woven interconnectedness.

The *infinite* nature of existence is an important part of the structural foundation that supports our entire *universe* and its amazing architecture of freedom, so naturally, it is expressed in many ways throughout this book. ***This vast extent: our infinite and intimately woven collection of universes, is called the multiverse.***

ANTIMATTER EXPERIMENTS

Antimatter represents another enormous gap in our understanding of the cosmos. *Antimatter* is fully physical, in that it has already been found to exist in our universe. Physicists have "seen" *antimatter* produced in high-*energy* experiments; traces from small and short-lived bits of this material have been detected after large, *energetic collisions* in *particle accelerators. Antimatter* is built just like normal matter, except all of its *particles* and components hold the exact opposite charge. *Antimatter* parallels *matter* in every way, and it is believed by many that all subatomic *particles* have their equivalents in *antimatter.* There could be entire *galaxies* identical to ours, but instead of being made of *matter,* they are made from *antimatter.*

When *matter* and *antimatter* come into direct contact with each other, all the *mass* of both is consumed in an enormous and complete release of pure *energy* that leaves nothing behind. Since these interactions free all the bound *energy,* such encounters would be magnitudes more powerful than even *atomic* and *thermonuclear reactions.* So far, these experiments have only produced very small *particles* of *antimatter* but, as I write, physicists are very busy trying to generate larger quantities in *supercolliders;* and yet they have no absolute certainty about what will happen when they do. It is noteworthy that, similarly, scientists also lacked certainty about what would happen when they exploded the first *atomic bomb.* ***In the early 1940s, some physicists even speculated that there was a theoretical possibility of starting a chain reaction from the bomb that could spread and annihilate the entire Earth, yet they still went ahead with their project! The possibility of world annihilation does not seem to be an effective deterrent for curious or driven human beings. Maybe this is because, deep down, we all understand that this physical world is just an illusion, so we are not so concerned with the "small" details, such as the annihilation of an illusion.***

Based upon what we understand today, *antimatter* and *dark matter* (see science appendix) are not directly related. *Antimatter* is physical, real, and detectable, and we understand how it reacts with *matter. Dark matter* is theoretical, yet it seems to be much more plentiful than even *matter.*

IT IS ALL ONLY INFORMATION

When physicist Brian Greene asked John Wheeler what the most important topic of physics will be as we deepen our explorations, Wheeler's immediate response was *"information!"* At the most basic level, he saw the very real possibility that the most elementary thing in the universe is simply *information.*

This *information* then expresses itself as the very *particles* that organize themselves, through patterns of *vibration* spanning multiple dimensions, to form the entire physical structure of our universe. **Patterns and overtones of vibrational information freely communicate and interact through all dimensions, thus creating the entire order and structure of our universe. In our three-dimensional reality, we do not see or experience this raw vibrational information; we only experience the "shadows" that eventually reach our three-dimensional realm, and only after information has been "distilled" from eleven or more dimensions into our familiar three dimensions.**

HUMAN PAIN AND SUFFERING

While all humans attempt to avoid pain and suffering, because of the biology of our nervous systems, every sentient *being* will experience pain. Pain is a direct physiological reaction to a physical stimulus, a warning sign that our bodies may be in danger. Suffering is different because it is not biological; it is a state of mind, a secondary human reaction or response; it is, therefore, something that we can regulate.

Even with our biological reaction to pain, by changing our attitude and learning how to regulate our inner environment of tension, we can shift its severity and influence. During two recent trips to undeveloped regions of South America, I interacted with two individuals who had badly deformed legs. Both were horribly damaged and clearly experiencing pain, yet both of these people lovingly engaged with me and were going about their days, smiling and acting as if nothing were wrong; one was even on a five-mile mountainous walk to a market, and my conversation with him was light and jovial, with not a word or indication about any pain, suffering, or his obvious physical problem. These two individuals were able to move beyond their pain using their natural ability to regulate it internally. Earlier I told a story about overcoming pain from my own life.

Suffering is a secondary psychological reaction and, for this reason, it is completely different from pain; it is an internal mental interpretation of external circumstances. We always have full control over how we respond to circumstances and this is what makes our suffering completely optional. As difficult as this may be to accept, personal suffering is always an individual choice. Except for with chronic conditions, which from my perspective always indicate deeper emotional or lifestyle issues, most pain usually recedes as the offending stimulus is removed. On the other hand, suffering can and often does continue long after the offense has ended. *Pain has a critically important biological purpose; it is the body's warning system, signaling the need for attention or quick action. Suffering is not biological; it is the secondary result of our thinking, so it always is a personal psychological response.*

When it is understood and used as nature intended, pain is a mid-level signal system that warns us of problems or imbalances that need to be addressed quickly. While pain is part of a healthy response, with chronic physical or *energetic* imbalances it only appears after the early more gentle signals have been ignored for some *time*. In our Western culture, our first response to any type of pain is to medicate. Avoiding pain is a natural human response but our culture takes this so much further by offering medications that completely numb and allow us to avoid many types of physical and emotional pain. In our quest to avoid any negative sensations, including physical pain, our natural protective biological response is shut down, so we completely miss our body's signaling of deeper problems. Then, whether repressed or suppressed, pain will no longer function as intended and signal that it is *time* to make important lifestyle changes. If we suppress our pain instead of allowing it to *flow* through us and communicate, our body's natural systems will fail us. When we resist this natural part of our healing process and continue to avoid our pain, we wind up locking our highly *energetic* pain within, but now in a different and more rigid form. Many of our techniques for avoiding pain just create more shadow and chronic imbalance which then leads directly to more pain and suffering.

With most human suffering, "victims" are likely to blame "others," including God, for their suffering. This common belief that "others" can be responsible for creating our "problems," is also known as "victim consciousness," which is, not surprisingly, one of mankind's most repeated misunderstandings. There are certainly many cases where it appears that this suffering is being created externally by war, slavery, torture, human-induced famine, terrorism, and other similar acts of human cruelty, but even in these disturbing cases, there is a significant degree of internal choice. In such difficult and terrible situations, when buffeted by the most brutal horrors that mankind can imagine and produce, it may seem impossible for an individual to make a better, non-fearful, and conscious choice, but even here suffering is still a personal choice. This has been demonstrated by some of our most celebrated historical figures who used atrocities, imprisonments, and disabilities as useful tools to sharpen their focus and help ferment ideas that later manifested as world-changing actions. Christ, Stephen Hawkins, and Nelson Mandela are notable examples. Beyond individuals, the mass mistreatment of large groups of people throughout history energized significant diasporas that completely changed human history. Attitude and action can, and do, transform what would otherwise be seen as hopeless suffering. Suffering, unlike pain, is an internal choice that can always be reversed once our internal power is recognized, realized, and fully illuminated.

THE END OF SUFFERING

It takes an extremely clear and broad awareness to see, recognize and understand the deepest hidden truth about suffering, that it is always a personal choice. This is one of the most difficult hurdles within any of our journeys towards personal freedom. At first glance, this viewpoint could seem absolutely cruel, cold, callous, and likely even perverse. However, through exploring our separation more fully, every "sufferer" eventually discovers that, despite appearances to the contrary, there are still free choices involved that can, and will, change the very nature of any type of experience, including our deep suffering. Understanding such impossible-seeming choices requires a profound change in attitude that includes a fearless courage and clarity that only comes from a deep trust of life. Only experience can build this trust.

Once we fully recognize that there are no "others," and every situation is perfectly choreographed for the growth and expansion of our collective *being*, then we also understand that there are never accidents or wrong-turns in life. We know that everything belongs to a "oneness" that is fully coordinated through constant intercommunication and understand that healing can only begin once we have the opportunity to work through our "victim consciousness": the "they did this to me" paradigm that traps us in endless cycles of separation and suffering. We must first be able to recognize that because there are no "others," what we are experiencing is only a reflection of ourselves. Understanding our awesome personal "responsibility is initially very frightening but ultimately it is completely empowering. Once we are fully ready through experience, the required change involves nothing more than a shift of our attitude.

Complete healing can occur instantly or in a moment of grace. It may take an entire lifetime, or it might unfold through many lifetimes as Buddha reminds us. Independent of our ego's desires, our *whole-ing* always unfolds with perfect timing for all involved; long or short in duration, it matters not because our healing is always progressing. From our position in *time*, the only unknowns are when and how much personal suffering will be experienced before our eventual and inevitable realization?

Once we have successfully made this bold internal leap, then after a period of integration and adjustment, our joy (our natural state of *being*) returns, even if the external conditions do not appear to change. We begin to experience life in a different way, not through a pleasure reliant on external circumstances, but through a deeper type of joy begot by opening to the full spectrum of life: light and dark, good and evil, all of existence *flowing* within and through our *being*. Along with this expanded awareness comes the deeper recognition of the perfection of human life. We then recognize how negative external-appearing forces are what shape the most profound and expansive parts of our journey; it is often through our most difficult challenges that we discover how the *infinite* possibilities for life extend far beyond what we had ever imagined. Through our difficult experiences, we discover that no words can describe our deepest nature; its form, extents, and most beautiful insights lie far outside the realm of human thought and imagination. Once the mind learns how to rest as a tamed servant of the heart, explanation is no longer necessary, for the empowered heart fully understands how to navigate this unseen, vast, *magical*, and always-available *space* of wonderment.

While our healing can't be rushed, the choice to end personal suffering remains completely in our control. Unlike our experience of pain, our suffering is always the emotional end-result of our internal thinking. If we can change our thinking, we can eliminate our suffering, regardless of any external circumstances.

"OTHERS" WHO ARE FAR LESS FORTUNATE

This discussion thrusts us directly into the extremely difficult and challenging conceptual hurdle of understanding the very real and profound human suffering observed throughout our world. This book's philosophy may be interesting as an abstract concept, but how can it possibly serve those in this type of dire need? It might be easy for someone with resources and a relatively healthy body to talk abstractly about "spiritual healing," but what about the billions of less fortunate who are born in war-torn impoverished countries or have a horrible untreatable condition or disfiguring disease? What does this philosophy mean for all the innocent children around the world who were innocently born into these types horrible conditions?

The deepest answer to these kinds of questions is multilayered, and opening to this level of wisdom requires the clear knowledge of "who we really are." On the surface, it is obvious to all who think about these types of human problems, that they are primarily created or exacerbated by the way we interact with each other on this planet; these most devastating human expressions are largely the direct result of human fear, greed, power, and war. These all are symptoms of our confusion about "who we are" and, as such, human suffering can be largely seen as a self-created problem built around our general misunderstanding of *self*. **If we were to fully recognize our deepest interconnection: that we are all smaller parts or aspects of a single being, then we would naturally treat each other very differently.** This is what Jesus was referring to when he commanded, "Thou shalt love thy neighbor as thyself. There is no other commandment greater than this." He relays the same message with many other statements such as, "As you did it to one of the least of these, my brothers, you did it to me."

From the deepest perspective, one where there are no "others," this unfortunate *being,* who is living with the less than ideal body, or in a terrible situation, can be better understood as an expression of some dissociated or rejected part of our greater collective *Self.* As each of us opens and becomes more aware of our deep interconnectedness and begins to experience less separation inside, we naturally recognize that this "other" is also a part of our own experience. With this adjustment, our entire way of living and relating to these "others" shifts.

Because many have not developed this awareness of connection, mankind continuously and unconsciously creates more suffering. Our compassionate response then becomes to try to "fix" or patch what we just broke. While all of our humanitarian relief efforts demonstrate love, hope, and care for others, and some may also temporarily relieve a degree of short-term suffering, they never can provide the "fix" for the real problem: our deep inner confusion about *self*. **The only real and permanent solution is for each of us to learn how to step beyond the world of fear and separation and end the division within.**

Person by person, as we heal inside, the outer world responds. Ripples from this new *awareness* then transform the appearance of the outer lives of everyone, including those who currently fill the "roles" of the "less fortunate."

As our relationship to the world, which is initially understood to be "out there," shifts and heals within, then, because we are all so completely interconnected, there will also be less separation, isolation, and suffering appearing "outside." ***The end of the war and division within will also mark the end of war and division without. As the inner landscape changes, so will the outer.***

The suffering of "others" can be seen as a vivid and instructive spotlight that illuminates our hidden, inner, *energetic* conflicts. The visible world is only the dramatic illumination of what is going on deep inside; nothing creates more clarity then our direct external experience of our inner landscape. When we witness and experience the suffering of "others," it means that our *soul* is still divided and suffering. If we ignore or hide from this suffering it means that we are hiding from our own. ***To better understand our shadow world inside, we need to look no further than the visible outer condition of our world.*** Some part of our collective *being* is *projecting* this difficult experience so that we may have, once again, another opportunity to witness, gradually understand, evolve, and become more aware, whole, and compassionate. Only when wholeness and balance are restored inside, will they become reflected throughout the outer world.

Through experience, our collective *soul* deepens and increases its understanding about our profound interconnectedness and our *attachment* to that which is *impermanent.* We unravel our true relationship to "others" only when we learn to love and integrate all the isolated and rejected parts of our *being.* These separated parts (*projected* as "other" individuals) can only become healed and whole through each of us learning how to live our lives within a recognition of the *oneness* of all *Being.* Only when that day arrives where there is no visible suffering in the "outside" world can we be certain that our inner world has been healed and completely integrated.

The outside world may seem harsh and cruel from the perspective of our temporary individual separation, but it is always seeking balance, completion, and perfection through all its natural processes. We can understand how storms, earthquakes, and disease are natural, and even necessary parts, of a balanced ecosystem. Similarly, the existence of human problems with many individuals facing extreme challenges, serves to shape the ecosystem for evolving our consciousness. ***All parts of our collective experience are projecting exactly what is needed in every moment to create more balance and deepen our awareness.*** Because we are so intimately interconnected, we ultimately will all benefit and evolve together. ***Life is seen***

as cruel and cold only if one is deeply attached to the illusion of time, our bodies, and this particular physical manifestation of life.

Our most challenging and difficult experiences also produce the greatest opportunities for our growth. This applies to both the individual and the collective. Steven Hawking, the famous cosmologist, expressed deep gratefulness for his extremely reduced physical condition. He fully credits his lifelong disease, ALS, for his focus, achievements, and even his satisfaction with life. The unique focus that we gain through our extreme hardships becomes the rich catalyst for a tremendous amount of growth and productivity. However, sometimes deep within the most profound opportunities for growth and opening, it even becomes necessary to cast off our physical body. Within the *time*-bound existence of all bodies, there always will arrive a *time* when the temporary body no longer serves the deepest needs of our *soul*. We gain inner peace needed for making this transformation comfortable when we know and remember that we are so much more than just these physical bodies.

THE HUMAN CONDITION – KNOW THYSELF"

It is so easy and natural to become depressed and despondent when contemplating the human condition, both present and historical. The astonishing degree to which mankind repeatedly acts against its own best interest is disheartening and overwhelming, and it is even more disturbing to many that this type of self-abuse continues to this day. Today, even with all of our increased awareness, it still appears that we are marching, at an accelerating rate and in lock-step, to destroy our beautiful planet and its life-giving ecosystem.

Of course, this book concludes that the best and only permanent solution to our being overwhelmed by our "fears of separation, and our pending certain death" is to become more aware of the "eternal *being* who we really are" and, subsequently, release all our attachments and any identification with these "temporary bodies." We naturally relax when we finally recognize and understand the illusionary nature of our identification with "our body." However, because this truth lies so far outside our normal awareness, it can't be quickly "understood" in any conventional sense. Our "understanding" requires the dramatic, illuminated, and well-timed revelations that our experiences within the "outer" physical world provide. **Life unfolds as it does, just so each of us can peer into the deeper truth of who we are.**

This inner integration of our divided parts does not happen quickly or easily. Once a new level of *truth* is recognized, the next period of integration typically becomes a life-long process; from our *"time*-bound" perspective, this process of Integration always takes time, possibly even many lifetimes. However, we now also understand that *time* is just a product of mind, bound within the limits of our three-dimensional *space*, an awkward artifact that is not yet understood. Ultimately, we also discover that our concept of and sense about *time* exists for only one purpose: to give us the *"time* and *space"* that is necessary for our integration and the expansion of our awareness.

When this same process is witnessed from a broader perspective – the *multidimensional overview* – our missteps can be understood as simply new opportunities to experience more of the many different aspects and possibilities available in the unfolding drama of our three-dimensional physical universe. Ultimately, there is no problem at all, regardless of how any of us respond. Our full experience on this plane requires all of these diverse reactions and experiences because only when taken all together, do they create balance and wholeness; even the most destructive-appearing ones provide critical momentum for eventual balance. In this *dualistic* theatre, we require tragic-appearing destruction as an integrated partner to inspired creation; there always exists a perpetual balance in all things, and in this physical theatre, this balance must be built upon *contrast*.

At this deeper level of understanding, it is also understood that none of this is "real" in the sense that it is actually "happening." What we experience is only the *projected* "movie-like" *images* emerging from the deeper levels of our energetic existence; so to "change" the entire "movie," all that needs to change are our resistances to the fundamental energies that would otherwise move each of us through a *flowing* demonstration of grace. Our exterior world, just as it appears, is the outward *projection* of our inner truth, made visible for our benefit, so that we can "know thyself" and heal. **This is why, despite all outward appearances to the contrary, everything is actually perfect, and all is well in our world.**

CONCLUSION – THE NEW BEGINNING

OUR WORLD IS NOT WHAT IT SEEMS

How could our universe possibly be the way that I propose, not real, solid, and fixed, but instead an instantly changeable secondary *image* similar to a movie? How could *time* be something that we create internally, or our deepest type of healing require nothing but internal change? How could our entire existence be so radically different from what we have believed for so long?

We knock on a wall and it feels solid. If we try to walk through a glass door, we injure ourselves. We "know" that we can't walk on water. Each day, as we look into mirrors, we see new signs of aging that validate advancing *time*. These experiences are all very "real" seeming. Our interpretation of a fixed and solid world seems clear and rock-solid; the world appears to be so much more logical and familiar than this new, strange, mutable, fluid, and instantly changeable, vibrationally responsive universe that has been described throughout this book.

The common, rational, and very normal reaction, right after our first exposure to the radical vision of this book, is to regard it as only the product of a wild and indulgent imagination or pure science fiction. Such disbelief is completely natural because this is the only "rational" or "logical" way that our three-dimensional minds can reason and respond to such a different vision. Our ability to think and reason is always, and entirely, based upon what we already think we know.

However, hundreds of years ago, the "real world" arguments of the "Flat-Earthers" were equally convincing but were eventually found to be completely wrong. In the same way, our current physical paradigm will eventually fall to some version of this new and very strange, but much more dynamic and interconnected, model of our existence.

Inspired visionaries – from our latest scientists to our greatest mystics, modern and ancient – have been slowly introducing us to these strange possibilities, including the powerful idea that our concepts and expectations have a "curious" way of determining our actual experiences. It has long been understood that our brains must learn new things by comparing new experiences to those which we already have processed; new insights are rooted in previous understandings—meaning that our thinking is naturally programmed to remain within its existing paradigm. But now our scientists are confirming what many have long-felt; that the universe works in a very different way than previously thought, involving an impossible-seeming interconnectedness that operates outside our *conceptual horizon*, beyond what our human thinking can rationally comprehend. *It is, therefore, not even possible to "understand" this completely different way of viewing life, at least with our minds, for they are a type of instrument that only build ideas upon existing and already-understood concepts. This means that for this next great shift in human consciousness to fully unfold, for each of us, the critical first step must involve our willingness to let go of anything and everything that we think we understand about our world and our lives.*

To cross this next critical threshold, we first must be willing and able to "let go" of our powerful attachment to the familiar and "known." Once we are able to successfully "step over," we will then also understand that we don't need to abandon the things we love in our lives; instead, they will be seen and integrated with a completely new light and *transmuted* so that their old meanings and significances shift naturally, easily, and fully contribute to our newfound awareness. In this type of expansionary growth, nothing is ever discarded except our old limiting thoughts and ideas; all of our past experiences can be seen and understood differently, as everything that appears in our lives becomes recognized and celebrated for their part in helping to illuminate the shadowed and isolated aspects of our wholeness.

As this new awareness emerges, it will naturally initiate powerful, unexpected, but also deeply revealing, new experiences. To fully benefit from this gift, we must practice staying completely open and available to all the new and different viewpoints revealed; they *resonate* so powerfully only because they are exactly what is required for expanding our capacity for life, love, and connection.

THAT WHICH WE SEE AND EXPERIENCE

Within the physics community, it has been clear for many years that the *classical physics* of Newton, which codified our *deterministic* machine-like world view for so long, is no longer the most accurate description of existence. Scientists have been gradually uncovering, developing, and exploring a brand new vision of reality: one that involves our opening to once-unimaginable possibilities including a completely different understanding of our relationship with each other and our universe; a physical expanse that we once thought of as "external" to our *being*.

Our expectations have been shown to be critical determinates of our experience, and we are beginning to understand that we are part of a continuum that includes everything in existence. Our world seems to be co-created in ways that we have yet to understand; we each, in intimate interaction with all others and all "things," somehow determine just how our world appears. In this incredible soup of interconnected "oneness," there are no "others" and yet, somehow, each of us holds the key to how our world appears.

Many *quantum physicists* might say that there is no such thing as an *objective reality;* or in the recent words of one, "The probability of an *objective reality* that exists independent of the *observer* is very small." The everyday three-dimensional world we experience is more like an ephemeral *illusion* shaped by a type of deeper, invisible, energetic *information.* Not coincidentally, this very strange conclusion of modern physics just happens to agree with many of our most celebrated spiritual traditions, yet we are just beginning to fathom the meaning and importance of this new vision, especially the critical role of our *observation* and participation.

However, because of the restrictive way that our minds and language must work, it is also from within this same *illusion* that we are attempting to explain these odd experimental results. Because we are only seeing the shadows or *images* of something else that is completely invisible to us, when we try to explain our universe from our current *viewport*, we always fall far short. We cannot explain or think about the universe in our normal, limited, three-dimensional terms and expect to be able to "understand" these *quantum* experimental results. In the depths of our *multi-dimensional* universe, *time,* and *space* do not divide experiences or events; it is only our minds that separate everything. Beyond our thinking minds, *time, space,* and things have either very different meanings or no meaning at all. In these depths where everything is fully inter-connected, life unfolds in ways that, today, with our type of thinking, we can only begin to imagine.

As we gain the broader level of "understanding" that experience provides, we learn how to relax our habitual need to explain all things using our logical and *time*-ordered system of thought. We also practice being *present* by learning how to let go of our habitual need to "do," so that we can then more fully witness and enjoy the experience of this great mystery. Three words of Christ describe this important process: "Become a passerby." By itself, this simple, but radical, perspective of just watching and witnessing is a very effective form of meditation, and through this practice alone, over *time*, a deeper type of understanding gradually forms within.

Through our new awareness, supported and reinforced by our "out-of-the-box" experiences, spiritual heritage, and the newest physics, we start to imagine the possibilities of a different type of universe: one that only exists "outside" of *time*, *space*, and all ingrained human concepts. A primary purpose of this physical experience that we call life is to better understand ourselves in order to become more whole. Here, through physical drama created by the deeper energetics, we are able to witness and respond to aspects of our *being* that otherwise might remain hidden. While we do this naturally just by living, *whole-ing* is a more conscious method for uncovering, exploring, and integrating the *shadows* that divide our *being*. *Whole-ing* is a wonderful practice that helps reduce our suffering to allow us to lead more enjoyable lives.

It is important to remember that the discoveries and ideas of *quantum physics* and *relativity* are quickly evolving, and many things will likely change as our new paradigm evolves. **The ideas in this book are discussed with the ever-present caveat that any explanation created from our three-dimensional conceptual mindset and language will always miss the mark because we are exploring something that originates and abides far beyond our current ability to discuss or understand. Our truest and most life-changing discoveries must be experiential; this is the deeper nature of our beautiful journey.**

OURS IS A NEVER-ENDING STORY

As our lives unfold in our familiar realm of *time*, we are in the midst of an ongoing process that involves an unlimited amount of mystery to discover, explore, and experience. When the idea of *"oneness"* is viewed from our *time*-bound realm, our awareness and understanding will always appear to be expanding as we experience and explore our universe of *separation*. However, because this "oneness" originates beyond the realm of *time*, formed in a different type of *space* where the alpha (beginning) and the omega (end) are one and the same, we too are forever engaged in this: a *timeless* process. In the depths of our *being*, where death can't exist, ours is truly a never-ending story.

APPENDIX ONE – THE SCIENCE

INTRODUCTION

The triad of our science, experience, and spiritual traditions form the common foundational ground where these strange-sounding ideas all originate. This section summarizes the history and meaning of the most relevant underlying physics to help the reader better understand this corner of this triangular foundation. With recent advances in our knowledge and technology, scientists are now able to peer further out into *space*, deeper into the composition of the tiny things that make up matter, and backward through *time*. What they have been witnessing is astounding, and the implications are beyond life-changing: they are fully transformational.

The following section is a very brief summary of some of the main points from the extensive science sections in my two earlier books, *The Architecture of Freedom* and *A Path to Personal Freedom*. Readers that reflexively shy away from any discussion involving math, physics, or science need not fear this chapter because it is not about the actual mathematics or science; it is mostly a simple history. Knowing something about the history of this science will help some readers to better understand the ideas and architecture that I am describing, as well as why it has taken so long for our culture to integrate the deepest implications. If, on the other hand, the reader desires a deeper explanation of the physics and how the physics support this vision for existence, then please refer to my previous books and the bibliography in this book. This is a fascinating journey that can dive to any depth that one's interest may desire.

I will describe first the highlights from the last one hundred years of physics and cosmology because this is the science that has completely changed how we understand our universe and existence. I also include some older, relevant scientific history to add context. While science has been a trustworthy guide for my process, understanding this science or its meaning is not necessary for our expansion or evolution—it will unfold, regardless.

THE PHYSICS OF THE LAST 100 YEARS

INTRODUCTION

This book will only start to make sense once the reader entertains the big idea that **the universe is infinitely larger, deeper, and more interconnected than we generally imagine; and much of what we consider to be the**

limits or bounds of our existence is simply the by-product of the limits of our conceptual thinking, senses, tools, and culture. Once we fully understand this principle, the rest of the book should *flow* and make complete sense.

Having only studied undergraduate physics and then taught high school physics for two years, I am far from an expert in *relativity* or *quantum physics*. I do not live and breathe the theory, experiments, and mathematics like many practicing physicists. Therefore, I must trust the theories, experimental analysis, and more technical aspects to those who are fully immersed in this science. **My lifelong driving interest has always leaned towards the philosophical so that I could better understand just what these astonishing, but very strange, discoveries are saying about our day-to-day lives.** Freed from the tedious calculations and dependence upon grants and professional peer review (and inherent politics), I have been mostly as a kid in a playground, joyfully exploring what is possible, using my friends and myself as the experimental subjects. It was largely through this type of "life-play experimentation" that I began to identify the more practical applications of this physics. This playful method eventually evolved into a heartfelt and complete vision of a universe throbbing with purpose and meaning. Infused into this vision is a powerful and growing awareness of the *infinite* possibilities for this wonderful adventure that we call life.

THREE SETS OF PHYSICS

Physics is the most basic scientific study of how physical things work in our world. Today's physics is rapidly evolving, but finds itself in a difficult quandary: divided by three somewhat independent sets of theories or laws. First, physicists still rely on the extremely functional set of laws derived from Newton. This is the physics that we now call *classical physics;* this set includes Newton's laws of motion and all the additional physics that it spawned. It does a great job of describing things in our own size range – objects that we can readily see and touch. There is also a second newer set of rules that describes very large things, high speeds, and great distances, such as *galaxies, light,* and *gravity.* This is the physics of Einstein's *relativity.* At the same *time* that *relativity* was being discovered by Einstein, a third new type of physics called *quantum physics* was also emerging. It describes very small things, such as *atoms* and subatomic *particles,* and the *energy* and *forces* that are associated with them: the realm of the very small or minuscule.

Classical physics, which still accurately describes all that is found in the middle, between the very big and the very small, is built upon that which we can see, experience, and measure directly in our three-dimensional "solid" world. For almost 400 years, this physics made a wonderful, logical sense to our rational minds. Derived from the direct observation of our three-dimensional world, it was developed over many years through repeatable experiments that historically did not rely on extremely complex or expensive equipment. Our "common sense" is largely tied to this physics.

The two newer branches, *quantum physics*, describing the workings of the very small *subatomic particles*, and *relativity*, seeking to understand the mechanics of the vast cosmos, illuminate many ideas that do not make the same type of "good sense" to our three-dimensional minds. When discussing *relativity* and its meaning, we find that we must venture outside of our long-understood, familiar, and "safe" three-dimensional paradigm. Because *relativity* works mathematically with *classical physics,* many physicists understand it as simply a revision or addition to *classical physics.* These physicists would probably say that our physics is divided into only two different parts—*quantum physics* and a combination of *classical* and *relativity*. However, because two hundred and fifty years and a paradigm shift involving "*time* and *space*" separate *relativity* and *classical physics*, I will continue to treat them as distinct and separate branches; this is only a personal choice. Even though *quantum physics* has been unbelievably successful and thoroughly tested for almost a century, it does not yet integrate as logically or mathematically with the other two branches. *classical physics* and *relativity* are studies of exactness and certainty, while *quantum physics* is about possibilities and probability. To our three-dimensional and logical minds, these two approaches make little or no sense together.

The advances in *quantum* and *relativistic physics* have directly led to the development and recent deployment of an abundance of new, powerful, and very functional technical devices. The strange mathematical predictions, the subsequent exploration of the behavior of subatomic *particles*, and our deepening knowledge of the cosmos have led directly to a wide array of electronic gadgetry that generates, drives and moves the vast amounts of *information* that today's world depends upon. We have so fully integrated these strange-sounding theories into the machines and tools of our contemporary lives that not only are we are entirely dependent upon them, but we even take them for granted. We occupy a very special and transformational place in the rapidly changing paradigm of human understanding. If most of us understood the deeper implications of the actual science behind our computers, iPods, automobiles, electrical grids, GPS systems, weather satellites, phone systems, or weapons, we would be in constant awe. We often hear that "the proof is in the pudding." It now appears that we have more than enough "pudding" to be convinced.

Quantum theory radically changes how we must look at our world; if it were simply an untested theory, then this entire book could, and should, be described as science fiction. However, every test of the quantum theory during its almost 100 years of existence has resulted in its absolute confirmation; no other major theory in physics has ever had this success rate. The majority of our national Gross Domestic Product (GDP) is dependent upon or built from products based upon this theory. The high-tech products from this theory have become as much a part of America as apple pie. Quantum theory, which has become a very critical and integrated part of our lives, is at least as real as anything else is in our world!

Today, we use these three sets of rules and observations to describe everything that we understand about our universe. *Classical physics*, formalized by a young Sir Isaac Newton more than 400 years ago, works best with objects sized for the human scale: things such as baseballs, bugs, airplanes, and bridges. It is relatively simple, intuitive, and practical; most of us have a gut-level understanding of much of this physics, even if the math looks complicated. We know that if we throw a ball into the air it will come down and if we hit a brick wall we will quickly stop moving. *Relativity*, Einstein's contribution that describes the very big objects we encounter at the cosmological scale, provides us with tools to begin understanding and describing things like *gravity*, the speed of light, and our galaxy's origin in the *big bang*. **Much of relativity is counter-intuitive to our three-dimensional mindset; its incorporation requires hard work and a great deal of imagination. In comparison, the other new theory, quantum physics, the study of the very tiny, can only be described as completely "weird and crazy." Understanding its deepest meaning and implications requires adopting an entirely new perspective on the nature of the universe, and even on life itself.**

The fact that we still need more than one type of physics to fully describe our world is a clear sign to most people, and virtually all physicists, that we do not yet really understand our universe. Many, if not most, physicists believe that there must exist a single, final, *unified field theory* that would incorporate and connect everything that we know to be true: a vision that would *unify* the four known *forces* (*gravity, electromagnetic, strong nuclear* and *weak nuclear*) into a single elegant theory. The expectation is that when found, it could be expressed as a simple equation or a set of clear equations describing everything known.

To date, *unification* has completely eluded the efforts of our very best scientific minds. As a result, many scientists acknowledge a general level of discomfort as we continue to divide the physical world into three somewhat disconnected parts. A *unified field theory* would change this, but in the effort to reach this important milestone, many scientists have devoted entire lifetimes without a clear breakthrough. Einstein spent most of his life and career trying! His three major theories, *special relativity, photoelectric effect,* and *general relativity,* came very early in his career; much of the rest of his life was devoted to his attempts at integrating *quantum theory* in order to realize the *unification* of the four known *forces.*

The deeper problem is that *quantum physics* and many of the implications of *relativity* can't be explained or understood from the limited framework of our conceptual thinking. Our mindset was shaped by this older, three-dimensional *classical physics that was built upon rational thought.* Today, scientists are consistently observing relationships, interactions, and movements that are so unexpected and profound that meaning, understanding, or visualization is impossible, even for the most talented physicists and philosophers. Even the meaning of the once-simple concept of "observation" has been thrown into deep re-questioning.

Remember that it was not too long ago—less than 500 years—when virtually all humans were convinced that the world was flat and ended with a distinct edge over which one could fall. Back then, everyone believed that the Earth was the physical center of everything in the universe and the Sun moved across its sky once every day. They were taught that the Earth was the most important part of the entire universe: everything else moved around it, making it the focal center for all of God's creation. From within that rigid mindset, no one could imagine the next great scientific leap: the understanding and cultural integration of the three-dimensional geometry of our revolving solar system, a geometric truth that Copernicus uncovered and Galileo later refined.

From their ancient flat-Earth mindset, adjusting to this new Copernican vision required a dramatic conceptual leap: the understanding and acceptance that our planet was just one small part of a much larger system, the solar system, with the Sun, not the Earth, as the central element. Mankind began his shift towards the *Heliocentric* (Sun-centered) view more than 400 years ago. We slowly and very gradually abandoned the old *Geocentric* (Earth-centered) idea, and then, through the years that followed, we further refined this new paradigm by discovering that our solar system was only one of many such systems within a much greater system called the *Milky Way Galaxy*. Gradually we realized that even our galaxy, which contains some two hundred billion stars, is just a small part of a much larger physical universe, containing at least hundreds of billions of other *galaxies* of similar or greater size than the *Milky Way*. Our understanding of the vastness of our observed physical universe has shifted our cultural understanding so far that it now seems quaint and amusing to think that people once believed in a Flat-Earth. ***Humans have a long history of slowly, but successfully, vaulting through dramatic paradigm shifts. We are deep in the process of doing this once again.***

Throughout the last six hundred years of the human journey, our physics, astronomy, and mathematics have been critical harbingers, guiding scientists towards new directions for exploration, often providing the first insights and clues. Today, looking back, we see this repeating pattern of successfully following the best mathematics, and then science, deep into what was once unknown and very strange territory. Eventually, as the very strange-seeming discoveries become more integrated, these new ideas begin to even seem normal. ***As a culture, we have shown that we can and do dramatically change the way we think; but it always involves a complex and extended process—one that can now be guided by our best theoretical math and science.***

For all of us, born into a world that was carved and framed by older rational concepts, to even begin incorporating the meaning of the last hundred years of quantum and relativistic research requires another complete paradigm shift— one that seems to require us to move beyond even our need to understand. Today, from within our conceptually dependent realm, this shift can only be experienced through an emerging new type of awareness that many now call flow.

THE NEED FOR EXTRA DIMENSIONS

There is general agreement among many physicists that we are moving closer to uncovering the "Holy Grail of physics"—the *unified field theory:* the single theory that combines *quantum mechanics, classical physics,* and *relativity.* Many physicists are extremely excited about the newest modifications to *quantum gravity* and *string theory* and their possible role in this *unification.* Now almost fifty years old, *string theory* is a mature sub-theory of *quantum physics* that involves ten or eleven dimensions. Ten or eleven dimensions, while impossible to understand or visualize, are relatively easy to work with when using the tools of mathematics. Once we include these mathematical dimensions, the three sets of physical laws begin to merge in a much more elegant way. However, at the same *time*, these added dimensions continue to be completely impossible for any of us to visualize or understand, and because they involve a different type of untestable *space*, they are likely to remain unprovable through our traditional and rational scientific methods. *Quantum gravity*, the most recent approach to *unification*, currently holds the greatest possibility for producing scientific proof. Built upon the more approachable *relativity*, it is attractive to many physicists because it does not require reaching into the unprovable depths of ten or eleven dimensions.

In this discussion, I explore several different independent lines of reasoning that all lead to the clear conclusion that our universe must be built upon extra dimensions. The addition of these extra dimensions also changes everything that we think we understand about our universe. We must first let go of old ideas and concepts to even begin to grasp what a universe constructed in this way might mean. Before we learn about dimensions and wade into this deeper water, let us back up and examine the history of modern physics in more detail.

CLASSICAL PHYSICS

OUR ANCIENT FATHERS

The ancient Greeks were largely responsible for embedding the Earth-centered model of the universe into our consciousness; they gave us two of our longest lasting paradigms: Euclidian three-dimensional geometry and the "flat-Earth" model of the universe. Euclid, who walked ancient Greece about 300 BC, is the father of our geometric three-dimensional mathematical language, and we honor his contribution by naming this mathematics *Euclidean geometry*. His contemporary, Aristotle, was the first person of record to declare that there are only three dimensions to the universe. He said, "The line has a magnitude in one way, a plane in two ways, and the solid in three ways, and beyond these, there is no other magnitude because three are all." Several hundred years later, Ptolemy proposed a mathematical proof of only three dimensions, based upon not being able to draw a single line that is perpendicular to all three axes of a three-dimensional object. For a long *time*, this seemed like the end of the story because it had been declared by the finest minds that *"three are all!"* For the next two thousand years, we found ourselves deeply locked into this rigid geometry, one that was shaped and easily understood by our senses and logical minds. Except for a few isolated and almost forgotten "fringe-thinkers," no one seriously imagined that we were missing a big piece of the picture until the middle of the eighteenth century. ***"Curved" or multidimensional space was not described, documented, or even thought possible until about 150 years ago.***

PHYSICS OF NEWTON

Throughout the Dark Middle Age of humanity, there was little *time* or *energy* for pure, rational science. Instead, through "trial and error" people learned how to strengthen castle walls, build higher cathedrals, make stronger armor and swords, and develop more effective and lethal weapons for warfare. Few of these activities were approached scientifically, and the church resisted any real attempts at scientific progress. Pure science was sidelined for more than a thousand years.

Starting in the 1300s, our Western culture began to be slowly shaped by a new mindset that eventually birthed the *Renaissance,* and this burst of *energy* and creativity (possibly fueled by the newest European import, coffee) eventually led to Newton's description of our physical world. With his *Laws of Motion (1687),* we entered a long period of mankind imagining the world to be like a mechanical machine governed only by these laws. A new belief grew amongst scientists and philosophers that, since many physical observations could be analyzed, predicted and repeated, eventually this logical system would unlock a world where everything could, and would, be predictable. The rational philosophy of Descartes meshed perfectly with this new attitude, and together they intertwined to shape a new paradigm that was built upon *determinism.* God still existed, but many people began to believe that his job description was reduced to "just controlling the gears." Anomalies, whenever they popped up, were usually dismissed as poor or inaccurate science. It was believed that absolutely everything would eventually be figured out because everything was predictable, except when God occasionally intervened by "adjusting the controls." Newton's declaration that *"Objects at rest tend to stay at rest, and objects in motion stay in motion unless acted upon by an outside force,"* along with his famous companion equation, *"force* equals *mass* times *acceleration* (F=ma)," became the unmovable foundation upon which this new paradigm was built.

For the next 300 years, *classical Newtonian physics* seemed to have the nearly flawless ability to answer almost every question thrown in its direction. Then just before the time of Einstein, as new scientific and mathematical tools emerged, the Newtonian model started to reveal some extremely worrisome and paradigm-threatening faults, and it became clear to some that very large pieces of the puzzle were still missing. To better understand the context of this change, one that we are still embroiled within, let us now examine the history of our exploration of *matter* and the cosmos.

SIZE OF THE ATOM

Ancient Greek Philosophers predicted the *atom* as the elemental building block of matter, but it was not until the seventeenth and eighteenth centuries that we began to understand the actual chemistry of the *atom*. The *atom* is extremely small; a half-trillion of them (500 billion), packed tightly edge to edge, would form a line only one inch long. If you were to take a marble and blow it up to the size of the Earth, its *atoms* would then become about the size of marbles. Stated in a slightly different way, the size of an *atom*, compared to the size of a marble, is the same as the size of a marble compared to the size of the Earth.

Today we also understand that *atoms* are far from "solid." They mostly contain empty *space*, with smaller *particles* (*protons, neutrons,* and *electrons*) that help define the borders of their *space*. How much empty *space* is in each *atom*? The nucleus (*protons and neutrons*) of an *atom* has a diameter more than 10,000 times smaller than the *atom* itself. The only other "things" in the *atom* are *electrons*, which are so small and flighty that they do not take up any real *space* at all. This means that even *atoms* are built from almost entirely "empty *space*."

A few additional analogies can give us a better sense of how much empty *space* is found inside *atoms*. If we inflated the smallest *atom* (the hydrogen *atom*) to the size of a football stadium, the nucleus would be the size of a pea in the middle of the fifty-yard line, and the single *electron* of hydrogen would be smaller than a speck of dust in the stands. The only actual matter in a region the size of a large football stadium is about the size of a pea. **What we experience as "empty space" makes up the vast majority of every solid-appearing atom.**

This means that the *atom* is certainly not the solid thing we assume it is when, for example, we hit a brick wall with our fist. Also, as we peer deeper into the "solid" appearing nucleus to study *protons* and *neutrons*, we discover that these very small *particles* that make up the "solid" parts of these *atoms* are similarly constructed— they also follow the same "mostly empty *space*" pattern. As we continue to discover new ways to "see" deeper into the subatomic world and off into deep *space*, it seems that this pattern repeats itself, on and on in both directions, towards both the smaller and the larger.

The universe seems to be constructed almost entirely from what we might consider to be "empty *space*." If they even exist at all, "solid things" are a very small component of the physical universe. However, as we explore in more detail, we discover that this "empty" *space* itself is not empty at all. Data from very recent *space* probes indicate that "empty" *space* contains most of the gravitational material in the universe, and it may hold the key to the ultimate destiny of our physical, three-dimensional universe.

SIZE AND ARRANGEMENT OF "THINGS"

THE SMALLEST "THINGS"

The smallest "thing" we can understand and work with today is Planks Constant, which is 6.626 X 10^{-34} j-s. This is a fixed number that describes the amount of *energy* in a *photon* and how it varies with *frequency*. Because of the equivalence of *energy* and *mass*, it is also a measure of *mass*, but it represents a unit of *mass* so seriously small that if a marble were to be blown up to the size of the entire universe and this Planks unit grew by the same amount, it would still be many orders of magnitude smaller than the size of an *atom* and completely invisible to our eyes and instruments.

Next in increasing size would be *strings*, which *string theory* describes as the smallest indivisible unit of matter. (*String theory* is discussed in more detail later.) After these are some relatively big players: the *particles* defined by the *standard model* of *quantum physics*. These include *electrons*, *neutrinos*, *muons*, *neutrons*, and the recently studied *Higgs boson*. Following these, we then jump to a relative giant, the *atom*, and this, in turn, is followed by *molecules*, *compounds*, and, eventually, all the human-sized *matter* that we actually can see.

OUR WORLD IS NOT SO SOLID

Since the *atom* is mostly "empty" *space*, and everything that we encounter in our day-to-day world is made of *atoms*, if we were small enough—say, the size of a *nucleus*—we would be able to see and directly experience all the "empty" *space* that is within and between the *atoms*. From this tiny *viewport,* we would easily understand that "solid" things are not very solid after all. They just seem solid to our senses because of our size: our particular and unique perspective. **If we were the size of a nucleus or smaller, observing the space between the particles would feel much the same as when we, at our normal size, look out at and ponder the space between distant stars.**

Physics has defined four primary forces that shape our universe: *electromagnetic, strong nuclear, weak nuclear,* and *gravity. Electrons* and *protons* are all held together by *electromagnetic force,* while the *strong nuclear force* holds the similarly charged and therefore repelling *particles* of the *nucleus* together. We have also identified a *weak nuclear force associated* with *radioactivity,* and of course, there is also our old friend, *gravity.* Einstein demonstrated that our experience of *gravity* as a *force* is only an artifact created by the extra-dimensional geometry of *spacetime.* It seems likely that one day we will understand these other *"forces"* in a similar way.

What we perceive as "solid" material is mostly "empty" *space* with a microscopic amount of "solid-appearing *particles*" that are shaped by these attractive and repulsive forces (*energy*). **Our bodies are made up of atoms, so we, too, are constructed from space and energy. Everything that we call matter, including our bodies, is only organized energy.** As we begin to understand this concept, the expected outcome from two "solid" appearing objects bumping into each other becomes less certain and less predictable. At some point, it becomes almost easy to imagine how the tiny *particles* of one object could pass through the enormous voids of the other. If the *energetic* conditions were right, our bodies should be able to pass right through brick walls.

A NEW PHYSICS EMERGES

EUCLIDEAN GEOMETRY HAS ITS LIMITS

Euclidean Geometry is the geometry that we all instinctively rely upon to find our friend's house, fit furniture into our rooms, throw a baseball, read a map, build a wedding cake, and determine if we can safely jump to the next rock. Even if someone fails high-school geometry, they still have an intuitive understanding of this geometry; it is built so deeply into our bodies and brains that we use it constantly, even though we never think about the math. However, this every-day, three-dimensional geometry, which has served us so well and for so long, has now been discovered to be incomplete, only approximate, and, possibly, even wrong in certain situations.

CLASSICAL PHYSICS FALLS SHORT

By the turn of the twentieth century, unexpected and very strange experimental results began to raise new and deep questions about the infallibility of the classical mechanistic model. What exactly is *gravity*? Why don't Newton's equations precisely describe the motion of the planets? What is light? Why don't *electrons* get pulled into the nucleus of the *atom*? As we learned more, the physics that once promised to explain everything not only left these and many other questions unresolved, but it also began to raise many new questions.

Relativity and *quantum physics* changed everything. Evolving almost independently but at the same time, these two new theories forced scientists to take a fresh look at these questions. ***As the years passed and both theories proved extremely successful, scientists and philosophers started to imagine that these theories must also be describing a radical new way of viewing our universe and life itself.***

Understanding and fully integrating this growing body of revolutionary knowledge will ultimately require a greater cultural adjustment than even that of our last great shift, the one that took us from the "Earth-centered" to the "Sun-centered" *viewport*. *Relativity* and *quantum physics* have allowed us unprecedented access to the very large, very fast, distant, and very small parts of our universe. ***Newton's laws still work, but now we realize that they only work when limited to certain sets of conditions or reference frames involving a specific range of size, speed, and time: objects having a size, mass, and velocity that are related to the scale of our bodies.*** However, when we analyze "things" so small or fast that we cannot observe them directly, or "things" so large, distant, or energetic that we have no frame of reference, the 400-year-old Newtonian laws of physics are revealed to be only working approximations.

EINSTEIN

REFERENCE FRAMES

Of the four groundbreaking papers Einstein published in 1905, the third was named *On the Electrodynamics of Moving Bodies;* only later was it renamed *special relativity*. In one sense, this theory was a modification to the long-standing agreement within *classical physics* about what scientists and mathematicians call *reference frames*. His theory begins by reinforcing the existing classical view that all uniform motion is relative to the motion of the *observer* and that there are, therefore, no privileged or special *reference frames*. Also, because everything in the universe is in constant motion, any reference point must also be in motion. Anything observed is, therefore, always relative to that motion. This first part of his theory is still describing *classical physics*, so it causes no major conflicts or problems.

Speed of Light

Einstein then introduced the entirely new idea that the speed of light was the same for all observers, regardless of their reference frame. Treating light uniquely was an entirely new idea, and with this new realization, much of our understanding about light, space, gravity, and time was suddenly and dramatically transformed! Once his math was understood and analyzed, it was realized that all our observations and measurements of *time* and distance depended on how fast we are traveling. *Time and space were now discovered to be variable, instead of fixed or absolute.*

Einstein needed to add another *dimension* to our coordinate system to make sense of these unexpected predictions. He showed that *time* was not separate from three-dimensional *space* and, instead, we exist in a unified *spacetime* coordinate system—a four-dimensional blend or *continuum* that combines our three-dimensional *space* with *time*. *Once Einstein deduced that we were living in this multidimensional combination of time and space, the real meaning of both "time" and "space" changed forever. However, today we have yet to integrate this dramatic and now 110-year-old realization into our general culture. More than a century after our existing understanding of time and space has been shattered, we are still living deep within our old paradigm.*

Einstein's first *relativity* theory also predicted some very odd and unexpected behavior for light involving the concept of speed. The measurement of speed is, as he illuminated, completely linked to how we perceive *time*. This is why, in a very real sense, this theory completely changed our understanding of *time*. *Light* behaves very differently from other common *wave phenomena*, such as sound, which someone can catch up to and even physically pass. Jets do this often, creating a *sonic boom* as they zoom past their own moving sound wave. Light is different; it obeys a unique set of speed rules but it requires much faster speeds for us to even notice the difference. The effects become most dramatic at speeds approaching the *speed of light*. At slower common speeds, the fact that light still seems to behave in a familiar way is only the result of our inability to discern the difference: it is an approximation. For example, flying on an airplane at 500 miles-per-hour you might see another airplane out of the window. If the planes fly close to each other and parallel, you can wave and even get a normal timely response from people on the other airplane. Because the planes are moving together at a relatively low speed, nothing unexpected is detected. But unlike sound *waves*, Einstein demonstrated that you can never catch up to a beam of light because no matter how fast you travel, light still moves away from you at the *speed of light,* which is about 186 thousand miles per second, or close to 670 million miles per hour.

Einstein's theory also predicts that if we were to travel at these very high velocities, very unusual things would begin to happen to us as our speed increased and approached the *speed of light. Outside observers* are persons or instruments watching us move at these great speeds, while not moving with us. *Outside observers* would notice that as we approached the *speed of light,* we would start to change shape. To the eyes or instruments of any *outside observer,* as we moved faster, our body would dramatically flatten or compress in the direction of our movement. An *outside observer* would also notice that our movements would slow down so much that we would appear to almost stop; it would seem like the force of our *gravity* had become much stronger. As we approach the speed of light, the outside observers might notice that the watch on our wrist almost stops and, compared to them, we age more slowly as our *time* slows down. These are some of the changes that would be seen from the perspective of any *outside observer* sitting in his lab back on earth. **However, (and this is where this gets even more strange) from the perspective of a passenger traveling with us at this high speed, nothing at all would have changed. To someone else in our own fast-moving reference frame, including ourselves, everything about our moving world would still look and feel quite normal.**

These observations are almost impossible to grasp or visualize when considered through the "common sense" of our conceptual minds, which are designed for efficient navigation through and around our three-dimensional world, but at much lower speeds. Despite these strange-seeming results, this physics, Einstein's *relativity,* has proved to be sound and durable, having been thoroughly tested and explored for more than one hundred years. The more we explore and discover, the more credible his theory becomes.

$E=MC^2$

Einstein's fourth publication from 1905, *Matter-Energy Equivalence,* taught us that *mass* is only *energy* that is being expressed in a different form. In this paper, he demonstrated that *energy* and *mass* are interchangeable forms of the same thing; *matter* is just *energy* that has been "frozen" and stored in a different state; it can always be converted back to *energy* under the right conditions. In this "frozen" state, *matter* embodies an enormous amount of *energy.* This relationship is described by his famous equation $E=MC^2$, or "*energy* equals *mass* times the speed of *light* squared." Einstein determined that to calculate the amount of *energy* contained in an amount of material *mass,* one needs to multiply the amount of *mass* by the speed of light and then, multiply that result by the speed of light, once again—a very big number multiplied by a very big number resulting in an enormous number. In our *gravity field,* we can think of *mass* as *weight,* and this equation means that a very small amount of anything contains an unbelievably large amount of *energy.*

A handful of material under the right conditions can release enough *energy* to destroy a large city, and this is exactly what occurred when the nuclear bombs were dropped on Hiroshima and Nagasaki at the end of World War II. The special condition, which allowed the *mass* to convert quickly to *energy* in such a dramatic way, was the aggregation of a *critical mass* of *fissionable material.* An uncontrolled *fission* process leads to a violent release of *energy* —the atomic bomb! Through this clear and dramatic demonstration of how a small amount of *mass* produces an enormous release of *energy,* this weapon, unfortunately, became the experimental proof of Einstein's theory. This particular method of proof was also his lifelong regret. Fortunately, when this release of *energy* is slowed and controlled, as it is within *nuclear reactors,* this same *process* can be used productively to generate heat and electricity.

Today, we universally recognize that *matter* is just another form of *energy.* This realization that everything in the universe could be reduced to *energy* entirely reversed the old classical view that *matter* and *energy* were two very different things. This awareness also leads us directly to one of this book's most critical concepts: ***There are no actual "things" in the "physical universe." There is only energy in all of its various forms. Everything ever experienced can be reduced to energy and energy exchange. Understanding this concept can help us let go of our old ideas of a rigid, solid, and mechanical universe, to make room for new and more fluid possibilities.***

GENERAL RELATIVITY

Ten years after publishing his *special relativity,* Einstein further modified this work with the publication of his *general relativity* theory. With this update, there was finally an explanation for *gravity,* which had always been the unspoken "elephant in the room" of *classical physics.* Einstein demonstrated that *gravity* is how we experience the temporary warp or deformation in the fabric of *spacetime,* caused by the presence of a large *mass.* This effectively replaced the idea developed by Newton that *gravity* was a *force* that causes masses to attract each other.

Suddenly, what was once described as a "mysterious" force could be recognized as the direct result of an unseen, but very real, geometry and beginning with this fresh understanding, the "mystery" became more about understanding the deeper levels of geometry. Gravity, which we rely on every day and all intuitively "understand," turns out to be a function of the architecture of our universe – its geometric shape and structure. We are not able to see this deformation of spacetime with our normal senses because our senses are only equipped to see the world in three dimensions. Instead we "feel," and then interpret this four-dimensional deformation as a force that we call gravity.

This is exactly what was being illustrated in "Flatland," the Victorian era novel discussed earlier. Our interpretation of *gravity* as a *force* is only an "artifact," caused by the "dimensional problem" that results from "dimensional reduction." *Similarly misinterpreted is our old understanding of time marching along a one-directional path. One day we will finally realize that this perspective of time is only because of our limited perception of something that is an integral, geometric part of an extra-dimensional architecture.*

WHAT RELATIVITY MEANS

Relativity signaled the erosion of our existing cultural paradigm in multiple and dramatic ways. It demonstrated that we all live in a space of more than three dimensions: Einstein's spacetime. It described how spacetime is actually curved and in local areas how this curvature creates the illusion of a force that we call gravity. Relativity also showed that gravity affects the rate of the unfolding of time, and we are only beginning to understand what this might mean. General relativity also established the mathematical basis for black holes, wormholes, and other strange phenomena that today's cosmologists are excitedly exploring. Most importantly, this relativistic physics demonstrates that many of the very things that we observe and use every day in our lives, including all things that involve gravity and time, are only our neurologically interpreted secondary impressions of this hidden geometry; one that is formed and held within a type of space that requires more than our normally-perceived three dimensions.

For more than one hundred years, we have been aware that our universe is constructed from more than the familiar three dimensions that were first described by Euclid. We learned that *gravity* and *time* were only trickle-down shadows, artifacts, or *projections* cast upon our *three-dimensional viewport* from this much larger multidimensional *space*. **This multidimensional geometry, which extends through regions and realms that we cannot directly see or sense, shapes the more fundamental space that contains our universe.** Even though we have seen evidence of this expansive structure for several generations, our culture still has not integrated this life-changing knowledge into our everyday lives and worldview.

Curved Spacetime and Artifacts

One of the bigger conceptual surprises from the mathematics of *relativity* is that lines that look straight to us, in our three-dimensional *space*, are actually "curved" because *spacetime* itself is *curved*. *General relativity* demonstrated that *gravity* is how we experience the distortion or *curvature* of the *spacetime continuum* created by massive objects such as the Sun or planets. *Curved* is a word we understand well in three dimensions, but in four dimensions *curved* has a different meaning, one that we cannot easily comprehend. Since we cannot directly experience or understand *spacetime*, *curved* becomes the closest three-dimensional concept to describe this new idea about the *shape* of *space*.

Gravity appearing as a *force* is just an *artifact*, one of many caused by *dimensional reduction,* a term that describes the way that four-dimensional phenomena appear within our three-dimensional *viewport*. We are not able to accurately perceive or describe multidimensional *space* because we have evolved and been conditioned to understand and function in just our three dimensions (refer to the discussion of the novel *Flatland*).

We all understand that the ground under our feet is the surface of a curved sphere, our planet Earth, even though it still feels and looks flat to our senses. Similarly, spacetime seems "flat" when viewed in three dimensions, even though it is mathematically curved.

Mathematics as a Guiding Light

While *spacetime* is impossible to visualize, using mathematics, we can still describe and manipulate its multidimensional geometry with relative ease. Mathematics is a language that often transcends our ability to visualize and even our need to logically "understand." The reader might now ask, "Is it fair or reasonable to be equating mathematics to things that have real meaning in our world?" History demonstrates that the answer to this question is unquestionably affirmative, for *time* and *time* again our mathematical predictions turn out to have important real-world implications. Our history demonstrates that mathematics has been a clear and accurate predictor and a faithful guide. ***Mathematics has served us by providing an extremely powerful searchlight to help illuminate new paths through the unknown.***

Mathematics is a language. Since at least the days of ancient Greece, there has been a lively philosophical debate about whether mathematics exists in nature and we just discovered it, or whether it is our creation. At the deepest levels, for reasons discussed later, we will discover that these two perspectives are actually one and the same. Regardless of its origin, mathematics has always provided us with expansive glimpses into the unknown, and these have completely changed our understanding of the universe. Like microscopes and telescopes, our mathematics is another powerful tool that allows us to see beyond the normal limits of our senses.

Of course, any attempts to provide human meaning to these mathematical results are completely subject to the limits of our *viewport* and current cultural paradigm. As a result, potential meanings will also shift and change as we reach new levels of individual, cultural, and scientific understanding. The more we learn, the clearer it becomes that there are vast parts of our universe that we simply do not understand or of which we are not even aware. Mathematics is a tool that can help to provide us with theoretical clues and vistas into these otherwise invisible worlds.

It was mathematics, alone, that gave us our first glimpse of extra-dimensional space about 150 years ago. This expanded view of space is now the key to an entirely new way of understanding our relationship with our universe. Mathematics can lead us to things and places that, otherwise, are hiding beyond our imagination.

VISUALIZING A MULTIDIMENSIONAL GEOMETRY

In the next major section, we will begin our journey into the strange world of *quantum physics*. *Relativity* required us to expand our dimensional paradigm one more step: from three dimensions to four. However, as we attempt to integrate *quantum physics*, even this four-dimensional coordinate system becomes woefully inadequate.

About the *time* of Einstein's birth, several mathematicians introduced the idea that the geometry of the universe required one additional dimension, making it four-dimensional. Today, most physicists realize that a four-dimensional *spacetime* is an absolute minimum; a fair number think existence requires many more dimensions. A significant number of physicists who have been exploring *string theory, superstring theory* and *m-theory* are predicting that as many as ten or eleven dimensions might be involved.

If we seek to understand the real mechanics and structure of our existence, we must first come to the place where we can accept and recognize that existence is built upon more than three dimensions, even though we do not have the direct ability to perceive or visualize them. These extra dimensions are not just for mathematicians, for they are necessary to form and maintain the familiar structure of the physical world that we all move through every day; these invisible dimensions shape our entire experience. *The reader's willingness to accept that our universe is built upon a multidimensional geometry is a prerequisite for many of the ideas discussed in this book. This hidden geometry is infinitely expansive in its ability to allow for new possibilities, including its ability to simultaneously connect everything in the universe in ways we cannot possibly imagine with our three-dimensional minds. It is currently being demonstrated that our universe is intimately interconnected and inter-responsive, despite the vast distances of space and the billions of years that appear to separate everything in our universe. It is this multidimensional architecture that suddenly makes the "impossible" very real!*

In this multidimensional *space*, all additional dimensions fit and work together in a way in which all dimensions are of equal and critical importance to the maintenance of the entire architecture; every level is equally important because if one level were to be lost, the entire structure would collapse. Like the foundation, walls, roof, and utility systems of a house, all the parts must function well and work together; they all depend on each other. Our universe is also *holographic,* meaning that all dimensions and parts also can be understood as simultaneously "inside" and "outside" of each other; there are no dimensions that are "more important" and can be described as "higher" or "lower." Ultimately we discover that our old ideas about "inside or outside" and "higher or lower" are artifacts from our limited *viewport. Existence is entirely interconnected and interactive so all dimensions contribute equally to shape the whole. To speak of "higher" or "lower" dimensions as more (or less) important is equivalent to thinking that our hands are more important than our feet simply because they sit higher on our bodies.*

Quantum Weirdness

Relativity was the beginning of our journey into the strange world of modern physics. While *relativity* might sometimes seem illogical when viewed from our old "common sense" perspective, we are about to discover how *quantum physics* takes "strangeness" to a completely new level.

Bohr Atom Model

In junior-high chemistry classes, my generation was taught that the *atom* was the *elemental* building block of all *matter*. We were instructed to visualize these *atoms* as having a solid center, called the *nucleus,* with *electrons* circling, creating a form that is similar to our *solar system*. This was a simple model and it seemed logical for the *atom* to be constructed like our *solar system* since nature tends to repeat itself. *Families* of *atoms* were identified, and we were taught that most of the differences between these *families* were due to the number of *electrons* in the outer rings of the *elements*. Interactions between *atoms* were determined by every *atom's* strong "desire" to have its outer ring filled with its ideal number of *electrons*; this allowed the *element* or *compound* to exist in a more stable, *lower-energy* state. To someone like myself, who thinks in pictures, this *image*, called the *Bohr atom*, was clear, simple, and satisfying.

Unfortunately, this model of the *atom* turned out to be crude, approximate, and even quite wrong. My formal studies of physics began in the 1960s, and back then, very few high school physics and chemistry teachers understood the rapid changes that were revolutionizing their fields. Starting in the 1920s, many working physicists had begun to realize that this once-convenient model had several serious, even fatal, problems. It still worked to describe many interactions between *atoms*, but it could no longer be recognized as an accurate physical representation. It typically takes many generations to shift a powerful cultural memory.

Ironically, the old and obsolete model of the *atom* has been named the *Bohr atom*, after Niels Bohr, the central figure in the development of *quantum physics*. Ernest Rutherford had earlier developed this now obsolete model of the *atom* that was passed to my generation, but later, Bohr modified it to include aspects of the new and rapidly developing *quantum mechanics*. This new, modified model was named the *Rutherford/Bohr atom;* but then, in one of the great ironies of science (maybe because his name was easier to remember), Bohr's name became wrongly and forever tied to the older form of the model that completely pre-dates his paradigm-changing contributions.

ELECTRONS AND CLOUDS OF PROBABILITY

One of the biggest misconceptions from the old *Bohr atom* model concerns the character and quality of *electrons* themselves. As physicists learned more about these extremely small but critical components, they began to realize that *electrons* are not actual "things" spinning around a core center, like the planets around our Sun. In fact, they are not "things" at all, most closely being described as hazy "clouds of possibilities" that only "exist" in a strange *indeterminate state* with a *potential* for physical expression. This *potential* for physical expression was described mathematically by Schrödinger's *probability wave function,* the most famous equation from the early days of *quantum physics.*

For almost a century, experiment after experiment has produced data most easily understood by accepting the strange idea that *electrons and other small particles* do not even physically exist until the moment that an *observer* becomes involved. The act of *observation* (or awareness), by itself, is what causes what was once only a *probability wave* to *collapse* and manifest as an actual physical thing. **Today, the most agreed-upon explanation is that electrons, photons, neutrinos, and other small particles do not exist at all in our physical realm until they interact with something else, or until someone or something is observing them. Only with interaction of some kind, often involving conscious awareness, does one of the many possibilities for their expression become real and physical.**

These tiny (but not fully existing) "pre-things" behave like potential ideas that have yet to form. **At the same time, these shifty particles are what make up all matter, including ourselves. We are constructed from particles that appear to have no existence until someone, or something, else observes them.** As weird and bizarre as this conclusion sounds, this is only the beginning of strange phenomena that we continue to encounter as we venture deeper into the *quantum* world. In recent years, as the experimental apparatus improved, physicists have directly observed this *quantum behavior* in larger and larger objects. This strange behavior has been found in very large carbon-based *molecules*, and some scientists are predicting that the observation of *quantum behavior* in baseball size and even larger objects is just around the corner.

Electrons are largely responsible for all chemical reactions in nature, most of which are the very processes that allow our bodies to function. **The bigger things in this world, such as our bodies, are regulated and controlled by the behavior of these tiny quantum particles, such as electrons, which** *essentially do not even exist, are extremely elusive, and always exhibit strange behavior. Since we are built from these same mysterious components, we also are not what we seem to be. Because we are made from these particles, we are also much more mysterious than we think and have an awe-inspiring and unlimited potential for forms of expression that are unimaginable today.* As Niels Bohr was very fond of saying, *"Anyone not shocked by quantum mechanics has not yet understood it."*

THE ELECTROMAGNETIC SPECTRUM

Visible Light is only a very small part of a much larger type of *energy* that we call *electromagnetic radiation*. This is a type of *radiation* found in a wide range of different *frequencies* that have been isolated and named, but all together they describe what is called the *electromagnetic spectrum*. When we speak about *visible light, we are* referring to only those *frequencies* or colors that we can see with our eyes, a very small and limited part of this much larger family of *radiation*. *Visible light* consists of only those *wavelengths* that fall between "heat-producing, *near-infrared"* on the *low frequency* or *long-wavelength* end of the *spectrum,* to the much more energetic *"near-ultraviolet"* on the higher end. There are many other *frequencies* or "colors" that humans can't see, but other creatures can sense readily. Above and below this narrow range that our eyes and nervous system register is an extensive continuum of higher and lower *frequencies* of *radiation*. While these *frequencies* may be invisible to our eyes, they still are extremely important to our lives.

At one end of the *electromagnetic spectrum,* we find the lowest *energy,* longest *waves,* and lowest *frequencies,* which we call *radio waves.* As the name implies, we use some of these *frequencies* as *carrier waves* for radio broadcasting. As the *frequencies* get higher and *wavelengths* correspondingly shorter, we have, in order: *microwaves, infrared light, visible light, ultraviolet light, x-rays,* and *gamma rays.* As we move through this list, the *wavelengths* shrink but the size of the *energies* involved increases dramatically. We now better understand the dangers of these higher *energy* forms of *radiation,* so we try to avoid exposure to high- *energy gamma rays, x-rays,* and even *ultraviolet.* Too much exposure to any of these higher *frequencies* is known to be damaging to our bodies. The Earth's *magnetic field* and atmosphere naturally shield and protect us from these more dangerous *frequencies* of *electromagnetic radiation.*

(As a cautionary note, we are increasingly bathing ourselves in a sea of human-produced *radio waves* and even *microwaves* from wi-fi, body scanners, electric grids, and cell phones. Some of these lower *frequencies* may also turn out to be less than completely safe—we might remember that *x-rays* were once thought to be harmless. As a general precaution, we should try to limit our exposure to these low-level *microwave frequencies* until we understand more about their long-term effects.)

Without *visible light,* we would not be able to see. Other species can "see" different parts of this *spectrum;* plants, birds, and bees "see" and use frequencies such as *infrared* and *ultraviolet* which fall above and below our visible range. In sonic frequencies a similar thing happens where dogs and other animals hear higher and lower-pitched sounds then we hear. ***Once again, we are reminded that so much, even in our known three-dimensional universe, remains hidden to our senses. Our inventions and instruments extend this range but they also have limits. We can only directly experience a small part of our known universe; vast amounts remain hidden and therefore mysterious.***

V*isible light* is responsible for much of *photosynthesis:* the critical beginning of our food chain. Plants can also use other parts of the *electromagnetic spectrum* that are completely invisible to us, such as high *ultraviolet,* which is outside our human range of vision. ***Photosynthesis in plants, an essential process that builds carbon-based matter, is another powerful reminder that we are all constructed from energy.***

Electromagnetic radiation is critical for physical life; we completely rely on it for many essential living processes. Without this *light energy* from the Sun, there would be no life on Earth, at least as we understand it. While *visible light* contributes some of the heat that fuels our planet, most of Earth's *heat* comes from the invisible *infrared* bands of *electromagnetic radiation* that fall below the lowest end of the *visible spectrum.* None of the plants and animals living on our planet could exist without this type of *radiation*; even deep-sea or cave creatures, which do not require it directly, still depend on it indirectly.

We can apply most of this book's discussions about *visible light* to all the other *frequencies* of the *electromagnetic spectrum.* The entire *spectrum* plays a critical role in our everyday lives, even if we are not consciously aware of this fact; it is responsible for transmitting and communicating *energy* and *information* at the *speed of light* throughout our planet and universe. Most of the *time,* when physicists refer to *light* they are speaking about the entire *electromagnetic spectrum.* I will do the same.

QUANTA AND THE PHOTOELECTRIC EFFECT

The first detailed mathematical description of the *electromagnetic spectrum* is attributed to James Maxwell in the mid-1800s; his groundbreaking physics paved the way for Einstein. Over the next fifty years, Faraday, Boltzmann, Plank, and several other physicists also studied the unusual nature of *electromagnetic radiation.* However, it was Einstein, in the first of his four groundbreaking research papers published in 1905—*The Photoelectric Effect*—who was the first to demonstrate that light travels in discrete particulate packets.

This *photoelectric effect* occurs because small discrete "packets" of *light,* which he named *quanta,* act like *particles* to then strike *electrons* within the material sending these *electrons* into motion. The easiest way to visualize this phenomenon is by thinking about what happens when the cue ball strikes other balls in a game of billiards. When *energy (mass)* is released from *matter,* it is always grouped in these discrete bundles or quantities, called *quanta (quantized* amounts of *energy).* The name *quantum physics* refers directly to Einstein's discovery that *light* and all other forms of *electromagnetic radiation* exhibit this property.

Einstein's paper on the *photoelectric effect,* along with others (including his two papers that later became the *theory of relativity),* established this short fifteen-year *time* span as the single most productive period of his life. However, one of the greatest ironies in the history of science is that the name *quantum physics* came directly from Einstein's research, even though he was not a *quantum physicist.* He spent much of his later life fighting the extended implications of his own discovery and never fully accepted *quantum physics.* His famous quotations, *"God does not play dice with the universe,"* and *"Spooky action at a distance"* were expressions of his criticism and distrust of *probability or superposition* and *non-local action* in *quantum theory.* Quantum weirdness seemed too strange even for him.

Einstein's work on the *photoelectric effect,* an important piece in the *quantum* puzzle, earned him the Nobel Prize in 1921, some sixteen years after its publication. Today we recognize that he deserved another Nobel Prize for *relativity,* but at that *time relativity* was seen as such a dramatic departure from anything known that the Nobel committee was not prepared to "stick their necks out" and award him their prize. The *classical physics* paradigm was rapidly dissolving as the old, once-reliable laws of physics were being turned upside-down from multiple directions, so it was completely natural that the established scientific community would be extremely confused. ***It was during***

this period of rapid discoveries that our once-certain world began to be recognized as a much more uncertain world, but also one full of infinite, new, and breathtaking possibilities.

WAVE-PARTICLE DUALITY AND OTHER STRANGENESS

The *photoelectric effect* also proved to be the key to uncovering one of the strangest principles of *quantum physics* —the dual or *wave-particle* nature of *light*. Einstein's work demonstrated that *light* could behave like a beam of *particles*, which he called *photons*. This *particle nature* of light is what causes the *photoelectric effect,* and this collision and the resulting movement of *electrons* is what creates the electricity in our solar battery-chargers, daylight sensors, solar lighting, and all our other *photovoltaic* systems. *Light particles* (*photons*) hit the *photocells* and dislodge *electrons*, causing the loose *electrons* to move (or at least move their charge) and generate electricity. **The fact that light can behave as if it were a beam of particles is now a fully integrated part of our scientific understanding and contemporary lives.**

However, through experiments that began in the early part of the twentieth century, physicists and chemists discovered that light could also behave like a wave. **When we look at the characteristics of light experimentally, we sometimes see light acting as a stream of particles and sometimes like a wave**. After a few well-analyzed "happy accidents" in laboratories, along with many clever, well-designed experiments, where all produced the same strange results, **physicists reached the difficult realization that light exhibits both wave and particle behavior, but never both at the same time.** At any given moment, *light* behaves as if made from **either** *particles* or *waves*. Physicists call this "not-yet-determined" state of "multiple possible expressions," *quantum superposition. The Architecture of Freedom* has detailed descriptions of the specific experiments that first revealed this strange and split personality of *light*.

As we investigate further, this split behavior gets even stranger because what determines whether light acts as waves or as particles seems to depend upon what the observer expects or anticipates. Because we are the observers, it is our expectations that determine how this electromagnetic radiation behaves or appears. The physics only gets weirder as we add the equally strange realization that before the moment of observation, these particles only existed as a probability wave; they were not yet "real" physical "things." They were, instead, in some strange kind of indeterminate state where they only had the potential to exist. They require some act of *observation* before their "wave of potentiality" can *collapse* to allow the *particle* or *wave* to appear as "real" and measurable.

In other words, in some unknown way, we create our 'real' world with our expectations. Here, we discover that our act of conscious participation will influence or determine the physical outcome. This is revolutionary because it means that our best science is telling us that our expectations directly influence the eventual expression of our physical world!

Particles also constantly "jump" to different locations, but they seem to do this in an unexpected and peculiar way: without passing through the *space* in-between. When an *electron* jumps to an entirely different *energy* state, it never appears anywhere in the middle, either physically or in the measured values. It is as if it just disappears from one place and then reappears in another.

The wave-particle decision and the "jumping" seem to happen instantly, meaning that the transfer of this information appears to happen even faster than the speed of light. These extremely small *particles* do not seem to understand *time* in the same way that we do. *Individual particles striking a target will behave like a single coherent wave, even if vast amounts of time separate the release of these individual particles; the passage of time does not appear to change their behavior at all.* Another astounding realization is that *these small particles can also appear to be present in two or more places at the same time.* (It is again informative to remind ourselves that our bodies are also made from these same tiny but very strangely behaving *particles*.)

One of the most important *quantum* principles uncovered is that we cannot measure both the *position* and the *momentum* of subatomic *particles* at the same *time*. If we measure *position*, the very act of measurement interferes with the *particle* in a way that makes accurate measurement of *momentum* no longer possible. The reverse is also true, so measuring *momentum* means we cannot accurately locate the *particle*. This is the *uncertainty principle*, first proposed by Werner Heisenberg, which states that we will always be uncertain about either the *position* or *momentum* of *quantum particles*. *Again, we are reminded that whenever we interact, we will always change the very thing that we are observing.*

One of the most interesting *quantum* properties is the ability of *subatomic particles* to communicate with each other instantly over *infinite* distances. Physicists call this characteristic *entanglement* and describe the *particles* as acting *non-locally,* which means that their location in *space* does not seem to matter or have an impact on their interaction. Einstein referred to this phenomenon as *"spooky action at a distance,"* since this was one of the implications of the experimental data from *quantum physics* that did not sit comfortably with him. *Non-local interaction* is an extreme understatement for a quality that is completely astounding and paradigm-altering. To understand *non-local interaction* requires that we radically transform our entire view of *space*, the universe, and even of life itself. Timothy Ferris, the former Berkley professor and the author of *The Whole Shebang,* describes this phenomenon using one of my all-time favorite quotes about the strange *quantum* world: ***"It is as if the quantum world has never heard of space—as if, in some strange way, it thinks of itself as still being at one place at one time."*** These small *particles* behave as if the vast distances of our three-dimensional universe are meaningless.

For over half a century, *entanglement* was considered interesting but quirky, and not of any real value to physics or engineering. This started to change in the 1970s as *lasers* came into labs and *entanglement* could be accurately tested (using Bell's inequality). In 2019, the world witnessed the first real-world experimental demonstration of this property. *Entanglement* turns out to be highly valuable for transmitting extremely secure network *information* (*quantum encryption*), and many other potential practical uses that are being developed as I write. This new *quantum* world is strikingly different from the old world we once thought we understood so well; it makes "real" so many previously unimagined possibilities.

MODERN QUANTUM RESEARCH

Quantum physics is the physics of extremely small things such as *particles,* like *photons, electrons,* and *quarks*, which are all much smaller than the things we can actually "see" with today's equipment or instruments. Since our eyes or instruments cannot directly see them, the behavior of these invisible *particles* must be explored indirectly by mapping the trails or traces that they leave on various detectors or screens. Unraveling this *quantum* world has been a primary focus of physicists for almost one hundred years.

Due to the high *energies* and speeds involved in *particle physics*, the equipment required for this type of research has been getting larger and larger as we discover new ways to peer into the smaller and smaller. While scientists conducted the early experiments with devices that easily fit on tabletops or in rooms, the newest piece of equipment for this type of study, the *CERN Large Hadron Collider*, an accelerator in Switzerland, is seventeen miles in diameter.

In Texas, construction began for an even larger and more powerful accelerator that was to be more than fifty miles wide. Named the *"Super-conducting Supercollider,"* its construction was halted at the halfway point in 1993 because of politics and budgetary concerns. The abandoned tunnels are now being used for growing mushrooms.

Using this expensive equipment, along with simpler experimental devices and modern astrometrical measurements from *space*, scientists have now collected a substantial amount of data and knowledge about this mysterious world of the ultra-small.

THREE BRANCHES OF QUANTUM THINKING

As *quantum physics* started to mature, and philosophers and physicists tried to understand and explain the unexpected and bewildering experimental results, several distinct schools of thought emerged.

COPENHAGEN INTERPRETATION

The original, and for many years the dominant, *quantum* explanation is called the *Copenhagen interpretation.* Niels Bohr held the position as its champion and chief interpreter until he died in 1962. His was a tightly encamped group that believed that *quantum* weirdness was best explained by accepting the fact that these small *particles* only exist as *waves of probability* or *potential* until the actual act of *observation.* (*Superposition* is their name for this indeterminate state of *matter* that exists before actual *observation.*) **At its most extreme, this view implies that there is no objective reality because it is only the actual act of observation that nails down all particles, and, therefore, our reality.** The Copenhagen camp went even further to declare that if something is not *observable,* then it does not even exist and is, therefore, not even worth talking about. **In other words, reality does not exist without observation.**

MANY WORLDS INTERPRETATION

In 1955, Hugh Everett proposed a very interesting, alternative explanation called *The Many Worlds Theory.* In a single bold leap, he introduced the radical idea that every *time* a decision is made or a path is chosen, our "world" actually splits or divides. With this split, another new world is instantly created and, even though we consciously observe only one world at a *time*, all the other possible outcomes exist from that moment on. Every *time* we make an act of observation or a decision, our world splits again and creates new versions that represent all the possible outcomes. All these new worlds then have the same *logically consistent* history, but now multiple new worlds exist, together expressing all paths to every possible future outcome. This process looks like an ever-branching flow chart or the branches of a tree. Another way of stating Everett's theory is that **all possible outcomes do occur in a very real and concrete sense, and every possible outcome results in an entirely new branch of history.** The movie "Sliding Doors" portrays how this might unfold within our lives.

"MANY WORLDS," BUT NO TIME-MY PERSPECTIVE

Everett's original *Many Worlds* view proposes that all possible outcomes are being "created" at the moment of decision or *observation*. This means that our three-dimensional sense of *time* is once again involved and responsible for ordering the events. However, I believe that if we can step beyond the *"time-ordering"* of our conscious minds, it could better be said that all of these possibilities have always existed, exist now, and will always exist in a *space* that is "outside" of *time*. **Therefore, these "many worlds" are not being manifested with a collapse of the wave at the moment of our observation or decision. Instead, every possible universe already exists but only in a realm that is beyond time. From our "time-bound" perspective it would appear that they already and always exist in a woven web-like continuum.** Our decisions and observations only determine where in this continuum we travel. We only perceive this as a moment in *time* due to our persistent illusion of a discrete moment of *time* that is generated by our brains. We then become aware and somewhat *attached* to one of these "pre-existing" worlds, while another individual's awareness may follow a different path. We all live in different worlds of our own making.

BOHM'S ENFOLDED UNIVERSES INTERPRETATION

The third major interpretation of *quantum* experimental results is David Bohm's *implicate order* or *enfolded universe* theory. While it was proposed a few years after Everett's *Many Worlds*, it was not until the 1970s, when *entanglement* experiments began to confirm the fundamental cornerstone of this theory (the *nonlocal* nature of *particles*) that it started to receive the attention it deserved.

This interpretation of *quantum mechanics* implies a "swirled" order to the universe, where every part is in direct contact with every other part. A demonstration, using the common salad dressing of peppered oil and water, provides one of the best ways to visualize *enfolding*. While the dressing is sitting still on the shelf, the three layers are very separate; but when shaken, suddenly the oil, water, and pepper touch each other everywhere.

Enfoldment must involve more dimensions than three because it allows for "direct contact" to occur over any "distance." This type of spatial interconnection can't occur with just three dimensions. However, through the nature of the "impossible to understand" *space* provided by extra dimensions, this unexpected interconnectedness suddenly becomes a possibility. *Nonlocal interaction* is only made possible by the type of multidimensional *space* that exists beyond our familiar and comfortable three dimensions. **Because of the deeper enfolded nature of existence, every part of the universe is always intimately connected to every other part of the universe. Nothing can ever be hidden. Everything in the universe is always fully interconnected, communicating, interactive, and responsive. Every thought, observation, or action is communicated throughout the entire universe and contributes to the unfolding of existence. Every thought, word, or action by anyone or anything has meaning for everyone and everything.**

Bohm, and many others since, compare this vision of the universe to a *hologram*, where every piece of *information* is stored everywhere in such a way that we can see a record of all existence within even the smallest pieces. Discussions about our *holographic universe* and its incredible implications are scattered throughout this book.

More and more physicists are taking the idea of *enfoldment* a step further by theorizing that, at some deep level, the reason for this *non-local behavior* and instant interconnectivity is that **all things that appear to be separate are only one thing!** As the careful reader already knows, I believe this represents a critical and fundamental truth about our existence. **Every thought or action from everyone and everything fully contributes to this incredible experience of life, an orgy of expression that is the visible manifestation of our "oneness."**

MOST RECENT INTERPRETATIONS

As physicists continue their quest for *unification,* two different and currently distinct areas of focus have emerged—*string theory* and *quantum gravity theory*; the former derives from *quantum physics,* while the latter builds upon *relativity*. Both directions already overlap and will most likely merge because they are both seeking to answer the same questions. Of the two approaches, *string theory* pushes the old paradigm further because it requires adding at least six more dimensions to our model of the universe.

ELEMENTARY BUILDING BLOCKS

String theory emerges directly from mankind's search for the smallest *elementary* building blocks. Many ancient cultures believed that air, water, fire, and earth were the *elementals*, the basic building blocks of *matter*. Then beginning with the ancient Greeks, the *atom* was recognized as the primary physical building block of *matter*. For 2,000 years, we believed that *atoms* could not be subdivided; they were the smallest indivisible *particle*. However, in the early twentieth century, we discovered that the *atom* itself was constructed from *protons, neutrons,* and *electrons*. Most physicists viewed this simplification as great progress since three small *particles* had replaced a list of more than one hundred larger *elements (atoms)*. However, those who felt that we had finally found the smallest pieces of *matter* with these subatomic *particles* were soon surprised when, in the 1960s, *particle physicists* discovered *quarks*, the even-smaller building blocks of *protons, neutrons,* and *electrons*.

These new *subatomic particles* behaved just as predicted and strictly adhered to the rules of *quantum mechanics*. At the *time quarks* were discovered, these *quantum* rules were called the *standard model*. Unfortunately, even with all these discoveries of small and then even smaller *particles*, there still was no acceptable way to combine *relativity* and *quantum theory*. For most physicists, the continuation of this *unification* problem indicated that something was not quite right.

EARLY STRING THEORIES

In the 1980s, a new theory called *string theory* re-energized the *unification* movement by presenting an entirely new interpretation for the *quantum* experimental results. This theory proposed that even *quarks* could be subdivided, and this *time* the smallest elemental pieces were no longer *particles*, rather more like tiny vibrating *strings*. According to *string theory,* everything physical manifests from these extremely small, vibrating *strings* that exist in various shapes, some even forming loops. S*trings* are always *vibrating*, and the precise way that they *vibrate* determines specifically which types of *particles* they manifest. **Even more interesting, the math of string theory indicates that these strings can only exist in a minimum of ten dimensions.**

String theory and its spin-offs (*super-string, super-gravity,* and *m-theory*) are mathematical and theoretical; there presently is no experimental proof of their existence. Since these theories are based on many extra dimensions, any experimental proof will likely elude us for a long *time,* for we have yet to conceive of a way to conduct physical experiments for testing theories that involve these extra dimensions. However, even without this experimental proof, this mathematics has created enormous excitement amongst physicists. It works relatively well, and the new forms revealed make a special kind of sense to those most familiar with this type of mathematical exploration.

The sticky problem of *gravity* not being mathematically describable from within *quantum mechanics* was addressed when *string theory* was further "modified" to allow for the existence of the *graviton,* the theoretical *particle* that obeys all *quantum* rules and helps to explain *gravity.* With the addition of this *graviton,* many physicists saw new potential for *unification* and called this sub-theory and its variant *super-string* and *super-gravity,* respectively. Before long, four other fully consistent *super-string* theories that described other possible arrangements were hypothesized, but there was a big problem because they all seemed to compete directly with each other.

M-THEORY

Finally, in the mid-1990s a *string theorist* named Edward Witten, who some feel is the most brilliant living mind on the planet (and not just for his contributions to physics), determined that the five different, originally-proposed *string theories* were simply different views of the same theory (again-*artifacts*), and could be combined into one theory if one more dimension were added to *string theory.* **M-theory, which is his further refinement of string theory, therefore requires eleven dimensions instead of just ten.**

According to *m-theory,* the source of all existence seems to be a *membrane* functioning like a multidimensional drumhead. The *strings* from ten-dimensional *string theories* still exist; but since they are built from one less dimension than these eleven-dimensional membranes, they can be thought of as dimensionally-reduced slices of this *membrane*—a relationship that is similar to when the sphere is being seen as a circle in the 1884 novel *Flatland.*

In *m-theory,* these multidimensional vibrating *membranes (also called m-branes, or just branes)* are the source of all material creation. In our three-dimensional *viewport,* all that we can ever see or experience of these *membranes* are the final *projections* (artifacts or shadows) that eventually manifest as the visible "objects" of our reality. **The particles that we encounter in our three-dimensional universe are only the dimensionally reduced artifacts, shadows, or projections that ultimately reach us from this unfathomable vibrating eleven-dimensional symphony.**

One of the most exciting potentials of *m-theory* is that it provides a pathway to combine *relativity* and *quantum mechanics* in a mathematically elegant way; this quality alone has attracted the attention of many in the physics community. **While m-theory does explain more of the mystery and connects relativity and quantum mechanics, it continues to bother many physicists because it is completely untestable and will remain so until we find a reliable method for peering into and experimenting with extra-dimensional space.**

Since these *strings* or *membranes* are said to *vibrate*, as they *vibrate* together they would then create music-like *resonances* and harmonies. These *vibrational* symphonies would then pass through the eight layers of dimensions (filtered from their eleven dimensions down to our three) to manifest as the small physical *particles* that are the building blocks of all *matter* in our realm. **According to the various string theories and m-theory, all matter begins with this type of vibration.**

In the beginning, there is this "song," which is originally performed in an eleven-dimensional concert hall! That with which we interact directly is only that projection (artifact) of the symphony, translated from dimension to dimension until it finally reaches our three-dimensional world as the physical expression that we experience.

QUANTUM GRAVITY

The impossibility of proofs for theories built upon eleven dimensions and the persistence of the sticky problem of gravity led theoretical physicists to focus their search closer to home. *Quantum gravity* theorists are attempting to incorporate *quantum observations* into a more understood and proven theory that already includes *gravity*: *Einstein's relativity*. *Quantum gravity* theories all are built upon *relativity* and operate in a better-understood realm, where proofs are still workable and *gravity* is already incorporated. Here, rather than probing the world of tiny *particles* and attempting to incorporate *gravity* into *quantum* theory as *string theorists* do, physicists are exploring the *quantum effects* that are fully visible in the greater cosmos. Most of these explorations *quantize relativistic spacetime* at the scale of the *Plank's constant*, a unit that is so minuscule that it lies far beyond our experimental capabilities.

Unfortunately, this may mean that this approach is, also, completely untestable. The most recent theory of this family is *loop quantum gravity* which describes the universe as a mesh of *quantized* points in time and *space*; a form that closely resembles what I have been calling the "Web of *Infinite* Possibilities." *Loop quantum gravity* is also a theory where the intersection with *string theory* becomes more apparent. One of my favorite resources for a better understanding of *quantum gravity* is Carlo Rovelli, an Italian physicist with a great gift for clear explanations in English.

UNIFIED FIELD THEORY

As discussed, the dominant and pressing problem within modern physics has been the inability to find a way to combine *quantum theory* and *relativity*. Spread between these two types of physics, we have useful descriptions of all four of the fundamental *forces*: *gravity, electromagnetic, weak nuclear,* and *strong nuclear*. The first of these, *gravity,* is explained only by *relativity,* while the other three are described quite well through *quantum theory,* yet most physicists agree that any complete description of our universe must include all four of these *forces*. Many talented physicists have devoted their entire lifetimes to this project, but up until now, all of their attempts to combine these two theories have been repeatedly thwarted. This *unified field theory* has been the Holy Grail of physics, but it was not until the 1970s, with the discovery of *string theory,* that anyone saw real possibilities for *unification*. I just discussed the other recent *unification* theory, *quantum gravity*. It also holds some promise for *unification* because it derives directly from *relativity,* meaning that it is built upon more-understandable four-dimensional *spacetime*. Because it is limited to the four more understandable dimensions, it may not be as far-reaching as the string theories, but this characteristic also makes it much more likely that a proof will be found.

When we include eleven dimensions into our calculations, as *m-theory* requires, suddenly *quantum mechanics* and *relativity* start to fit together more easily. Calculations become mathematically elegant as the equations balance and the pesky problem areas begin to self-resolve. With these extra dimensions in the mix, *quantum physics* and *relativity* start to make mathematical sense together. Although currently unprovable, *M–theory* may still be the best trailhead for guiding us to the long-sought-after *unified field theory*.

COSMOLOGY

Cosmology, the study of the origin and development of our universe, involves every branch of science. Modern *cosmology* began with the paradigm-shifting work of Kepler, Copernicus, Brahe, and Newton, but the last 100 years of *relativity* and *quantum physics* have completely revolutionized the field. Just as *classical physics* helped usher us out of the "flat-Earth" paradigm, *relativity* and *quantum physics* have stretched the limits of imagination by introducing many new ideas about the potential nature and *shape* of our universe. **Since the biggest things are made from the smallest, we are now discovering that there is far more connection between "out there" and "in here" than we ever imagined.** Because of recent discoveries, *cosmology* now includes dramatic new approaches to issues such as *gravity, space, time,* consciousness, and the *observer.* In addition, a steady stream of "outside" discoveries uncovering mysterious things such as *black holes* and *dark matter* are rapidly, and thoroughly, changing our old ideas about the nature of our universe.

DARKNESS WITHIN OUR UNIVERSE—DARK MATTER

One of the most unexpected discoveries in modern *cosmology* has unfolded as new technologies and methods allowed scientists to more accurately analyze the relative motion of the various objects of our known universe. It has become clear that there is not enough visible *mass* present to produce the relative motions of the stars and *galaxies*, account for rotational speeds of *galaxies,* and explain the *gravitational bending* of light from distant stars. The amount of *gravitational matter* necessary to explain these inner workings of the universe was not just a little bit off. It was not even off by double—it was off by at least a factor of five! This means that the visible *matter* is less than one-fifth of what is necessary to produce the relative motion of everything we can see or measure. By some other calculations, we only see and understand about one-tenth of the *gravitational material* that must exist.

One of two things is probably occurring: either all physicists are making a colossal mistake in theory or calculations, or between eighty and ninety percent of the *gravitational* "stuff" that makes up our known universe is completely invisible to our senses and technology. We have no idea what this invisible *gravitational material* is, but we know that it affects and interacts with everything else that we can see.

Since this mysterious "stuff" emits no light and we cannot see it directly, we call it *dark matter*—meaning it is "dark" to our tools and senses. We now include this *dark matter* in all contemporary calculations about the cosmos, even though its existence only became fully recognized in the 1990s.

To further complicate this picture, it was recently discovered that dark matter is not the only hidden "dark" stuff in our universe. There is also dark energy, of which there is even more than dark matter. Before we examine *dark energy* in more detail, we first need to understand what is being described when cosmologists talk about the *shape* of our universe.

SIZE AND ARRANGEMENT OF OUR UNIVERSE

Looking at the extents of our known physical universe and wondering about how big it is, scientists imagine that there are five basic possibilities. The first possibility is that the universe could be *finite*, alone in existence, and have an edge somewhere beyond the limits of our present view. The second possibility is it could be alone, but be *infinite* and extend forever. The third is an interesting hybrid of the first two: the universe could be *finite* and *closed* in a bubble-like fashion, but it might be only one of an *infinite* number of these *finite* "bubble" universes. The fourth possible combination is that it is *finite* and one of a *finite* number of other universes; and the fifth combination is that the universe is *infinite* and one of an *infinite* number of other *infinite* universes.

The first possible configuration, the lone, *finite* universe, is the easiest for most of us to understand. With this configuration, the entire universe would simply end somewhere beyond our *cosmic horizon,* which today is a distance of about 40 billion *light-years* in any direction. A light-year is the distance that light travels in one year, which is almost six trillion miles. The light from the most distant stars that we can see today originally left those stars about 15 billion years ago, back when these stars were much closer to us. Since then, the universe has been rapidly expanding, so our visible universe has now expanded in size from 15 billion *light-years* to about 40 billion *light-years* in every direction. Since this particular configuration of the universe has been deemed to be *finite*, somewhere beyond the limits of our vision, the physical universe must somehow end. (To me, this sounds very similar to the way our "flat-Earth" ancestors imagined their flat pancake world to have an outer edge.) If this particular arrangement was the full extent of our universe, it would be *finite,* but it would still be enormous and full of breathtaking possibilities. Today, we have observed hundreds of billions of *galaxies* and each one contains hundreds of billions of stars. Even without additional regions or dimensions, this type of universe would still be large enough for the existence of other worlds similar to our own.

To understand the other four possible arrangements, we will first need to become familiar with a few new terms and concepts. When cosmologists speak of the *shape* of the universe, they are referring to a specific mathematical quality that lies beyond our three-dimensional ability to visualize. We now understand from Einstein's *relativity* that a large *mass*, such as a star, deforms four-dimensional *spacetime* locally, near the large *mass,* and we call this deformation *gravity*. All the *matter* and *energy* in the universe, because they are two forms of the same thing, also work *non-locally* and in unison to deform, or bend, the entire universe. Scientists named this larger deformation of the universe its *shape*. To simplify our discussion, we will use well-understood, three-dimensional forms such as ball, pretzel, saddle, and tabletop to describe these *shapes*, while always remaining aware that these are only three-dimensional approximations of the actual four-dimensional forms, a concept used to satisfy our way of thinking since four-dimensional forms are impossible for us to visualize. Our human ideas about "shape" are quite different from any actual *shapes* that would exist in a *multidimensional* universe. When we speak about the *shape* of our universe, we are referring to its shape within *spacetime*, which is a realm that we cannot see or even imagine because *spacetime* is built from more than our familiar three dimensions, and we have no spatial concepts that allow us to describe or visualize it. These mathematical *shapes* lie outside our three-dimensional *viewport,* beyond our *conceptual horizon.*

Cosmologists also describe the three potential ways that the universe might curve: it could have a *positive curve*; it could have a *negative curve,* or it could be *flat* with a *zero curve*. There are other variants and possibilities, some of them very specific, but these three *shapes* are the most important ones to understand for this level of discussion. The direction of *curvature (positive, negative,* or *flat)* is determined by the total amount of attractive *mass* and *energy* that the universe contains within a given amount of volume—its density. A higher density (more *matter* and associated *energy)* will produce a *positive curvature* – a ball *shape* is the closest three-dimensional equivalent. A lower amount of *energy* and *mass* will produce a *negative curvature,* with a saddle *shape* being the best three-dimensional equivalent. It takes just the right amount of gravitational material to produce a zero curvature or *flat shape;* in three dimensions, a tabletop describes this shape.

Knowing its *shape* helps us understand whether our universe is *infinite* or *finite*. A *positive curvature* will lead to a closed system, which will always be *finite*. A *negative curvature* usually, but not always, results in an *infinite* universe, while a *zero curvature* (*flat*) means that the universe **must** be *infinite.*

There is another possibility that the universe could *curve* back on itself and twist in some unexpected way that resembles a four-dimensional racetrack, a variant that we will call a pretzel *shape.* Then, instead of looking deep into eternity, we might be looking at some of the same objects multiple times, and from different angles, due to multiple loops and twists around the twisted track. This idea evolves directly from Einstein's work. If the universe *curves* in this way (convex or ball), it results in a very big universe, but not an *infinite* one; this type of universe has been called a *bubble universe.* If ours turns out to be a *bubble universe,* it could be all alone, one of a *finite* number, or one of an *infinite* number of these bubble universes.

However, a substantial amount of new but very reliable data is surprising many cosmologists because, if correct, it means that our universe has exactly the right amount of mass, including all dark energy and dark matter, to make the universe flat. This also means that, to the best of our current scientific knowledge, the universe must be infinite.[2]

It is also important for scientists to understand whether the universe will continue to *expand,* accelerate its expansion, or eventually start to *contract.* This, after all, will be a critical factor determining any "future" for our Earth! The mathematics of a *flat universe* is particularly elegant because, in a *flat universe,* the total overall *energy* will be zero. Therefore, the *flat* universe is also one form that could have emerged from a net-zero *energy* condition during a "creation" process. Analyzed mathematically, we find that this *flat* universe, without any other input, will continue to *expand,* but over *time* the *rate of expansion* will gradually slow down.

DEEPER DARKNESS – DARK ENERGY

When Einstein presented *relativity* at the turn of the last century, he was personally quite certain that the universe was stable and *non-expansive*; but to his dismay, he discovered that his equations demonstrated otherwise. In the late 1920s, Edwin Hubble confirmed by observation that the universe was

---

[2] This new data was compiled from the WMAP, *Wilkinson Microwave Anisotropy Probe* satellite, which was launched in 2001. Its seven-year mission was to make fundamental measurements of *cosmology,* including measuring *cosmic microwave background radiation.*

indeed *expanding*, but he could not yet tell if this expansion was slowing down and might someday reverse.

If constant or *accelerating* expansion is occurring, then some powerful *force* must be continuing to push out: one that is more powerful than the opposing contraction caused by *mass*-induced *gravity*. It is relatively easy to understand why this is so. If we throw a ball straight up in the air, at first it travels quickly, but then it gradually slows down (*de-accelerates*) until it eventually stops and falls back to Earth at an ever-increasing rate of speed. Because gravity pulls it back, the thrown object eventually falls unless there is some additional *energy* added into the system, such as the firing of a rocket engine, to overpower the gravitational *force* pushing in the opposite direction. A single impulse event such as the *big bang* is analogous to the act of throwing a ball up into the air. If the universe began with a single *big bang* with no additional *energy* added later, then due to *mass* attraction (*gravity*), the outward movement would eventually stop and reverse itself. The universe would then collapse back into itself.

Through the years cosmologists have learned that the *expansion of the universe* is not slowing down as quickly as would be expected from a single impulse *big bang* kind of event. By using Earth-bound telescopes and instruments, satellite observation, and, finally, the WMAP satellite, they recently determined that some unseen additional *force* must be working against *gravitational mass attraction* and is responsible for adding *energy* to the expansion of the universe. Without this extra *force* pushing out, the universe would eventually start to contract. From the best current data, most cosmologists are quite sure this contraction is not going to occur, and there is no question that some repulsive *force,* opposing *gravity,* must be adding to and reinforcing, the initial *big bang.* **This means that there exists a mysterious type of repulsive energy that we cannot directly see or measure.** This *energy* has been named *dark energy*. Like *dark matter,* it is not visible to us, but we need it to explain the types of movement we are observing in our universe.

While Einstein demonstrated that normal *matter* and *energy* are equivalent, it is clear that *dark matter,* which makes up at least 80 percent of the *gravitationally attractive mass* of the universe, and *dark energy* are very different things. Our naming this newest unknown influence "energy" was somewhat unfortunate and inconsistent because *dark energy* is not just another form of *energy*. Normal *energy* is convertible to *mass* and is, therefore, attractive or *gravitational.* Similarly, *dark matter* produces an attractive *"force"* so it behaves just like other matter: it *gravitationally attracts.* On the other hand, *dark energy* counteracts the normal *gravitational* action of *matter* and *energy*; it produces a *repulsive "force,"* which makes it completely different, even opposite, from "normal" *energy.*

We have absolutely no idea what this outward "force" is, how it works, or where it comes from. However, cosmologists have calculated that it represents 73 to 76 percent of the total energy and matter in the universe. We have only named this *force "dark energy"* because, once again, we cannot see it. *Dark matter* and *dark energy,* while very different, do have three things in common. First, they both affect the size, shape, and expansion of the universe in significant ways; second, we can't "see" them; and third, we know almost nothing about them.

Let us take a closer look at the astounding breakdown of the composition of our "known" universe. **The most recent WMAP III analysis indicates that the universe is 76 percent dark energy, 22.5 percent dark matter, and only 1.5 percent baryonic matter.**

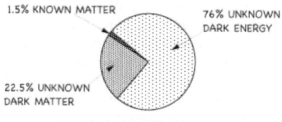

1.5% KNOWN MATTER

76% UNKNOWN DARK ENERGY

22.5% UNKNOWN DARK MATTER

OUR "KNOWN" UNIVERSE

(Baryonic matter is normal physical matter, such as cosmic dust, planets, stars, asteroids, etc.). When we examine the *atoms* of this very small percentage of *baryonic matter,* we find that even this, which makes up such a small part of "everything that we think we understand," is mostly empty *space.* On top of that, this *space* within the *atoms* is also filled with *dark energy.* **The actual amount of physical "stuff" that we (and all the stars, planets, asteroids, and cosmic dust) are made of is, at the most, a minuscule proportion of the observable parts of existence. We know little or nothing about what gives form to our universe and, therefore, form to our bodies! Everything we think of as matter is some form of energy, and the majority of the energy shaping our universe is completely mysterious and invisible to our senses and our devices. In addition, when we observe our universe, we are only looking at a tiny part of existence because we have not yet addressed any of it that is farther**

away than 15 billion light-years, smaller than a quark, or existing beyond three dimensions.

Over the last one hundred years, we have made enormous advances in our understanding, but we have also learned that the deeper we peer into the nature of the cosmos, the more we uncover a much larger and rapidly growing mystery. With each deeper probe, the number of discoveries that cannot be rationally explained seems to be expanding to occupy a much more significant part of our scientific awareness. The deeper we explore, the more we continue to discover new unknowns and perspectives that clearly must exist, yet lie beyond our comprehension. **As we learn more, the unknown parts of the mystery only seem to grow.**

This discussion has focused on just the visible or measurable three-dimensional portion of our universe. Even if our universe were only three dimensional and found to be finite, it appears that in some of its strange and extended pockets, our universe has room for a never-ending vastness: a wondrous and expansive playground for the expression of existence. However, as discussed, a *finite* universe is not likely because, through just our three-dimensional exploration, we are discovering multiple avenues to the *infinite*. **If our universe turns out to be finite, then due to its enormous size and extents it is possible, even likely, that other "worlds" (some may be just like ours) exist within its vastness. If, however, the universe is found to be infinite, there can be no real debate among physicists and mathematicians about the likelihood of other worlds exactly like ours. If it is infinite and also "outside" of marching time, then this very strange and unbelievable outcome is an absolute given.**

Why this is so all hinges on the meaning of our concept of *infinity*. Since there is likely to be confusion about the mathematical concept of *infinity*, I will discuss and explain it in more detail before we proceed.

UNDERSTANDING INFINITY

> "If the doors of perception were cleansed, everything
> would appear to man as it is, infinite."
> *William Blake*

Our minds require relationships to function; we are constantly referencing or relating new experiences and ideas to older concepts that we already know and understand. Our minds tend to work more easily within a certain range of numbers and sizes. Easily understood are numbers like one, ten, one thousand, or even fractions, such as one-quarter. A bit more difficult to understand, at least until we have practice, are quantities such as one million dollars (six zeros), the Earth's eight billion inhabitants (nine zeros), 22 trillion dollars of U.S. debt (twelve zeros), or, on the other end of the scale, very small numbers, such as the width of a human hair (100 microns or 4/1000 of an inch). While these numbers may be quite large or small, they still have a very direct relationship to our lives. However, when we start looking at distances between ends of the visible universe (the enormous) or size of the *particles* that make up *matter* (the minuscule), our ability to comfortably understand these numbers falls short because we have nothing in our direct experience to relate to them. We can give these quantities creative names such as *googol* which is a "one" followed by a hundred zeros, or a *googolplex,* which is a "one" followed by a *googol* of zeros; but we still cannot really "understand" their size. Even those who work with these numbers daily cannot fully fathom what numbers of this scale mean.

Any fixed number, minuscule or enormous, does not represent infinity. Infinity cannot be described with a number—it is a concept that lies beyond numbers. If we can assign any number or amount to something, then it is not infinity.

Infinity is a mathematical idea that everyone can understand and visualize because it is a concept that we can understand fully in the three-dimensional realm. Actually, we only require one dimension to explain the concept of *infinity* using simple examples such as a line or just counting.

If we pick a number and add one to it, and then take that new number and again add one to it, we have initiated the common process of counting, which does not ever need to end. We will never reach a last and final number because no matter how many times we repeat this process we can always add one more to the count. This process demonstrates a simple example of *infinity*. If at any *time*, we stop the process of counting, no matter how large the number, we are no longer describing *infinity*.

Another way to imagine *infinity* is to pick a line of any length and divide it in two. Take one of the pieces, divide it, and keep repeating the process. At any point, the line may look extremely short, but there will always be a midpoint to every line, no matter how short the line. This is also an *infinite* process because it will go on and on, forever. Actually, any line can be divided an *infinite* number of times and, therefore, any line contains an *infinite* number of points. *Infinity* can be fully understood using only this one dimension—a line.

An *infinite* universe means that no matter how many *atoms*, cells, organisms, planets, solar systems, galaxies, or universes we have encountered, there will always be more to be found. Our discovery of new aspects or regions will continue forever and never end.

Our Universe Is Infinite

I have briefly outlined several lines of reasoning that all lead to the conclusion that we live in an *infinite* universe. *So far, we have arrived at this conclusion using only three-dimensional concepts. Once we add more dimensions to this structure, any remote chance that our universe is "finite and alone" will quickly disappear because, with these added dimensions, the additional pathways to the infinite become endless.*

While a finite universe, based on fixed and limited bounds, might be more easily understood and, therefore, comfortable for us, it has become impossible to avoid the pressing weight of evidence demonstrating that we live in an infinite universe. The only real questions now seem to be: in what ways is the universe infinite and exactly what does an infinite universe mean to our lives?

Other Worlds Exactly Like Ours

Even though the universe, as a whole, is *infinite*, our physical Earth is still *finite*. There is an enormous number of *atoms* that make up our Earth, but that number can still be determined. A calculation, based on the *mass* of the Earth, estimates the number of *atoms* that make our planet to be near 10^{50} (or the square root of googol). This, of course, is a very big number, but because it is an actual number it is still *finite*. All these *atoms* can be moved around and rearranged in many different ways, but eventually, we will run out of new possible arrangements because we are rearranging a *finite* number of *atoms*: *the number of possible arrangements of the Earth's atoms, while enormous, is still finite.* This is not a theory or speculation—this is a completely rational and long-understood mathematical fact.

Imagine a simple example. Let us say we have three shirts, two pairs of shoes, and one pair of pants, and we want to calculate the total number of different fashion combinations. After a little exploration, we will discover that this number is three (number of shirts), multiplied by two (number of shoes), multiplied by one (number of pants), or six total different possible combinations (3 x 2 x 1= 6). Test this at home. You can arrange a fixed number of articles of clothing in only a certain fixed number of ways because eventually, you run out of options. The number of *particles* that make up Earth

(10^{50}) is also fixed, so they can only be arranged in a fixed (*finite*) number of arrangements (10^{50} x 10^{49} x 10^{48}, etc.). **In other words, a finite number of things can only be arranged a finite number of ways.**

In an *infinite* universe, the same types and number of *particles* or *atoms* that are found on Earth must eventually reappear somewhere else in the universe, not only once, but again and again, forever. There will be an *infinite* number of every type of *particle,* and they will collide, interact, and arrange themselves in different groupings an *infinite* number of times. **In an infinite universe, all the possible combinations for arranging any finite collection of particles will have been tried, so somewhere in this infinite universe, these arrangements must start to repeat. This will happen over and over; and so eventually, somewhere, an exact copy of Earth will be formed. Then, as impossible as this seems, because the universe is infinite, this exact duplication of Earth will also continue to reoccur again and again.**

As astounding as it seems, if time is not a factor, it is a fundamental mathematical and logical fact that an infinite number of exact copies of Earth will exist in an infinite universe. Even when this clear and simple logic if fully understood, this fact is very difficult to accept. However, this is a critical concept, so I will state it in a slightly different way. **In an infinite universe, where time is no longer a limit, every one of the finite numbers of particles making up Earth will, by mathematical necessity, appear again and again, somewhere, over and over, forever. These particles will combine an infinite number of times in every possible arrangement; there is no end to the number of times these particles can combine into different combinations. Since the Earth only contains a finite number of particles, every so often the exact number and type of atoms that make up our world will be repeated, and the repetitions of the Earth's set of atoms will then occur in every possible arrangement. Sooner or later they will be arranged so that they exactly duplicate our Earth, along with everyone and everything on it! Then, this will process will repeat itself again and again.**

This means that somewhere, within the *infinite* expanse of our universe, every *atom* on this Earth will be duplicated and connected to all other *atoms*, in exactly the same arrangement as on Earth. This duplication will contain every *particle* and *atom*, arranged the same way as our world, duplicating everything in it. **In an infinite universe, this process goes on forever, so an infinite number of these exact copies of Earth will also exist in our universe! If the universe is just enormous, this may or may not happen. If the**

universe is infinite, then this impossible-to-conceive outcome must happen.

Most of us, when confronted by the awesome nature of this proposition, will logically reason that it is ridiculous, or that the odds of it happening are so small that it will never really happen! To our "dimensionally and *time*-bound" conceptual minds, the conclusion that there exists an *infinite* number of worlds just like ours sounds like nothing more than raw material for science fiction. It is useful to remember that the "flat-Earthers" once believed the same to be true of a spherical Earth. **However, it is becoming more and more likely that this is how our universe is built; and once we break through our normal (but inherited and cultural) constraints and begin to see our universe for what it really is (full of infinite possibilities), then everything that we think we understand about life will instantly, and forever, change.**

Some *cosmologists* argue that even if *space* is *infinite*, there is still a limit to the number of other worlds that would be like our own. They argue that *"physical evolution takes time and because of the amount of time that is required for the Earth's evolution, there are physical 'time' limits to the number and kind of duplicates of Earth, no matter what kind of spatial arrangements are mathematically possible and available."* This argument is entirely based on their perception of a *time* constraint. They argue that since the universe is only about 15 billion years old, there has not been enough *time* for all these other parallel Earths to have fully evolved, even if there were *infinite space* and material for an *infinity* of them. **The weakness of this reasoning is that it is based entirely on three-dimensional logic and linear time.** If our universe was only three-dimensional and *time* was linear and limited, these constraints would certainly exist; but we now recognize that our sense of a limited, one-directional, and linear *time* is internally generated and not generally accurate, for the greater cosmos treats time very differently. This argument fails to allow for new possibilities beyond those of our old three-dimensional *conceptual horizon,* and it does not even utilize our best current scientific understanding of *time*; it is reminiscent of the flat-earther's resistance.

EXTRAPOLATING INFINITY INTO EXTRA DIMENSIONS

Earlier I described infinity using only one dimension. The idea of *infinity* survives completely intact when it is expanded to two dimensions, and then, again, to three dimensions. Therefore, it seems likely that expanding the idea of *infinity* to four dimensions or more would not significantly restrict or reduce the meaning of *infinity*. When we extrapolate any concept into extra-dimensional *space*, it is generally understood that the number of new possibilities will, in fact, only increase. The concept of *infinity* already contains never-ending possibilities, so it easily survives such a transition intact: *infinity* expanded into multiple dimensions would still be *infinity*.

Even if we had determined that our three-dimensional universe was *finite*, any possibility for the conclusion that the rest of existence is also *finite* quickly disintegrates as we enlarge our *reference frame* to include extra dimensions. Because of the universe's *holographic* nature, these unseen dimensions can be thought of as simultaneously hidden "beyond" and deeply *enfolded* "within" our three-dimensional universe. Because this extra-dimensional perspective introduces many additional ways to imagine the *infinitely* vast "size" of our universe, it creates a lot of extra "room" for new possibilities. These extra dimensions also provide important mechanisms for explaining the deep, instantaneous connections between all the physically distant parts of our realm. This instantaneous interconnection throughout our universe, or *non-local* behavior, is another property of matter that seems mysterious only because it cannot be understood using three-dimensions alone.

We can also describe the universe as *infinite* in two different directions: "going out" through never-ending three-dimensional *space*, and "going in" through the unseen dimensions. *Infinity* is expressed at multiple levels throughout existence. As recent discoveries continue to illuminate the *infinite* nature of our universe, we are just beginning to recognize what this realization might be saying about our lives.

Much of this section is devoted to helping the reader understand and become comfortable with this conclusion about the *infinite* nature of existence. Some readers will resist this leap because, even though this is the arrangement that now seems most likely, it has not been scientifically proven and is likely to remain unproven for some *time*. **However, also unproven is the opposite conclusion: that the universe is finite. Finite systems are much easier to understand, quantify, and describe than infinite systems; therefore, it is much easier to prove that a system is finite. The fact that the universe has not yet been proven to be finite is, by itself, very revealing. Taken alone, this lack of proof of a finite universe throws tremendous weight towards the conclusion that our universe is infinite!**

AN INFINITE MULTIDIMENSIONAL UNIVERSE

I am far from alone in my conclusion that our universe is *infinite* and *multidimensional* because many, probably most, contemporary cosmologists and physicists also believe that this is the true form of our universe. **Much of what this book proposes would still be true in an infinite three-dimensional universe. However, the addition of extra dimensions, enfolded in ways that we cannot see or understand, creates a plethora of additional possibilities of interconnectedness, timelessness, and unity.** A multidimensional universe can help to explain the strange experimental results and the hidden mystery in our lives. The added dimensions permit a kind of interconnectedness that is not possible in only a three-dimensional universe, as it also allows room for an *infinite* amount of mystery and a never-ending *evolutionary* process.

Even if someone with extraordinary talent could visualize or understand four or more dimensions of *spacetime*, they would still be incapable of communicating this meaning to the rest of us. Our cultural imagination reaches its limits somewhere on our side of the veil that hides four-dimensional *space* from our senses and consciousness. This inability to communicate about "glimpses of extra-dimensionality" was precisely what happened to the main character in the Victorian novel *Flatland,* when he tried to describe his three-dimensional experience to his two-dimensional peers. He discovered that he had no effective words or ideas to communicate his encounter, even though he fully understood everything while it was happening. **He even found that once he tried to use words and concepts, his clear understanding became muddled and confused as the limiting nature of language and concepts effectively eroded his deeper "vision."**

We simply do not have any cultural reference for what conscious life utilizing this *infinite space* would be like. Any shift in the meaning of *infinity,* as we expand it into more dimensions, must allow for even more possibilities and connections than those we might rationally deduce. *Infinity* in multiple dimensions will include *qualitative* changes that go far beyond our present ability to imagine.

BUBBLE UNIVERSES

If cosmologists have made significant miscalculations, the universe might still close back on itself like a sphere or a pretzel. If this turned out to be our true *shape*, our universe would then be a *finite* "bubble" universe. However, even then it might be only one of an *infinite* number of other very different, similar, and identical "bubble" universes. If there are an *infinite* number of these, then we already know that, even though each universe is *finite,* there will still be *infinite* room for every possible arrangement. Within this *infinite* series of *finite* bubble universes, there would still exist an *infinite* number of full bubble universes that are identical to ours in every aspect.

EXPANDING UNIVERSE

I described how our universe is structured from mostly empty *space*, which is shaped through *energetic* interactions, and how this same type of arrangement repeats itself when we examine smaller or larger systems. We see this pattern in our *atoms* and subatomic *particles*, and we see it in the planets of our solar system, the stars of our Milky Way galaxy, our region of *galaxies,* and the known universe with its billions of *galaxies.* All of these forms follow a similar, repeating pattern of vast amounts of what appears to be empty *space* and very little, if any, actual "solid" material, with everything being held in place by *energetic* interactions. Solid-appearing *matter,* if it exists at all, is, at most, a minuscule component of our universe; it is universally recognized that our known universe is almost entirely constructed from empty *space* and *energy*. **Since we are made of atoms, we too are almost entirely empty space and energy. However, scientists have recently discovered that space is not actually empty. As I write, fresh insights into the nature of "empty" space continue to lead us towards an evolving understanding that space is instead filled with something extremely interesting, but not at all understood.**

As mentioned, when Einstein first realized that his theory of *relativity* predicted an expanding universe, he became greatly disturbed. This caused him to modify his theory by adding the now-infamous "fudge factor." The *cosmological constant* was an adjustment to his math that allowed it to predict a *stable* universe, which better fit his sensibilities. However, less than a dozen years later, lawyer-turned-cosmologist Edwin Hubble discovered that the universe was indeed *expanding*. After Hubble's discovery, Einstein began to describe his own "after-the-fact" alteration of the original *relativity* theory as the greatest mistake of his career, but it was not until much later, in the 1970s, that *cosmologists* began to uncover a fuller picture. **As then determined, not only was our universe expanding, but the average velocity of expansion was about three times the speed of light.** In the *time* that it took for the light to reach us from the farthest stars—just over 14 billion *years*—the outer edge of this visible universe expanded outward in every direction by an additional 30 billion *light-years*. **This means that in the last 14 billion years since the "big bang," the width of the known universe has more than tripled.**

Wait! You might think that nothing can travel faster than light. As it turns out, the stars, themselves, are not moving at that incredible rate into some empty void. Instead, *space* itself is expanding at that rate. While things contained within *space* cannot move faster than light, *space* itself seems to be able to do anything it wants. As mentioned above, there seems to be at least one other thing that can also move faster than light. ***Since non-local interaction is instantaneous, somehow information also communicates between entangled particles faster than the speed of light.*** Again, we are reminded that we understand so little.

When we think about the stars moving away from us through expansion, we naturally imagine three-dimensional *space* and not *spacetime*. Three-dimensional *beings* can't visualize this expanding *space* because we have absolutely no understanding of the structure or a "container" into which this *space* is expanding. What contains this three-dimensional *space* and allows it to expand?

VISUALIZING THE EXPANDING UNIVERSE

The common classroom demonstration of expanding *space* involving the instructor blowing up a balloon is not an accurate model because the balloon is only expanding into three-dimensional *space*, a realm that we already understand. We have yet to find a way to illustrate extra-dimensional *space*. However, while limited, a balloon can still be a useful tool to help explain the principle.

Blow up a balloon partially, but don't tie it. With a marker, make two or three dots on the surface of the balloon and measure the distance between the dots. Imagine that these dots represent galaxies. Think of the surface of the balloon as a two-dimensional surface, just like the surface of the Earth; both feel "flat" to any inhabitants, even though we know the surfaces are actually "curved." Now inflate the balloon fully and again measure the distance between the dots. This distance is now greater because the "flat surface" expanded as the balloon was inflated. What is this two-dimensional "flat surface" of the balloon expanding into? It is expanding into three-dimensional *space*. Now, in our minds, we can scale this analogy up one more dimension to provide a better understanding of our expanding universe. ***We inhabit the seemingly "flat surface" of three-dimensional space, and when this surface expands, it is expanding into four-dimensional spacetime!***

EMPTY SPACE AND VIRTUAL PARTICLES

As just mentioned, *space* is not actually "empty" at all. It was recently discovered that *space* between *quarks* is filled with an endless succession of very short-lived *particles* that fluctuate in and out of existence. Because these *particles* only exist for the briefest of moments, we usually refer to them as *virtual particles*.

Looking closely at the most massive part of *atoms*: *protons*, physicists now know that ninety percent of their *mass* is found in this "once-empty" *space*. Some physicists theorize that the *mass* must come from these *virtual particles*. Since we are made from *atoms*, our bodies also are made from these same *virtual particles* that continuously move in and out of existence. **Nothing seems to be fixed or solid, everything is always in flux and motion, including at least ninety percent of our bodies.**

There were good arguments for relating *virtual particles* to *dark energy* since "empty" *space* contains both. In an attempt to establish this relationship, the *mass* of *virtual particles* was calculated, but this calculated *mass* was found to be far too large to explain *dark energy*. This calculated amount was not just a little off; it was off by the unbelievably huge factor of 10^{120}. This enormous discrepancy was so dramatic that some have called this the worst prediction in the entire history of physics. Based on this analysis it seems clear that *virtual particles* are not directly responsible for *dark energy*.

Even though the equations of physics describe both *particles* and *virtual particles* equally, physicists had not been able to see, touch or measure any of these *virtual particles*, despite the existence of many observable physical phenomena that result from, and even require, the involvement of *virtual particles*. Today, their "virtual" nature is being re-assessed because one form of these *particles*, *virtual photons*, has just been "captured" like "real" *photons* and then used to create *light*. The shifting of other *subatomic particles* from *virtual* to *real* is now seen as imminently possible due to this recent discovery involving *virtual photons*.

Virtual particles are just one more "thing" that we do not understand, but they are of special interest because they seem to be "located" somewhere near the cusp, or outer-edge, of our current understanding. *Virtual particles* could be very well positioned to become the next new window into a treasure-trove of discovery that lies just beyond our three-dimensional *viewport*.

WHEELER, PENROSE AND BLACK HOLES

John Wheeler, who taught at the University of Texas after many productive years at Harvard, and Sir Roger Penrose, of both Cambridge and Oxford, developed much of the physics that describe *black holes, wormholes,* and other unusual topological features within *spacetime.* They, and now others, have assembled an enormous body of work describing unusual features that all derive directly from *relativity* and can be described mathematically. While *wormholes* are still only theoretical, physicists have substantial physical evidence of actual *black holes,* and they are constantly discovering new ones scattered about our universe. In 2017 cosmologists found several new, enormous *black holes*, one of which is 800 million times the *mass* of our Sun. By February of 2018, they discovered even more of these monsters, some of them over ten billion times the *mass* of our Sun.

Black holes, often found near the center of *galaxies,* can be described as enormous deformations of *spacetime* caused by an ultra-dense *gravitational mass*, usually a collapsed star. Massive, black holes are physically very small in terms of three-dimensional *space* and this makes them extraordinarily dense and *gravitational. Gravity* becomes so great in these *cosmological events* that nothing, not even light, can escape its *gravitational* pull. On the other hand, it is thought that the still theoretical *wormholes*, might be bridges between two remote areas of *spacetime*. It has even been seriously proposed that *wormholes* might function as physical doorways to *parallel universes*.

Due to their enormous size and nature, these powerful structures involve extremely large and, therefore, extremely interesting *energies*. They act as big cosmic engines for our universe, functioning like the typhoons, hurricanes, and tsunamis that churn our planet's weather. Along with the tiny *virtual particles*, these enormous structures are providing a unique and revealing window for understanding our universe.

A recent and very interesting theory about *black holes* proposes that *holographic information* describing our entire universe could be mapped onto their sides; a *black hole* might function like the nerve center or brain for our physical universe.

APPENDIX TWO – GLOSSARY

- *Advaita Vedanta* –an ancient and modern branch of Hindu philosophy that focuses on self-inquiry for gaining spiritual realization.
- *amygdala* – an almond-sized mass of gray matter at the bottom of each cerebral hemisphere involved in experiencing emotions.
- *Assisted Whole-ing Experience (AWE™)* – a therapeutic method for two or more to experience a shared space of oneness
- *atoms* – the smallest unit of ordinary matter that still constitutes each element.
- *attachments* – people, things, and ideas that have become important to our sense of self-identity.
- *avatar*–a temporary representation (icon or image) of a person that is used in video games.
- *Being* – (when capitalized) the deepest collective essence of life.
- *being* – an aspect of Being observed as separate and individual in space and time.
- *Buddhism* – a religion that evolved from the teachings of Buddha.
- *cognitive dissonance* – the mental conflict that occurs in the human mind when beliefs are contradicted by new information.
- *coherence* – the quality of forming a unified whole.
- *coherent light* – light beam consisting of one frequency in which a predictable phase relationship is maintained for a long time. Coherent LASER light does not spread rapidly.
- *conceptual horizon* – the outer limit of our human ability to use rational thought and logic.
- *core vibratory resonance (CVR)* – the vibratory information describing an individual's current state of being and potential for interconnectivity. Similar to soul.
- *(A) Course in Miracles (ACIM)* – a channeled update of Christianity recorded in the latter half of the 20th Century by a Colombia University research psychologist, Helen Schucman.
- *déjà vu* – a new situation or encounter that has somehow already been experienced fully (already seen).
- *doppelgänger*–an identical copy of oneself.
- *dualistic* – (1) the simultaneous existence of our physical separation and our deeper oneness (2) all things in a physical world are constructed from polar opposites.
- *electromagnetic field* – a field of force created by moving or stationary electric charges–a combination of magnetic and electric fields.

- *electromagnetic radiation* – a type of radiation including light, radio waves, microwaves, gamma rays, and x-rays, in which electric and magnetic properties are intertwined.
- *electrons* – the stable subatomic particles in atoms that hold a negative charge.
- *Emotional Freedom Technique (EFT™)* – a mind-body technique that improves the *flow* of energy in the body–aka Tapping.
- *energy* – the capacity for doing work.
- *enfolded* – (in physics) – intertwined in a way where every part touches every other part.
- *enlightening* – the gradual process of opening and clearing emotional blockages to allow more of the "clear light of conscious awareness" to move through, animate and motivate you.
- *enlightenment* – What some believe is the ultimate state of being that is possible in physical life. I believe it is, instead, a continuous process in which every person is always engaged.
- *entangled* – particles that remain connected where actions or changes in one instantly affects the other no matter their physical locations; they can be many light-years apart.
- *epigenetics* – the new field of study of the heritable changes of DNA that can affect gene expression but don't involve the sequencing of DNA.
- *field* – a physical quantity represented by a number that has been assigned a specific and dynamic value for every point in *space-time*. *Fields* are physical in that they occupy space, contain energy, and preclude a true vacuum.
- *field theory* – in physics, the study of *fields*.
- *fight or flight reflex* – an automatic deeply programmed response contained entirely within the lower brain designed to protect the physical body.
- *flow* – the ability to let life unfold by no longer controlling, blocking or interfering with whatever appears in your particular life.
- *fundamental particles* – the most elementary building blocks of matter. Today, these are thought to be quarks, leptons, and bosons.
- *genotype* – the genetic makeup of an individual which is then expressed as the phenotype.
- *Gnostic Christian* – a Christian sect or individual that relies on direct experience of the transcendent instead of doctrine.

- *gravity* – once believed to be an attractive force pulling any mass toward the center of the earth, but now known to be the result of the deformation of spacetime.
- *hippocampus* – the elongated ridges on the bottom of each lateral ventricle of the brain. Thought to be the center of emotion, memory and autonomic nervous system.
- *hologram* – a three-dimensional image formed by the interference of light beams.
- *holographic* – relating to holograms, where each smaller piece contains information about the whole.
- *holographic projection* – an imaging technique using laser light to produce life-like 3-D images called holograms.
- *holographic images* – the image of a holographic projection.
- *holographic storage* – using laser to store computer data in three dimensions. The storage of information where each bit of information still describes the whole.
- *holographic plate* – the 2-D plate (glass) that a 3-D holographic image (hologram) is stored upon.
- *image* – a secondary representation of form – a "picture" of something.
- *infinite* – something that has no end or limit and continues forever.
- *information* – what is conveyed or represented by a particular arrangement of things.
- *intra-communication* – information shared internally within a single entity.
- *karmic* – an important personal or energetic relationship that extends beyond just the interactions or information of a particular lifetime.
- *light* – the electromagnetic spectrum – particularly that part which stimulates sight. Also, the spiritual illumination which facilitates awareness and moves deeper aspects of being.
- *limbic* – the lower brain system (amygdala-hypothalamus-hippocampus) responsible for emotions and motivation.
- *matter* – physical substance (particles)
- *magic* – that which while very real, we are unable to understand or explain because it unfolds beyond our *conceptual horizon*. Also *miracles*.
- *magnetic fields* – the region around something that is magnetic.
- *medial prefrontal cortex* – the region of the brain that plans, makes decisions, and deals with complexities – orchestrates thoughts and goals.

- *mind* – the fully subjective image formed by the brain, central nervous system, other stimuli, producing our highly-filtered interpretations of personal experiences.
- *Mind* – (capitalized) "mind-at-large." An open awareness which allows access to all the information, intra-connection and experience in existence.
- *molecules* – a group of atoms bonded together representing the smallest unit of a chemical compound.
- *multiverse* –a collection of many, or an infinite number of, universes. Many believe this is the true form of existence.
- *neutron* – the subatomic particle in the nucleus of an atom that is without an electrical charge.
- *parameter* – a numerical or some other measurable factor that defines a system.
- *particles* – a small localized object that can be ascribed a volume, mass, and density.
- *particle-wave duality* – see *wave-particle duality.*
- *phenotype* – the set of observable characteristics of an individual resulting from int interaction of the genotype and the environment.
- *photon* – a particle representing a quantum of light or quantum of other types of electromagnetic radiation.
- *proton* – the subatomic particle in the nucleus of an atom with a positive electrical charge.
- *post-traumatic stress disorder* (PTSD) – a condition of persistent emotional and mental stress occurring from severe psychological shock.
- *psychedelic-assisted therapy* – a professionally guided therapeutic technique that utilizes psychedelics to ease and deepen the exploration of personal trauma and other psychological disturbances.
- *quantum gravity* – a contemporary area of research within *quantum physics* that attempts to probe and integrate the aspects of quantum mechanics that can be found within *relativity.*
- *quanta* – the plural of quantum.
- *quantum* – the smallest discrete quantity of energy. Its size will be proportional in magnitude to the frequency of energy it represents.
- *quantum field theory* – the theoretical framework that combines *classical field theory* with *quantum mechanics* and *special relativity.* In this theory, all the particles that make up our universe only represent the excited state of the underlying field.

- *quantum foundations* – the discipline that seeks to understand and explain the counter-intuitive aspects of *quantum physics*. Traditionally associated with the *Copenhagen interpretation*.
- *quarks* – any of a number of subatomic particles carrying a fractional electrical charge that combine form the larger hadrons such as neutrons and protons.
- *radio waves* – electromagnetic waves of longest wavelength and lowest frequency and energy which are used for long-distance communication.
- *sadhu* – a holy man (feminine sadhvine) within Hinduism or Jainism.
- *Self* – (capitalized) the fully intra-connected self that is then living with an awareness of everything within existence – able to access Mind.
- *self* – the ego-created self-identity that we form in our minds. Entirely illusionary, the only place it actually exists is in our minds.
- *selfing* – Paul Hedderman's term for the human egoic tendency to always be building, shaping, and protecting our human concept of a divided self.
- *soul* – the deepest energetic quality that forms an individual's viewport. Also CVR (core vibrational resonance).
- *spacetime* – Einstein's concept of 3D space and time being fused in a 4D continuum.
- *spiritual materialism* – using spiritual knowledge, practices, or techniques to prop up or build human ego – a term coined by Chögyam Trungpa.
- *string theory* – a quantum sub-theory where tiny vibrating and interacting strings are believed to be the most fundamental source of matter.
- *Sufism* – the most mystic order of worship within Islam that can be found throughout many of its sects.
- *tantric* – traditionally relating to doctrines of Hindu, Jainist, or Buddhist tantras of meditation, yoga, mantras, and ritual that aim to transform the body into a temple. More recently interpreted in the west as a spiritual focus on intimate practices involving sexual energies.
- *transmutation* – the complete and radical change in the way something appears to us.
- *un-selfing* – learning how to stop or slow the persistent selfing process.
- *viewport* – the collection of individual and cultural filters that determine just which parts of existence each of us can participate within and

experience. By discovering and untangling our filters we can expand our viewport and experience.

- *vibration* – the oscillation of the parts of a fluid or elastic solid whose equilibrium has been disturbed. Traditionally described as fluctuations within time.

- *visible light* – the small part of a much larger electromagnetic spectrum of radiation that allows us to see the world as a collection of discrete appearing objects.

- *waves* – a disturbance at a point of a field such that the values oscillate repeatedly about a stable equilibrium point.

- *wave-particle duality* – the concept of *quantum physics* that every particle can also be observed as a wave but never as both wave and particle at the same time.

- *whole-ing* – the opening to all aspects of who, and what, we actually are.

- *x-rays* – the scientifically useful region of frequencies of the electromagnetic spectrum above ultra-violet but below gamma. This range of wavelengths is very energetic, penetrating, and dangerous to human life.

RESOURCES
(Please feel free to recommend other resources.)

PHYSICS, MATH AND QUANTUM PHYSICS

Albert Einstein
-*Essays in Physics,* 1950, Philosophical Library
-*Out of My Later Years,* 1956, Citadel Press

Joseph Schwartz and Michael McGuiness
-*Einstein for Beginners,* 1979, Pantheon Books

Francis S. Collins, head of Human Genome Project
-*The Language of God,* 2006, Free Press

Brian Greene
-*The Elegant Universe: Superstrings, Hidden Dimension, and the Quest for the Ultimate Theory*
-*The Fabric of the Cosmos: Space, Time and the Texture of Reality,* 2004, Knopf
-*The Hidden Reality: Parallel Universes and the Deep Laws of the Cosmos,* 2011, Alfred A. Knopf

Michio Kaku
-*Beyond Einstein,* 1997, Oxford University Press
-*Visions,* 1999, Oxford University Press
-*Hyperspace,* 1994, Oxford University Press

Carlo Rovelli
-*The Order of Time,* 2017, Riverhead Books
-*Seven Brief Lessons on Physics,* 2014, Riverhead Books
-*What is Time? What is Space?,* 2006, Di Renzo Editore
-*Reality is Not What it Seems,* 2017, Riverhead Books

Timothy Ferris
-*The Whole Shebang: A State of the Universe(s) Report,* 1997, Simon and Schuster

Brian Cleg
-*The Quantum Age: How the Physics of the Very Small has Transformed our lives,* 2014, Icon Books Ltd.

Paul Davies
-*About Time: Einstein's Unfinished Revolution,* 1995, Simon & Schuster

David Kaiser
-*How the Hippies Saved Physics*: *Science, Counterculture, and the Quantum Revival*, 2011, W.W. Norton &Company

Art Hobson
-*Tales of the Quantum*: *Understanding Physics*, 2017, Oxford University Press

Max Planck
-*The Origin and Development of the Quantum Theory,* 2012, Amazon Kindle Direct Publishing

Jim Al-Khalili
-*Quantum*: *A Guide for the Perplexed*, 2003, Weidenfeld & Nicolson

Bertrand Russell
-*The ABC of Relativity*, 1958, George Allen

Louisa Gilder
-*The Age of Entanglement*: *When Quantum Physics Was Reborn*, 2008, Alfred A. Knopf

Tom Siegfried
-*Strange Matters*, 2002, Joseph Henry Press

Lawrence Krauss
-Lecture "A Universe from Nothing", You-Tube

ASTRONOMY AND COSMOLOGY

William J. Kaufmann, III
-*Discovering the Universe*, 1987, W.H Freeman and Company

Carl Sagan
-*Cosmos*, 1980, Random House

ANTHROPOLOGY

Robert Ardrey
-*African Genesis*, 1961, Macmillan
-*Territorial Imperative*, 1966, Kodansha Globe

METAPHYSICS

Fritjof Capra
-*The Tao of Physics*, 1975, Bantam Books
-*The Web of Life*, 1996, Anchor Books-Doubleday
-*The Turning Point*, 1982, Bantam Books
- "Mindwalk," movie directed by Bernt Capra

David Darling
-*Equations of Eternity*, 1993, Hyperion
-*Zen Physics–The Science of Death; The Logic of Reincarnation*, 1996, Harper Collins

Frank J Tipler
-*The Physics of Immortality*, 1994, Doubleday

Frank Wilczek with Betsy Devine
-*Longing for the Harmonies*, 1987 W.W. Norton Co.

Fred Alan Wolf, Ph.D.
-*The Dreaming Universe*, 1994, Simon and Schuster
-*Eagle's Quest*, 1991, Touchstone-Simon and Schuster
-*Parallel Universes*, 1988, Touchstone

Paul Davies
-*Are We Alone*, 1995, Orion Publications
-*About Time*, 1995, Orion Publications
-*The Mind of God*, 1992, Orion Productions
-*Space and Time in the Modern Universe*, 1977,
Cambridge University Press

PHILOSOPHY OF SCIENCE

Gary Zukav
-*The Seat of the Soul*, 1989, Fireside
-*The Dancing Wu Li Masters*, 1979, Bantam

Robert M Pirsig
-*Zen and the Art of Motorcycle Maintenance*, 1974, Bantam

Ken Wilber
-*A Brief History of Everything*, 1996, Shambhala
-*The Holographic Paradigm*, 1982, New Science Library

Eugene Pascal Ph.L.
-*Jung to Live By*, 1992, Warner Books

Krista Tippett -*Interviews*
-*Einstein's God-Conversations about Science and the Human Spirit*, 2010, Penguin

Adam Becker
-*What is Real: The Unfinished Quest for the Meaning of Quantum Physics*, 2018, Basic Books

J. Krishnamurti & David Bohm
-*The Ending of Time: Where Philosophy and Physics Meet*, 2014, Harper One

Steve McIntosh
-*Integral Consciousness*, 2007, Paragon House

PSYCHOLOGY

Stanislav Grof M.D.
-*The Holotropic Mind: Three Levels of Human Consciousness*, 1993, Harper-San Francisco

Antonio Damasio
-*The Feeling of What Happens-Body and Emotions In the Making of Consciousness,* 1999, Harcourt, Inc.
-*Self Comes to Mind-Constructing the Conscious Brain,* 2010, Vintage Books

Thomas Metzinger
-*The Ego Tunnel: The Science of the Mind and the Myth of the Self*, 2010, Basic Books

Judith Orloff, M.D.
-*Second Sight: An Intuitive Psychiatrist Tells Her Extraordinary Story*, 1996, Three Rivers Press

Brian L. Weiss M.D.
-*Same Soul, Many Bodies*, 2004, Free Press
-*Many Lives, Many Masters*, 1988, Simon and Schuster
-*Only Love Is Real*, 1997, Grand Central Publishing

RESEARCH SCIENCE

Dr. Zach Bush, M.D.
-Numerous interesting articles and interviews on web–human microbiome
https://zachbushmd.com

Rollin McCraty, Ph.D., Executive Vice President and Director
The Institute of HeartMath
-Numerous research papers
–Quoted in movie, *"I Am"*

Scientific American
-*The Secrets of Consciousness-From the Editors of Scientific American,* 2013,
Scientific American

Bruce H. Lipton, Ph.D
-*The Biology of Belief-Unleashing the Power of Consciousness, Matter and Miracles,* 2005, Hay House

Lewis Thomas
-*The Lives of a Cell,* 1974, Viking Press

J. Konrad Stettbascher
-Making Sense of Suffering, 1993, Meridian

MUSIC

Robert Jourdain
-*Music, the Brain and Ecstasy,*
1997, Bard Press

Daniel J. Levitin
-*This Is Your Brain on Music,* 2006, Plume

FICTION

Edwin A Abbott
-*Flatland,* 1952, Dover Publications

Michael Murphy
-*Golf in the Kingdom,* 1997, Arkana
-*Jacob Atabet,* 1977, Jeremy P. Thacher

NON-FICTION

Carol Riddell
-*The Findhorn Community,* 1990, Findhorn Press

Michael Pollan
-*How to Change Your Mind-What the New Science of Psychedelics Teaches Us About Consciousness, Dying, Addiction, Addiction, Depression and Transcendence,* 2018, Penguin

Tom Shroder
-*Acid Test-LSD, Ecstasy and The Power to Heal,* 2014, Penguin

Aldous Huxley
-*The Doors of Perception & Heaven and Hell,* 1954, 1956, Harper Collins

Albert Hofmann
-*LSD and The Divine Scientist-Final Thoughts and Reflections of Albert Hofmann,* 2011, Park Street Press

Elizabeth Gilbert
Eat, Pray, Love, 2007, Penguin Books

POETRY AND ART

Kahlil Gibran
-*Between Morning and Light,* 1972, Philosophical Library
-*The Prophet,* 1923, Alfred A Knopf

WESTERN SPIRITUALITY AND RELIGION

Elaine Pagels
-*The Gnostic Gospels,* 1979, Vintage Books

Jean–Yves Leloup
-*The Gospel of Philip,* 2003, Inner Traditions

Michael Wise, Martin Abegg Jr. and Edward Cook, *editors and translators*
-*Dead Sea Scrolls,* 1996, Harper: San Francisco

James M. Robinson, editor
-*The Nag Hammadi Library,* 1978, Harper and Row

Stephan A Hoeller
-*Jung and the Lost Gospels: Insights into the Dead Sea Scrolls* 1989, The Theosophical Publishing House

Kyriacos C. Markides
-*Riding with the Lion: The Search of Mystical Christianity,* 1995, Penguin

Mathew Fox
-*One River, Many Wells,* 2000, Tarcher/Putnam
-*The Coming of the Cosmic Christ,* 1988, Harper Collins

Rupert Sheldrake with Matthew Fox
-*The Physics of Angels: Where Science and Spirit Meet,* 1996, Harper: San Francisco

Eckhart Tolle
-*The Power of Now,* 1999, New World Library
-*The New Earth: Awakening to Your Life's Purpose,* 2005, Dutton Publishing

Anthony DeMello
-*Awareness,* 1990, Doubleday
-*The Way of Love: The Last Meditations of Anthony DeMello,* 1991, Doubleday

William Dych. S.J., *editor*
-Anthony DeMello writings, 1999, Orbis Books

Jack Kornfield
-*After the Ecstasy, the Laundry: How the Heart Grows Wise on the Spiritual Path,* 2000, Bantam Books

Sam Harris
-*Waking Up: A Guide to Spirituality Without Religion,* 2014, Simon & Schuster Inc.

Paul Hedderman
-*The Escape to Everywhere—Based on Early Talks,* 2015, James S. Cloud

Stephen Mitchell, *editor*
-*The Enlightened Heart,* sacred poetry, 1989, Harper and Row

Paul Ferrini
-*Reflections of the Christ Mind,* 2000, Doubleday

Nicole Gausseron
-*The Little Notebook: The Journal of a Contemporary Woman's Encounters with Jesus,* 1995, Harper San Francisco

Joan Borysenko Ph.D.
-*The Ways of the Mystic: 7 Paths to God,* 1997, Hay House

Sam Keen
-*To a Dancing God,* 1970, Harper

Helen Schucman and William Thetford
-*A Course in Miracles,* 1975, Foundation for Inner Peace

EASTERN SPIRITUALITY AND RELIGION

Thich Nhat Hanh
-*Peace Is in Every Step: The Path of Mindfulness in Everyday Life*, 1991, Bantam Books

Chögyam Trungpa
-*Cutting Through Spiritual Materialism*, 1973, Shambhala

Christmas Humphreys
-*Buddhism*, 1951, Penguin Books

Edward Conze, *editor and translator*
-*Buddhist Scriptures*, 1959, Penguin Books

Swami Prabhavananda and Christopher Isherwood, editors and translators
-*How to Know God: The Yoga Aphorisms of Patanjali*, 1953 & 1981, Vedanta Press

Christopher Isherwood
-*Vedanta for the Western World*, 1945, Vedanta Society of Southern California

Paramahansa Yogananda
-*The Autobiography of a Yogi*, 1946, Self-Realization Fellowship

J. Krishnamurti
-*Krishnamurti's Journal*, 1982, Harper and Row
-*Think on These Things*, 1981, Harper One
-*You Are the World*, 2001, Krishnamurti Foundation
-*Freedom from the Known*, 1975, Harper Collins

Stuart Holroyd
-*Krishnamurti: The Man the Mystery & the Message*
1991, Element

Lao Tzu
-*Tao Te Ching*, 1997, Wordsworth Editions

Birgitte Rodriguez
-*Glimpses of the Divine: Working with the Teachings of Sai Baba,*
1993, Samuel Weiser

Howard Murphet
-*Sai Baba Man of Miracles*, 1971, Samuel Weiser Inc.

Ramana Maharshi
-*The Spiritual Teaching of Ramana Maharshi*, 1972,
Shambala Publications

Osho
-*Intuition: Knowing Beyond Logic*, 2001,
Amazon

Adyashanti
-*The Way of Liberation*, 2012, Open Gate Sangha, Inc.
-*The End of Your World*, 2012, Sounds True

Brian Hodgkinson
-*The Essence of Vedanta: The Ancient Wisdom of Indian Philosophy*, 2006,
Arcturus Publishing Ltd.

Kahil Gibran
-*The Prophet*, 1979, Alfred A. Knopf
-*The Voice of the Master*, 1958, Bantam Books

Satyam Nadeen
-*From Onions to Pearls: A Journal of Awakening and Deliverance*,
1996, Hay House
-*From Seekers to Finders*, 2000, Hay House

Gopi Krishna
-*Living With Kundalini: The Autobiography of Gopi Krishna*, 1993, Shambhala

Paul Lowe
-*In Each Moment: A New Way to Live*, 1998, Looking-Glass Press
-*The Experiment Is Over*, 1989, New York

Shri Nisargadatta Maharaj
-*I Am That*, 1982, Acorn Press

Ram Dass
-*Be Here Now*, 1971, Lama Foundation
-*Still Here*, 2000, Riverhead Books
-*The Path of Service*, audiobook, 1990, Sounds True Recordings

Bubba Free John
-*The Knee of Listening*, 1972, Dawn Horse Press

Nirmala-Daniel Erway
-Nothing Personal: Seeing Beyond the Illusion of a Separate Self,
2001, Endless Satsang Press

Sri H.W.L. Poonja-Papaji
-THIS: Prose and Poetry of Dancing Emptiness, 2000, Samuel Weiser

Eli Jaxon-Bear, editor
-Wake Up and Roar: Satsang with H.W.L. Poonja, 1992, Gangaji Foundation

Thomas Byrom
-The Heart of Awareness: A Translation of the Ashtavakra Gita,
1990, Shambhala

CROP CIRCLES

Freddy Silva
-Secrets in the Fields, 2002, Hampton Roads

Steve and Karen Alexander
-Crop Circles, Signs, Wonders and Mysteries, 2006,
Arcturus Publishing
Website-www.temporarytemples.co.uk

GENERAL PHILOSOPHY

Immanuel Kant
-Observations on the Feeling of the Beautiful and Sublime
1764

PHILOSOPHY OF PAIN, HEALING, AND PTSD

Deepak Chopra M.D.
-Ageless Body, Timeless Mind, 1998, Harmony Books
-Quantum Healing, 1989, Bantam Books
-(Plus many others: one of the most popular writers on alternative healing.)

David Deida
-Enlightened Sex: Finding Freedom and Fullness Through Sexual Union,
audiobook, 2004, Sounds True

Peter Levine Ph.D.
-The Unspoken Voice: How the Body Releases Trauma and Restores Goodness, 2010, North Atlantic Books

Bessel van der Kolk
-The Body Keeps the Score: Brain, Mind, and Body in the Healing of Trauma, 2014, Penguin

Bill Moyers
-Healing and the Mind, 1993, Doubleday

Andrew Weil M.D.
-Spontaneous Healing, 1995, Knopf

Michael J. Tamura
-You Are the Answer, 2002, Star of Peace Publishing

Jill Bolte Taylor
-My Stroke of Insight, 2008, Penguin

Larry Dossey MD
-Meaning & Medicine, 1991, Bantam
-Reinventing Medicine, 1999, Harper
-Space, Time & Medicine, 1982, Shambhala
-Beyond Illness, 1984, New Science Library

HEALING SELF-HELP

Louise L. Hay
-You Can Heal Yourself, 1984, Hay House

W. Brugh Joy M.D.
-Joy's Way, A Map for the Transformational Journey: An Introduction to Potentials of Healing with Body Energy,
1979, Archer Putnam

Thorwald Dethlefsen and Rudiger Dahlke M.D.
-The Healing Power of Illness, 1990, Element Books

HEALING SELF-HELP THERAPIES

Chris Jarmey and John Tindall
-Acupressure, 1991, Gaia Books, London

Ben E. Benjamin, Ph.D.
-Listen to Your Pain, 1984, Penguin

Barbara and Kevin Kunz
-Complete Reflexology for Life, 2007, DK Publishing

Bonnie Prudden
-*Pain Erasure*, 1980, Ballantine Books

Moshe Feldenkrais
-*Awareness Through Movement*, 1972, Harper San Francisco

Pete Egoscue
-*Pain Free*, 1998, Bantam Books

MODERN SPIRITUAL SELF-HELP

Marianne Williamson
-*A Return to Love*, 1992, Harper Collins
-*Tears to Triumph*, Spiritual Healing for Modern Plagues of Anxiety and Depression, 2016, Harper One

Neale Donald Walsch
-*Conversations with God* (3 Volumes), 1995, G.P. Putnam's Sons

Rhonda Byrne
-*The Secret*, 2006, Atria Books

Debbie Ford
-*The Dark Side of Light Chasers-Reclaiming Your Power, Creativity, Brilliance, and Dreams,* 1998, Riverhead Books

Gary R. Renard
-*The Disappearance of the Universe*, 2002, Hay House

Charlotte Kasl, Ph.D.
-*If the Buddha Married: Creating Enduring Relationships on a Spiritual Path*, 2001, Penguin Compass

SHAMANISM

Michael Harner
-*The Way of the Shaman*, 1980, Harper San Francisco

Heather Cumming and Karen Leffler
-*John of God-The Brazilian Healer Who's Touched the Lives of Millions*, 2007, Atria Books

Guy Crittenden
-*The year of Drinking Magic-Twelve Ceremonies with the Vine of Souls,* 2017, Apocryphile Press

Peter Gorman
-Ayahuasca in My Blood-25 Years of Medicine Dreaming, 2010, Gorman Bench Press

Olga Kharitidi, MD.
-Entering the Circle, 1996, Harper San Francisco

Sandra Ingerman
-Soul Retrieval, 1991, Harper Press

ALTERNATE TECHNOLOGIES/ WORKSHOPS

Richard Bartlett, DC, ND
-Matrix Technology: The Physics of Miracles, 1990, Atria (Simon and Schuster)

Michael Brown
-The Presence Process: A Journey into Present Moment Awareness, 2005, Beaufort Books

Rasha
-Oneness: The Teachings, 2003, Amazon

Betty Edwards
-Drawing on the Right Side of the Brain, (workshops and book),
1979, Penguin Books

FILM

Tom Shadyac
-I Am, 2011, Shady Acres Entertainment

Rhonda Byrne
-The Secret, 2006, Prime Time Productions

William Arntz, Betsy Chase, Mark Vecente
-What the Bleep, 2004, Roadside Attractions

Bernt Capra
-Mindwalk, 1990, New Yorker Films

Louis Malle
My Dinner with Andre, 1981, New Yorker Films

PLACES AND TEACHERS IN AUSTIN, TEXAS

Ecstatic Dance Austin Community
Austin Texas 78702
www.ecstaticdanceaustin.net

Tribal Joy Ecstatic Dance
700 Dawson Rd. Austin Texas 78704
oscarmadera@att.net

Austin Body Choir
www.bodychoir.org

Five Rhythms — Gabriel Roth
Various cities around the country

Contact Improv
Various locations throughout country

Nia Space
Dance, Therapy, and Workout
Various Locations around the country

Casa de Luz
Macrobiotic Community Center and Meeting Place
1701 Toomey Rd., Austin TX 78704
512-476-2535
www.casadeluz.org

Amala Foundation
Transformational Programs and Service
1006 S. 8th St., Austin, TX 78704
512-476-8884
info@amalafoundation.org
www.amalafoundation.org

Vanessa Stone
Humanitarian and Teacher
Transformational Programs and Service
Texas, California, and Hawaii
www.vanessastone.org

About the Author

Tim Cross is an architect living in Austin, Texas, who studied undergraduate physics and is a trained ecologist. Before turning his attention to architecture, he taught high school physics and worked in the field of environmental science. He served in the Peace Corps and in the late 1970s was the *Environmental Advisor to the Government of Fiji*. However, throughout all of his adventures, his primary passion has always been to learn as much as possible about this incredible experience that we call life. This book series began as a simple letter to his two daughters, but quickly grew into something quite unexpected.

Although Tim is an architect, his books are not about the architecture of buildings. They are, instead, books about personal and cultural freedom, and how the invisible architecture of our universe fully supports each and all of our personal journeys towards freedom. He defines *freedom as letting go of those things in our lives that block us from experiencing our birthright, which is living fully within the flow of this infinite multiverse—a flow* that is shaped and interconnected with a deeper type of Love.

Tim is now at a place in his life where his only remaining desire is to allow for the deepening of his experience with Love. These books are living and evolving records that describe his process and what he continues to learn about living in this amazing and infinite universe.

Comments or corrections to:
tim@thearchitectureoffreedom.com